MULTI-ANTENNA SYNTHETIC APERTURE RADAR

MULTI-ANTENNA SYNTHETIC APERTURE RADAR

Wen-Qin Wang

CRC Press
Taylor & Francis Group
Boca Raton London New York

CRC Press is an imprint of the
Taylor & Francis Group, an **informa** business

MATLAB® is a trademark of The MathWorks, Inc. and is used with permission. The MathWorks does not warrant the accuracy of the text or exercises in this book. This book's use or discussion of MATLAB® software or related products does not constitute endorsement or sponsorship by The MathWorks of a particular pedagogical approach or particular use of the MATLAB® software.

CRC Press
Taylor & Francis Group
6000 Broken Sound Parkway NW, Suite 300
Boca Raton, FL 33487-2742

First issued in paperback 2017

© 2013 by Taylor & Francis Group, LLC
CRC Press is an imprint of Taylor & Francis Group, an Informa business

No claim to original U.S. Government works
Version Date: 20130220

ISBN 13: 978-1-138-07647-1 (pbk)
ISBN 13: 978-1-4665-1051-7 (hbk)

Library of Congress Cataloging-in-Publication Data

Wang, Wen-Qin.
  Multi-antenna synthetic aperture radar / author, Wen-Qin Wang.
  pages cm
  Includes bibliographical references and index.
  ISBN 978-1-4665-1051-7 (hardback)
  1. Synthetic aperture radar. 2. Radar--Antennas. I. Title.

TK6592.S95W36 2013
621.3848'5--dc23   2012050937

Visit the Taylor & Francis Web site at
http://www.taylorandfrancis.com

and the CRC Press Web site at
http://www.crcpress.com

Contents

List of Figures   xiii

List of Tables   xxi

Author Bios   xxiii

Preface   xxv

Abbreviations   xxvii

1   Introduction   1

   1.1   What is Multi-Antenna SAR 1
     1.1.1   Multichannel SAR 2
       1.1.1.1   Multiple Channels in Elevation 2
       1.1.1.2   Multiple Channels in Azimuth 3
       1.1.1.3   Multiple Channels in Azimuth and Elevation 4
     1.1.2   Multi-Antenna SAR 4
       1.1.2.1   SIMO Multi-Antenna SAR 5
       1.1.2.2   MIMO Multi-Antenna SAR 6
   1.2   Multi-Antenna SAR Potentials and Challenges 8
     1.2.1   Benefits of Multi-Antenna SAR 8
       1.2.1.1   Improved System Gain 8
       1.2.1.2   Increased Degrees-of-Freedom 8
     1.2.2   Application Potentials 9
       1.2.2.1   High-Resolution Wide-Swath Remote Sensing 9
       1.2.2.2   Ground Moving Targets Indication 9
       1.2.2.3   Three-Dimensional Imaging 10
     1.2.3   Technical Challenges 11
       1.2.3.1   Waveform Diversity Design 11
       1.2.3.2   Spatial, Time and Phase Synchronization . . 12
       1.2.3.3   High-Precision Imaging Algorithm 12
   1.3   Organization of the Book 13

2 Background Material   **17**

   2.1 Convolution and Correlation 17
      2.1.1 Convolution Integral 17
      2.1.2 Convolution Theorem 20
      2.1.3 Correlation Function 21
      2.1.4 Relations Between Convolution and Correlation 22
   2.2 Sampling Theorem and Interpolation 22
      2.2.1 Sampling 23
      2.2.2 Interpolation 24
      2.2.3 Aliasing Effects 25
   2.3 Linearly Frequency Modulated Signal and Matched Filtering 27
      2.3.1 Principle of Stationary Phase 27
      2.3.2 LFM Signal 29
      2.3.3 Matched Filtering 30
      2.3.4 Pulse Compression 32
   2.4 Radar Ambiguity Function 33
      2.4.1 Range Ambiguity Function 35
      2.4.2 Velocity Ambiguity Function 36
      2.4.3 Properties of the Ambiguity Function 37
      2.4.4 Example: LFM Ambiguity Function 38
   2.5 Basic Principle of Synthetic Aperture 38
      2.5.1 Synthetic Aperture Radar Imaging 39
      2.5.2 Remote Sensing Swath Width 42
      2.5.3 System Sensitivity 42
      2.5.4 Ambiguity-to-Signal Ratio 44
          2.5.4.1   Azimuth Ambiguity-to-Signal Ratio 44
          2.5.4.2   Range Ambiguity-to-Signal Ratio 46
   2.6 Point Spread Function 47
   2.7 Basic Image Formation Algorithm 49
      2.7.1 Two-Dimensional Spectrum Model 50
      2.7.2 Range-Doppler (RD) Algorithm 52
      2.7.3 Chirp-Scaling (CS) Algorithm 54
      2.7.4 Numerical Simulation Examples 61

3 Azimuth Multi-Antenna SAR   **63**

   3.1 Constraints on Resolution and Swath 63
   3.2 Displaced Phase Center Antenna Technique 66
   3.3 Single-Phase Center Multibeam SAR 68
      3.3.1 Azimuth Multichannel Signal Processing 69
      3.3.2 System Performance Analysis 72
   3.4 Multiple-Phase Center Multibeam SAR 76
      3.4.1 System Scheme and Signal Model 76
      3.4.2 Nonuniform Spatial Sampling 78

	3.4.3	Azimuth Signal Reconstruction Algorithm	81
	3.4.4	System Performance Analysis	85
	3.4.5	Numerical Simulation Results	87
3.5	Azimuth Scanning Multibeam SAR		89
	3.5.1	Signal Model	91
	3.5.2	System Performance Analysis	93
3.6	Azimuth Multi-Antenna SAR in GMTI		95
	3.6.1	GMTI via Two-Antenna SAR	95
	3.6.2	Three-Antenna SAR	98
	3.6.3	Simulation Results	103
3.7	Conclusion		109

4 Elevation-Plane Multi-Antenna SAR — **111**

4.1	Null Steering in the Elevation-Plane		111
4.2	Elevation-Plane Multi-Antenna SAR		114
4.3	Several Practical Issues		117
	4.3.1	PRF Design	117
	4.3.2	Ill-Condition of the Sensing Matrix	117
	4.3.3	Interferences of Nadir Echoes	118
	4.3.4	Blind Range Problem	121
4.4	Multi-Antenna Chirp Scaling Imaging Algorithm		123
4.5	System Performance Analysis		127
	4.5.1	RASR Analysis	127
	4.5.2	SNR Analysis	129
4.6	Numerical Simulation Results		129
4.7	Conclusion		131

5 MIMO SAR Waveform Diversity and Design — **135**

5.1	Introduction		136
5.2	Polyphase-Coded Waveform		139
5.3	Discrete Frequency-Coding Waveform		141
5.4	Random Stepped-Frequency Waveform		144
	5.4.1	Basic RSF Waveforms	145
	5.4.2	RSF-LFM Waveforms	147
	5.4.3	Phase-Modulated RSF Waveforms	148
5.5	OFDM Waveform		149
	5.5.1	OFDM Single-Pulse Waveform	150
	5.5.2	Ambiguity Function Analysis	154
5.6	OFDM Chirp Waveform		156
	5.6.1	Chirp Diverse Waveform	156
	5.6.2	OFDM Chirp Diverse Waveform	165
	5.6.3	Waveform Synthesis and Generation	169
5.7	Constant-Envelope OFDM Waveform		171

5.7.1 Peak-to-Average Power Ratio 173
5.7.2 Constant-Envelope OFDM Pulse 174
5.8 Conclusion 174

6 MIMO SAR in High-Resolution Wide-Swath Imaging   **177**

6.1 Introduction 178
6.2 MIMO SAR System Scheme 182
    6.2.1 Signal Models 183
    6.2.2 Equivalent Phase Center 185
6.3 Multidimensional Waveform Encoding SAR HRWS Imaging 188
    6.3.1 Multidimensional Encoding Radar Pulses 190
    6.3.2 Intrapulse Beamsteering in the Elevation Dimension . 192
    6.3.3 Digital Beamforming in Azimuth 194
    6.3.4 Range Ambiguity to Signal Ratio Analysis 195
6.4 MIMO SAR HRWS Imaging 196
    6.4.1 Transmit Subaperturing MIMO Technique 197
    6.4.2 Transmit Subaperturing for HRWS Imaging 200
       6.4.2.1 NTNR Operation Mode 200
       6.4.2.2 NTWR Operation Mode 202
    6.4.3 Range Ambiguity to Signal Ratio Analysis 203
    6.4.4 Image Formation Algorithms 207
    6.4.5 Numerical Simulation Results 210
6.5 Space-Time Coding MIMO SAR HRWS Imaging 214
    6.5.1 Space-Time Block Coding 214
    6.5.2 Space-Time Coding MIMO SAR Scheme 216
       6.5.2.1 Space-Time Coding Transmission in Azimuth 217
       6.5.2.2 MIMO Configuration in Elevation 218
    6.5.3 Digital Beamforming in Elevation 220
    6.5.4 Azimuth Signal Processing 223
6.6 Conclusion 226

7 MIMO SAR in Moving Target Indication   **227**

7.1 Introduction 227
7.2 MIMO SAR with Multiple Antennas in Azimuth 229
7.3 Adaptive Matched Filtering 231
7.4 Moving Target Indication via Three-Antenna MIMO SAR . 234
7.5 Moving Target Indication via Two-Antenna MIMO SAR . . 240
    7.5.1 DPCA and ATI Combined GMTI Model 240
    7.5.2 Estimating the Moving Target's Doppler Parameters . 243
    7.5.3 Focusing the Moving Targets 243
7.6 Imaging Simulation Results 244
7.7 Conclusion 249

**8 Distributed Multi-Antenna SAR Time and Phase
Synchronization**   **251**

8.1 Frequency Stability in Frequency Sources 252
  8.1.1 Oscillator Output Signal Model 253
  8.1.2 Frequency-Domain Representation 254
  8.1.3 Time-Domain Representation 258
    8.1.3.1 True Variance 258
    8.1.3.2 Sample Variance 258
    8.1.3.3 Allan Variance 259
    8.1.3.4 Modified Allan Variance 259
8.2 Time and Phase Synchronization Problem in Distributed SAR
  Systems 260
8.3 Impacts of Oscillator Frequency Instability 262
  8.3.1 Analytical Model of Phase Noise 263
  8.3.2 Impact of Phase Synchronization Errors 265
  8.3.3 Impact of Time Synchronization Errors 270
8.4 Direct-Path Signal-Based Time and Phase Synchronization . 275
  8.4.1 Time Synchronization 275
  8.4.2 Phase Synchronization 277
  8.4.3 Prediction of Synchronization Performance 281
    8.4.3.1 Receiver Noise 281
    8.4.3.2 Amplifiers 281
    8.4.3.3 Analog-Digital-Converter (ADC) 282
  8.4.4 Other Possible Errors 283
8.5 GPS-Based Time and Phase Synchronization 285
  8.5.1 System Architecture 285
  8.5.2 Frequency Synthesis 286
  8.5.3 Measuring Synchronization Errors between Osc_PPS
    and GPS_PPS Signals 288
  8.5.4 GPS_PPS Prediction in the Presence of GPS Signal . 290
  8.5.5 Compensation for Residual Time Synchronization
    Errors 293
  8.5.6 Compensation for Residual Phase Synchronization
    Errors 298
  8.5.7 Synchronization Performance Analysis 301
8.6 Phase Synchronization Link 301
  8.6.1 Two-Way Synchronization Link 303
  8.6.2 Synchronization Performance 304
    8.6.2.1 Continuous Duplex Synchronization 304
    8.6.2.2 Pulsed Duplex Synchronization 305
    8.6.2.3 Pulsed Alternate Synchronization 305
  8.6.3 One-Way Synchronization Link 306
    8.6.3.1 Synchronization Scheme 306
    8.6.3.2 Synchronization Performance 310

8.7 Transponder-Based Phase Synchronization 312
8.8 Conclusion 316

9 Distributed Multi-Antenna SAR Antenna Synchronization 317

9.1 Impacts of Antenna Directing Errors 318
    9.1.1 Impacts of Range Antenna Directing Errors 318
    9.1.2 Impacts of Azimuth Antenna Directing Errors 320
    9.1.3 Impacts of Antenna Directing Errors on Distributed
        InSAR Imaging 321
9.2 Beam Scan-On-Scan Technique 323
    9.2.1 One Transmitting Beam and Multiple Receiving Beams 323
    9.2.2 One Transmitting Beam and Flood Receiving Beam . 324
    9.2.3 Flood Transmitting Beam and One Receiving Beam . 324
    9.2.4 Flood Transmitting Beam and Multiple Receiving
        Beams 324
    9.2.5 Flood Transmitting Beam and Flood Receiving Beam 325
9.3 Pulse Chasing Technique 325
9.4 Sliding Spotlight and Footprint Chasing 326
    9.4.1 Transmitter Sliding Spotlight and Receiver Footprint
        Chasing 327
    9.4.2 Transmitter Staring Spotlight and Receiver Footprint
        Chasing 327
    9.4.3 Azimuth Resolution 328
9.5 Multibeam Forming on Receiver 328
9.6 Determination of Baseline and Orientation 330
    9.6.1 Four-Antenna-Based Method 330
    9.6.2 Three-Antenna-Based Method 332
9.7 Conclusion 334

**10 Azimuth-Variant Multi-Antenna SAR Image Formulation
Processing   335**

10.1 Introduction 336
10.2 Imaging Performance Analysis 340
    10.2.1 Imaging Time and Imaging Coverage 340
    10.2.2 Range Resolution 341
    10.2.3 Azimuth Resolution 342
    10.2.4 Simulation Results 344
10.3 Azimuth-Variant Characteristics Analysis 347
    10.3.1 Doppler Characteristics 347
    10.3.2 Two-Dimensional Spectrum Characteristics 349
10.4 Motion Compensation 353
10.5 Azimuth-Variant Bistatic SAR Imaging Algorithm 356
10.6 Conclusion 362

11 Multi-Antenna SAR Three-Dimensional Imaging   **363**

  11.1 Introduction 363
  11.2 Downward-Looking SAR Three-Dimensional Imaging 365
      11.2.1 Signal and Data Model 365
      11.2.2 Imaging Resolution Analysis 367
      11.2.3 Three-Dimensional Range Migration Algorithm 368
  11.3 Side-Looking SAR Three-Dimensional Imaging 371
      11.3.1 InSAR for Terrain Elevation Mapping 371
      11.3.2 Side-Looking Linear Array SAR 372
  11.4 Forward-Looking SAR Three-Dimensional Imaging 374
  11.5 Frequency Diverse Array SAR Three-Dimensional Imaging . 376
      11.5.1 FDA System and Signal Model 377
      11.5.2 Application Potentials in Target Imaging 381
      11.5.3 Several Discussions 386
          11.5.3.1 Waveform Optimization 386
          11.5.3.2 Array Configuration 388
          11.5.3.3 Optimal Array Processing 388
  11.6 Conclusion 390

Bibliography   **391**

Index   **433**

List of Figures

1.1   General stripmap and spotlight SAR geometries. 2

1.2   Typical configurations of the multiple antennas in monostatic multi-antenna SARs. 3

1.3   Example semi-active and fully active multistatic SIMO SAR systems. 5

1.4   Two typical azimuth-invariant bistatic SAR configurations. . 6

1.5   Illustration of a MIMO system with M transmit antennas and N receive antennas. 7

1.6   Illustration of a multi-antenna SAR-based GMTI processing. 10

1.7   Two typical multi-antenna SAR antenna arrangements for three-dimensional imaging. 11

1.8   Chapter organization of the book. 14

2.1   A time-invariant linear system. 18

2.2   Illustration of a function represented in terms of pulses. . . 18

2.3   Graphical representation of the sampling theorem. 24

2.4   Comparative graphical representation of the aliasing effects. 26

2.5   Effect of different different sampling frequencies when sampling a cosine signal. 27

2.6   Typical up-chirp and down-chirp LFM pulse signals. 29

2.7   Typical LFM signal time domain and frequency domain. . . 30

2.8   The pulse compression result of a LFM signal. 33

2.9   An example pulse compression for LFM radar system. . . . 34

2.10 The three-dimensional and contour plots of an LFM ambiguity function. 39

2.11 Geometry of a side-looking strip-map SAR. 40

2.12 Geometry of a simplified side-looking strip-map SAR. 41

2.13 Geometry of remote sensing swath. 43

2.14 Illustration of azimuth ambiguities. 44

2.15 Illustration of range ambiguities. 46

2.16 Image formation from the convolution of signal with the PSF. 47

2.17 Coordinate systems for SAR imaging processor. 50

2.18 Geometry of a strip-map SAR range model. 51

2.19 Block diagram of the range-Doppler algorithm. 53

2.20 Equalized range curvatures resulting from the chirp scaling algorithm. 58

2.21   Flow diagram of the chirp scaling algorithm.   61
2.22   The amplitude of the radar returns.   62
2.23   Imaging result of the chirp scaling algorithm.   62

3.1   Equalized range curvatures resulting from the chirp scaling
       algorithm.   64
3.2   Geometry of the phase center approximation.   67
3.3   Geometry of two different DPCA techniques: (a) single-phase
       center DPCA, (b) multiple-phase center DPCA.   68
3.4   Geometry of the SPCM SAR imaging.   69
3.5   Multichannel signal reconstruction algorithm in case of three
       channels.   71
3.6   Azimuth Doppler spectra synthesis for multichannel
       subsampling in case of three channels.   72
3.7   The detailed azimuth Doppler spectra synthesis.   73
3.8   The AASR of an example azimuth multi-antenna SAR as a
       function of PRF.   74
3.9   Illustration of the MPCM SAR systems.   77
3.10   Equivalent phase centers in case of three receiving
       subapertures.   78
3.11   Reconstruction for multichannel subsampling in case of three
       channels.   82
3.12   The equivalent reconstruction filter in case of three channels.   83
3.13   Reconstructing the azimuth signal with FFT algorithm. . .   85
3.14   Illustration of the original sepctrum with ambiguous elements.   86
3.15   The QPE as a function of azimuth slow time.   88
3.16   The QPE as a function of acceleration speed.   89
3.17   The impact of platform speed derivation on MPCM SAR
       imaging results.   90
3.18   The imaging results with the multi-channel reconstruction
       algorithm.   91
3.19   Illustration of the azimuth scanning multibeam SAR in the
       case of two beams.   92
3.20   Illustration of the ground covered by the two beams at
       different azimuth time.   92
3.21   Equivalent geometry model of the azimuth scanning
       multibeam SAR.   92
3.22   Spectra of the echoes received at two adjacent azimuth
       positions.   93
3.23   The reconstructed spectra for the two subswaths.   93
3.24   Illustration of azimuth scanning multibeam SAR azimuth
       ambiguities.   94
3.25   Geometry of an ATI SAR system.   96
3.26   Three-antenna DPCA-based GMTI scheme.   98
3.27   Geometry of a three-antenna DPCA-based GMTI system. .   99

3.28   Illustration of the SFrFT and DPCA combined parameters
       estimation algorithm.   104
3.29   Processing results before clutter cancellation by DPCA
       operation.   105
3.30   Processing results after clutter cancellation by DPCA
       operation.   106
3.31   FrFT domain of the return of the single moving target.   106
3.32   The focused image of the single moving target.   107
3.33   Example correlation coefficients as a function of
       clutter-to-noise ratio.   108

4.1   Two-element elevation-plane array allows for range ambiguity
       suppression.   112
4.2   Suppressing range ambiguity with the null steering technique.   113
4.3   Illustration of the elevation-plane multi-antenna SAR. . . .   114
4.4   Geometry of the elevation-plane multiaperture SAR.   115
4.5   Illustration of imaging holes in the farthest and nearest swath. 119
4.6   Comparative pulse compression results before and after
       suppressing the nadir echoes.   122
4.7   Illustration of the blind range suppression via variable PRIs.   123
4.8   The logical flow diagram of the image formation algorithm.   128
4.9   The sensing matrix condition number for different platform
       altitude.   130
4.10   The sensing matrix condition number for different number of
       subswaths.   131
4.11   Comparative RASR results between the elevation-plane
       multi-antenna SAR and a general single-antenna SAR. . . .   132
4.12   The comparative imaging results.   133

5.1   Comparison of correlation performance between the correlated
       optimization and entropy optimization.   142
5.2   An example DFCM waveform set.   143
5.3   An example RSF time-frequency grid.   146
5.4   An example RSF-LFM time-frequency grid.   147
5.5   Basic diagram of OFDM modulation and demodulation
       system, where ω_i is the subcarrier frequency.   150
5.6   Structure of an OFDM code matrix.   151
5.7   Spectra of an example OFDM pulse with five single-chip
       subcarriers.   152
5.8   The OFDM pulse frequency hopping patterns for $N_s = 5$. .   155
5.9   Wideband ambiguity functions of example OFDM pulse. . .   157
5.10   Spectrum of an example OFDM pulse transmitted by one
       antenna.   158
5.11   Different combinations between two subchirp waveforms: (a)
       $k_{r1} = k_{r2}$, (b) $k_{r1} = k_{r2}$, (c) $k_{r1} = -k_{r2}$, (d) $k_{r1} = -k_{r2}$. . .   160

5.12 Correlation characteristics of the different chirp combinations:
      (a) and (b) have equal chirp rates, (c) and (d) have inverse
      chirp rates.   161
5.13 Correlation results for different combination of chirp
      waveforms.   162
5.14 Example chirp diverse waveforms with 8 subchirp signals. .   163
5.15 The spectra transmitted by the MIMO SAR subantennas: (a)-
      (h) denote the spectra transmitted by the subantenna (a)-(h),
      respectively, (i) is the equivalent wideband transmission. . .   164
5.16 An example of matched filtering results of a subantenna for a
      single point target: (a) is the received spectra, (b)-(i) denote
      the separate matched filtering result of the chirp waveform
      (a)-(h), respectively.   164
5.17 The comparative matched filtering output resolution:
      no transmit beamforming is employed in the above one and
      transmit beamforming is employed in the below one.   165
5.18 Impacts of the frequency overlap and frequency separation: (a)
      there is a frequency overlap of $0.5B_r$, (b) there is a separation
      of $0.5B_r$.   166
5.19 Two example pulses of the designed OFDM chirp diverse
      waveform.   167
5.20 Comparative auto-correlation and cross-correlation of the
      OFDM chirp diverse waveform.   168
5.21 Ambiguity function of the designed OFDM chirp diverse
      waveform.   169
5.22 The amplitude and spectra of the designed OFDM chirp
      diverse waveform.   170
5.23 Block diagram of the parallel DDS-based waveform generator.
      Here only two orthogonal LFM waveforms are shown.   172

6.1   Illustration of the range ambiguity suppression with OFDM
      chirp diverse waveform. Different OFDM chirp diverse pulses
      are transmitted in different PRI.   181
6.2   Illustration of the transmission and reception relations in the
      case of three subpulses, where the horizontal axis denotes
      three subswaths.   182
6.3   Geometry mode of the MIMO SAR antennas.   184
6.4   Illustration of an example MIMO SAR system.   185
6.5   Virtual phase center of the uniform linear array.   187
6.6   Virtual phase centers of the nonuniform linear array.   188
6.7   Two typical virtual arrays for case B.   189
6.8   An example multidimensional encoding for the transmitted
      radar pulse.   190
6.9   Illustration of the intrapulse beamsteering in the elevation
      dimension.   192

6.10 Illustration of the MIMO SAR with multiple elevation antennas. 197

6.11 Illustration of an example transmit subaperturing array. Only two subarrays are shown, it is for illustration only. 198

6.12 Two different MIMO SAR operation modes. Only three beams are shown, it is for illustration only. 200

6.13 NTNR MIMO SAR receiving beampattern distribution. . . 201

6.14 NTWR MIMO SAR receiving beampattern distribution. . . 202

6.15 The equivalent phase centers of the two MIMO SAR operation modes, where P_{ij} ($i \in [a,b,b]$, $j \in [a,b,b]$) denotes the equivalent phase center from the i-th transmitting beam to the j-th receiving channel. 204

6.16 The geometry of range ambiguity. 205

6.17 Illustration of the NTNR MIMO SAR range ambiguities. . . 206

6.18 Illustration of the NTWR MIMO SAR range ambiguities. . 207

6.19 The geometry of range ambiguity. 209

6.20 The range-Doppler-based processing steps of the MIMO SAR imaging algorithm in the case of three transmitting beams and three receiving channels, where RMC denotes the range migration correction. 211

6.21 The MIMO SAR wide-swath imaging simulation results. . . 213

6.22 Comparative RASR performance. 214

6.23 Side-looking space-time coding MIMO SAR. 217

6.24 The STBC MIMO SAR with Alamouti decoder. 219

6.25 Separately matched filtering and multibeam forming on receiver in elevation for the STBC MIMO SAR. 220

6.26 Equivalent phase difference $\Delta\Phi(\eta_1, \eta_2)$ between two successive PRI transmission as a function of the instantaneous squint angle. 224

7.1 Geometry of MIMO SAR with multiple antennas in the azimuth dimension. 230

7.2 Illustration of the adaptive matched filtering algorithm. . . . 234

7.3 Comparative matched filtering results. 235

7.4 Scheme of MIMO SAR with three azimuth-antennas for GMTI applications. 236

7.5 Azimuth three-antenna MIMO SAR ground moving target imaging with double-interferometry processing algorithm. . 239

7.6 Illustration of the two-antenna MIMO SAR transmission and reception. 240

7.7 Signal processing chart for the two-antenna MIMO SAR GMTI algorithm. 245

7.8 The focusing results by a stationary scene matched filter. . . 247

7.9 The focusing results after processed by the clutter suppression algorithm. 247

7.10 The peaks in the FrFT domain from which the targets' Doppler parameters can be estimated. 248

7.11 The final focused moving targets SAR image via the CLEAN technique. 248

8.1   Ideal sinusoidal oscillator and practical oscillators. 253

8.2   Example phase noise density. 257

8.3   Distributed bistatic SAR signal model. 261

8.4   Impacts of various oscillator frequency offsets: (a) constant offset, (b) linear offset, (c) sine offset and (d) random offset. 263

8.5   Statistical time-domain phase noise generator. 264

8.6   Phase noise PSD of one typical 10 MHz local oscillator. . . . 265

8.7   Phase noise in 50 seconds. 266

8.8   Phase noise in 50 seconds after subtraction of a linear phase ramp. 266

8.9   Reconstructed phase noise PSD. 267

8.10 Residual phase noise in 15 seconds. 267

8.11 Simulation results of oscillator phase instabilities (noises) with ten realisations: (a) predicted phase noises in 10 seconds in X-band (linear phase ramp corresponding to a frequency offset has been removed), (b) predicted high-frequency including cubic and more phase errors. 269

8.12 Impacts of phase noise on bistatic SAR systems: (a) impact of predicted phase noise in azimuth, (b) impact of integrated sidelobe ratio in X-band. 271

8.13 Illustration of clock signal with jitter. 272

8.14 Illustration of impact of clock jitter on radar data. 273

8.15 Predicted time synchronization error in a typical oscillator. . 274

8.16 Impact of time synchronization error on SAR imaging. . . . 275

8.17 Illustration of the direct-path signal-based time and phase synchronization. 276

8.18 Range compression results for one single target: (a) results of range compression with time synchronization errors, (b) results of range compression with range alignment. 278

8.19 Illustration of the direct-path signal-based time and phase synchronization. 281

8.20 Predicted phase synchronization accuracy via direct-path signal with five realizations. 284

8.21 Block diagram of GPS-disciplined frequency source. 287

8.22 Illustration of the GPS-disciplined scheme using the PPS epoch. 288

8.23 Measurement of synchronization errors between GPS_PPS and Osc_PPS signals. 289

8.24 Example observation of GPS clock signal stability over two days. 291

8.25 Iterative processing procedure for predicting the *GPS_PPS* signals. 294

8.26 Estimation convergence of the estimator. 295

8.27 Predicted GPS clock bias while compared with the real data observed on May 18, 2008. 296

8.28 Estimation variance (sigma) as a function of statistical samples. 296

8.29 Block diagram of the proposed range alignment. 298

8.30 Prediction of timing jitter in the GPS PPS. 302

8.31 Typical phase synchronization compensation errors (are the difference in the two curves in Figure 8.30). 302

8.32 Illustration of the one-way synchronization link. 307

8.33 Time diagram for the synchronization pulses with spatial synchronization information including motion compensation information. 308

8.34 Model of a general PLL with additive noise. 311

8.35 Comparative results of output phase noise between synchronization channel and SAR channel. 313

8.36 The transponder-based phase synchronization, where VCA and DDS denote the voltage controlled attenuator and direct digital synthesizer, respectively. 313

9.1 Distributed SAR beam coverage problem. 318

9.2 A general bistatic SAR geometry. 319

9.3 Impact of antenna directing synchronization errors in range. 320

9.4 Impact of antenna directing synchronization errors in azimuth. 322

9.5 Sliding spotlight geometry definition. 327

9.6 Illustration of the antenna directing synchronization. 329

9.7 Passive antenna synchronization using two antennas on receive only. 330

9.8 Two three-antenna radar platforms and the possible combinations of navigation signals. 332

10.1 Semi-active (left) and fully active (right) bistatic SAR. . . . 337

10.2 Two typical azimuth-invariant bistatic SAR configurations. . 337

10.3 General bistatic SAR configuration with two-channel receiver. 338

10.4 General geometry of azimuth-variant bistatic SAR systems. 340

10.5 Range resolution results of the example bistatic SAR configurations. 344

10.6 Azimuth resolution results of the example bistatic SAR configurations. 346

10.7 Azimuth-variant Doppler characteristics. 348

10.8 Targets with the same range delay at zero-Doppler but that will have different range histories. 348

10.9   The constraints of the Loffeld's bistatic SAR model.   352
10.10   Illustration of the near-space vehicle-borne bistatic SAR
        motion errors.   353
10.11   Illustration of the mapdrift autofocus algorithm.   357
10.12   Azimuth-variant bistatic SAR geometry and its equivalent
        model with stationary receiver.   358
10.13   Range errors caused by the equivalent bistatic SAR model. .   359
10.14   The NCS-based azimuth-variant bistatic SAR imaging
        algorithm.   361
10.15   Imaging simulation results with the NCS-based image
        formation algorithm.   362

11.1   Illustration of a downward-looking SAR system.   366
11.2   Geometry of downward-looking SAR cross-track resolution.   368
11.3   Geometry of InSAR for terrain elevation mapping.   371
11.4   Geometry of a side-looking linear array SAR for
        three-dimensional imaging.   373
11.5   Geometry of a forward-looking linear array SAR for
        three-dimensional imaging.   375
11.6   A linear FDA with an identical frequency increment.   377
11.7   Comparative beampattern between FDA and conventional
        phased array, where $\Delta f = 30kHz$, $M = 10, N = 12$, $f_0 = 10$
        GHz and $d = \lambda/2$ are assumed.   379
11.8   Comparative beampattern between FDA and conventional
        phased array, where $\Delta f = 30kHz$, $M = 10, N = 12$, $f_0 = 10$
        GHz and $d = \lambda$ are assumed.   380
11.9   A linear FDA with an identical frequency increment.   381
11.10   FDA multibeam forming with coding technique.   383
11.11   Illustration of the proposed FDA radar. Here, only 18 scanning
        beams are shown. It is for illustration purposes only.   384
11.12   Comparative single-point target imaging results.   387
11.13   Chirp waveform implementation of an FDA system.   389
11.14   Example two-dimensional planar virtual array and
        three-dimensional cylindrical virtual array.   390

List of Tables

2.1   Simulation parameters.   61

3.1   Performance parameters of an example azimuth
      multi-antenna SAR.   76
3.2   MPCM SAR system simulation parameters.   88
3.3   MPCM SAR imaging simulation results.   88
3.4   Simulation parameters   104

4.1   System parameters using for simulating the sensing matrix.   132
4.2   System parameters using for simulating the imaging
      algorithm.   134

5.1   An example DFCW set optimized by the genetic algorithm.   145

6.1   The comparisons between conventional single-antenna SAR
      (case a), NTNR MIMO SAR (case b) and NTWR MIMO
      SAR (case c) in case of three transmitting beams and three
      receiving channels.   203
6.2   The system simulation parameters for the NTNR and NTWR
      MIMO SAR systems.   212

7.1   Simulation parameters   246

8.1 $S_y(f)$ region shape and associated frequency dependence. .   256
8.2   Phase noise parameters of a typical local oscillator.   265
8.3   Phase noise parameters of a typical oscillator.   270
8.4   Possible phase errors caused by ADC.   283

10.1 Example bistatic SAR configuration parameters.   345

Author Bios

Wen-Qin Wang received a B.S. degree in electrical engineering from Shandong University, Shandong, China, in 2002, and M.E. and Ph.D. degrees in information and communication engineering from the University of Electronic Science and Technology of China (UESTC), Chengdu, China, in 2005 and 2010, respectively.

From March 2005 to March 2007, he was with the National Key Laboratory of Microwave Imaging Technology, Chinese Academy of Sciences, Beijing, China. Since September 2007, he has been with the School of Communication and Information Engineering, UESTC, where he is currently an Associate Professor. From June 2011 to May 2012, he was a visiting scholar at the Stevens Institute of Technology, Hoboken, New Jersey. Currently he is a visiting scholar at the City University of Hong Kong. His research interests include communication and radar signal processing and novel radar imaging techniques. He has authored over 100 papers.

Dr. Wang was the recipient of the Hong Kong Scholar in 2012, the New Century Excellent Talents in University in 2012, The Young Scholar of Distinction of UESTC in 2012, the Excellent Ph.D. Dissertation of Sichuan Province in 2011, the National Excellent Doctoral Dissertation Award nomination in 2012, the Project Investigator Innovation Award from the Wiser Foundation of Institute of Digital China in 2009, and the Excellent Paper Award of the 12th Chinese Annual Radar Technology Conference in 2012. He is the editorial board member of four international journals, and the Technical Program Committee Cochair of the International Conference on Computational Problem Solving, Chengdu, in 2011.

Preface

Synthetic aperture radar (SAR) is a well-known remote sensing technique, but conventional single-antenna SAR is inherently limited by the minimum antenna area constraint. This book deals with multi-antenna SAR in microwave remote sensing applications, such as high-resolution imaging, wide-swath remote sensing and ground moving target indication (GMTI). Particular attention is paid to the signal processing aspects of various multi-antenna SAR from a top-level system description.

Multi-antenna SAR allows for simultaneous transmission and reception by multiple antennas, compared to conventional SARs with only a single antenna. This provides a potential to gather additional information and to benefit from this information to overcome the restrictions of conventional single-antenna SARs. Multiple antennas can be placed either in a monostatic platform or distributed platforms. The simplest multi-antenna SAR is bistatic SAR which can be extended to multistatic SAR by having more than two transmitters or receivers. Many different terms for multi-antenna SAR are used in literature. These include multistatic SAR, multi-antenna SAR, netted SAR, multisite SAR and distributed SAR. In this book, we use the term multi-antenna SAR as a "catch-all" to embrace all possible forms.

This book is a research monograph. Its backbone is a series of innovative microwave remote sensing approaches that we have developed in recent years. These approaches address different specific problems of future microwave remote sensing, yet the topics discussed are all centered around multi-antenna SAR imaging. By stitching these approaches together in a book, we are able to tell a detailed story on various aspects of multi-antenna SAR imaging within a consistent framework.

The subject of this book falls into the field of electrical engineering, more specifically, into the area of radar signal processing. The main targeted readers are researchers as well as practitioners in industry who are interested in microwave remote sensing, SAR imaging and radar signal processing. We have made a significant effort to present the materials in a rigorous and self-contained way, so that interested readers can learn the multi-antenna SAR framework thoroughly. Besides newly developed approaches, a considerable portion of the book presents tutorial-like materials; for example, in Chapter 2 we review the necessary concepts in a gradual and motivated manner. Little background knowledge is required beyond a basic understanding of signal and system theorem. Therefore this book can also serve as an introduction to SAR imaging.

Following Chapters 1 and 2, which contain the introduction and background material, the other chapters are structured into five parts. Chapters 3 and 4 are devoted to single-input multiple-output (SIMO) multi-antenna SAR imaging. The heart is to achieve high-resolution wide-swath remote sensing for azimuth multi-antenna SAR and elevation multi-antenna SAR. Chapters 5 to 7 deal with multiple-input multiple-output (MIMO) SAR waveform diversity, high-resolution wide-swath remote sensing and GMTI. Chapters 8 and 9 study time, phase and spatial synchronization for distributed multi-antenna SAR systems. Chapter 10 discusses azimuth-variant distributed SAR imaging algorithms. Finally, Chapter 11 introduces the multi-antenna SAR for three-dimensional imaging, where the frequency diverse array (FDA) is also introduced. Additional material is available from the CRC website:

http://www.crcpress.com/product/isbn/9781466510517.

With great pleasure I acknowledge many people who have influenced my thinking and contributed to my knowledge, in particular: Profs. Qicong Peng and Jingye Cai (University of Electronic Science and Technology of China), Prof. Xiaowen Li (Institute of Remote Sensing Applications, Chinese Academy of Sciences), and Prof. Hongbin Li (Stevens Institute of Technology). As always the support of my wife Ke Yang is gratefully acknowledged for her constant and gentle encouragement.

I also wish to acknowledge the support provided by the National Natural Science Foundation of China under grant No. 41101317, Doctoral Program of Higher Education for New Teachers under grant No. 200806141101, Fundamental Research Funds for the Central Universities under grant No. ZYGX2010J001, Chinese Post-Doctor Research Funds under grant No. 20110490143, Open Research Funds of the State Laboratory of Remote Sensing Science under grant No. OFSLRSS201011, and many other research funds.

Finally, I would like to Leong Li-Ming, Schlags Laurie and Gasque Ashley at CRC Press for their professional assistance in the preparation and publication of this manuscript.

Abbreviations

AASR azimuth ambiguity signal ratio

ADC analog-to-digital converter

AM amplitude modulation

ASR ambiguity-to-signal ratio

ATI along-track interferometry

CE-OFDM constant-envelope OFDM

CPI coherent processing interval

CS chirp scaling

DAC digital-to-analog converter

DDS direct digital synthesizer

DEM digital-earth model

DFCW discrete frequency-coding waveform

DFT discrete Fourier transform

DPCA displaced phase center antenna

FDA frequency diverse array

FFT fast Fourier transform

FM frequency modulation

FrFT fractional Fourier transform

GMTI ground moving target indication

GNSS global navigation satellite systems

GPS global positioning systems

HRWS high-resolution wide-swath

IF intermediate frequency

IFFT inverse fast Fourier transform

IMU inertial measurement units

INS inertial navigation system

ISAR inverse synthetic aperture radar

ISLR integrated sidelobe ratio

InSAR   interferometry synthetic aperture radar
LFM   linear frequency modulation
LO   local oscillator
LOS   line-of-sight
MIMO   multiple-input multiple-output
MPCM   multiple-phase center multibeam
MSE   Mean square error
MTI   moving target indication
NCS   nonlinear chirp scaling
NESZ   noise equivalent sigma zero
NTNR   narrow-beam transmission narrow-beam reception
NTWR   narrow-beam transmission wide-beam reception
OFDM   orthogonal frequency division multiplexing
PAPR   peak-to-average power ratio
PLL   phase locked loop
PPS   pulse per-second
PRF   pulse repetition frequency
PRI   pulse repetition interval
PSD   power spectra density
PSF   point spread function
PSP   principle of stationary phase
PSR   point scatter response
QPSK   quadratically phase shift keyed
RCMC   range cell mitigation correction
RCS   radar cross section
RASR   range ambiguity to signal ratio
RD   range-Doppler
RF   radio frequency
RMA   range mitigation algorithm
RSF   random stepped-frequency
SAR   synthetic aperture radar
SCR   signal-to-clutter ratio
SFrFT   simplified fractional Fourier transform
SIMO   single-input multiple-output

SNR	signal-to-noise ratio
SPCM	single-phase center multibeam
SRC	second range compression
STAP	space-time adaptive processing
STBC	space-time block coding
UAV	unmanned aerial vehicle
USO	ultra-stable oscialltor
VCA	voltage controlled attenuator
VCO	voltage controlled oscillator
WRI	waveform repetition interval
WVD	Wigner-Ville distribution

1

Introduction

Synthetic aperture radar (SAR), originating fin 1950s, is a form of radar imaging technology which utilizes the relative motion between an antenna and its target region to simulate an extremely large antenna or aperture electronically, and which generates finer spatial resolution than that is possible with conventional beam-scanning means [211]. It is usually implemented by mounting radar on a moving platform such as airplane or satellite, from which a target scene is repeatedly illuminated with pulses of radio waves. The radar returns received successively at different antenna positions are coherently detected and stored and then post-processed together to resolve elements in an image of the target region.

One of the main advantages for SAR imaging is the ability to operate in almost any weather, day or night, thereby overcoming many of the limitations of other passive imaging technologies such as optical and infrared. SAR can also be implemented as "inverse SAR" by observing a moving target over a substantial time with a stationary antenna. As single-antenna SAR is a well-proven remote sensing technique [51, 386], this book concentrates only on multi-antenna SAR systems and signal processing. The SAR basics can be found in several excellent books [73, 76, 353].

This chapter is organized as follows. Several basic concepts related to multi-antenna SAR imaging are introduced in Section 1.1, where several multi-antenna SARs and their system characteristics are discussed. The benefits of multi-antenna SAR, as compared to conventional single-antenna SAR, is investigated in Section 1.2. This is followed by the advances in monostatic and distributed multi-antenna SAR imaging in Section 1.3. Finally, the organization of this book is outlined in Section 1.4.

1.1   What is Multi-Antenna SAR

Conventional single-antenna stripmap and spotlight SAR systems (see Figure 1.1) are inherently limited in meeting the rising demands of future remote sensing missions. Innovative SAR imaging techniques are thus needed to be developed. The most promising of these techniques usually employ multiple antennas [161, 186, 187]. Multi-antenna SAR allows for simultaneous trans-

1

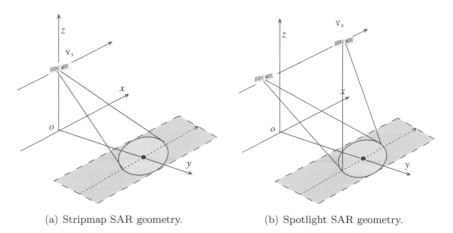

(a) Stripmap SAR geometry.   (b) Spotlight SAR geometry.

FIGURE 1.1: General stripmap and spotlight SAR geometries.

mission and reception by multiple channels, compared to conventional SAR systems with only a single channel. This provides a potential to gather additional information and to benefit from this information to overcome the restrictions of conventional SAR systems.

Multiple antennas can be placed either in a monostatic platform which results in a monostatic multi-antenna SAR system or in distributed platforms which results in a distributed multi-antenna SAR system. In the following, we introduce them separately.

1.1.1 Multichannel SAR

Multiple antennas can be either arranged in flight direction ("along-track") [77], perpendicular to it ("cross-track") [164], or in both dimensions [45]. They are illustrated, respectively, in Figure 1.2.

1.1.1.1 Multiple Channels in Elevation

The concept of multiple channels in elevation dimension proposed by Griffiths and Mancini [164] consists of an array antenna split in elevation dimension. The overall antenna dimension is smaller than that implied by the minimum antenna area constraint yielding a broad beam in elevation dimension that covers a wide swath (the ground width covered by the SAR beam) but at the same time gives rise to range ambiguous echoes. The range ambiguities can be suppressed by adaptively steering nulls in the antenna pattern in elevation to the directions of the ambiguous returns. In this way, a widened imaging swath can be obtained; however, in a monostatic SAR the swath may become no longer continuous because blind ranges will be introduced when the receiver is switched off during transmission. Such restrictions could be overcome with

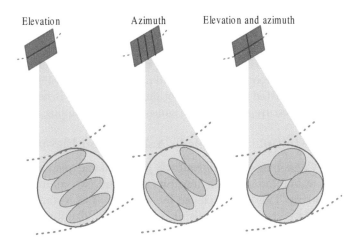

FIGURE 1.2: Typical configurations of the multiple antennas in monostatic multi-antenna SARs.

bistatic geometry which allows for simultaneous transmission and reception [217, 218] and hence, short compact antennas could be used for high resolution SAR systems with continuous wide area coverage.

In 2001, Suess et al. [361] proposed an innovative SAR which is built on an array antenna consisting of multiple elements in elevation. The proposed processing algorithm combines the echoes from the multichannel SAR in a way that forms a narrow beam in elevation which "scans" the ground in real-time in order to follow the returns of the transmitted signal. This enables a suppression of range ambiguous returns and ensures a high antenna gain. However, this technique requires knowledge of the observed terrain topography, otherwise a mispointing of the narrow elevation beam may occur, resulting in severe system gain loss [216].

1.1.1.2 Multiple Channels in Azimuth

In 1992, Currie and Brown [77] proposed the displaced phase center antenna (DPCA) in azimuth technique, which is based on dividing the receiving antenna in the along-track direction into multiple subapertures, each receiving, down-converting and digitizing the radar echoes. Hence, for every transmitted pulse the system receives multiple pulses in the along-track direction. This means that additional samples can be gathered, thus increasing the effective sampling rate. Consequently, either the azimuth resolution can be improved while the swath width remains constant, or the swath can be widened without increasing the azimuth ambiguities or impairing the azimuth resolution. That is to say, the system benefits from the whole antenna length regarding azimuth ambiguity suppression while azimuth resolution is determined by the dimension of a single subaperture, thus decoupling the restrictions on high-resolution and wide-swath remote sensing. This technique recovers the azimuth signal by simply interleaving the samples of the different receiving channels without any further processing, but it imposes a stringent timing requirement on the system regarding the relation between platform velocity, pulse repetition frequency (PRF) and antenna length [145]. These parameters have to be adjusted in order to obtain a signal that is equivalently sampled as a single-aperture signal at the same effective sampling rate.

1.1.1.3 Multiple Channels in Azimuth and Elevation

In 1999, Callaghan and Longstaff [45] proposed the quad array concept which is based on an antenna split into two rows and two columns yielding a four-element array. This approach can be understood as a combination of the two approaches described previously. It combines the advantages of gathering additional samples in azimuth to suppress azimuth ambiguities and simultaneously enabling an enlarged swath for a fixed PRF. This system may result in blind ranges in the imaged swath. Additionally, it requires also a stringent timing constraint to ensure a uniform spatial distribution of the sampled signals in azimuth dimension.

In addition, Classen and Eckerman [65] proposed an alternative concept which is based on steering multiple beams to different azimuth directions and assigning each of the corresponding footprints to a different slant range. In other words, the footprint of each beam steered to a different squint angle corresponds to a different subswath of the overall imaged region. As the antenna elevation dimension is larger than the imaged swath, range ambiguities can be well suppressed, thus allowing for a PRF high enough to suppress azimuth ambiguities. But, there will be coarsened resolution and impaired performance arising from the needed high squint angles [187].

1.1.2 Multi-Antenna SAR

The earliest radars were bistatic, which employed continuous-wave waveforms and separate transmitting and receiving antennas [446]. However, the bistatic form of radar was largely abandoned as a design approach after the invention of the duplexer. This allows the colocation of transmitting and receiving antennas thus simplifying the radar system design complexity. In recent years, the use of opportunistic illuminators, such as the global positioning systems (GPS) [360, 420] and broadcasting satellites [66], is considered as a main trigger for the significant growth of bistatic radars [165]. The books authored by Willis [446] and Cherniakov [60, 61] provide an excellent summary of bistatic radars. Multistatic radar extends the bistatic concept by having more than two transmitters or receivers.

Multistatic SAR extends conventional multistatic radars by employing aperture synthesis. It should be noted that many different terms for multistatic SAR are used in literature. These include multistatic SAR, multi-

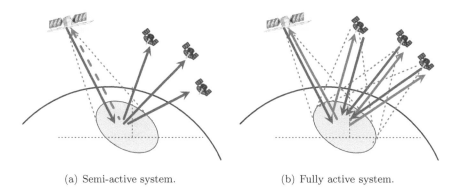

(a) Semi-active system. (b) Fully active system.

FIGURE 1.3: Example semi-active and fully active multistatic SIMO SAR systems.

antenna SAR, netted SAR, multisite SAR and distributed SAR. In this book, we use the term multi-antenna SAR as a "catch-all" to embrace all possible forms. Note that the term multistatic SAR has in the past been more typically used to refer to a distributed SAR network. We classify multistatic multi-antenna SAR into two types: single-input multiple-output (SIMO) SAR and multiple-input multiple-output (MIMO) SAR.

1.1.2.1 SIMO Multi-Antenna SAR

In multistatic SIMO SAR systems, only a single waveform is transmitted by one or more antennas. Multistatic SIMO SAR systems can be realized with different platform pairs. For example, future SIMO SARs could use "receive only" receivers mounted on airplanes [397], ground [86], unmanned aerial vehicles, near-space vehicles [402], satellites, or both the transmitters and receivers are mounted on airplanes [301] or satellites [286]. Multistatic SIMO SAR systems can be further divided into semi-active and fully active configurations, as shown in Figure 1.3. Semi-active configuration combines passive receiver with an active radar illuminator; fully active configuration uses conventional radars flying in close formation to acquire data during a single pass. Both of the radars have a fully equipped radar payload with transmit and receive capabilities.

The simplest and most representative multistatic SIMO SAR is the bistatic SAR. In this book, we further classify bistatic SAR into azimuth-invariant and azimuth-variant configurations. Two typical azimuth-invariant configurations are the pursuit monostatic mode (see Figure 1.4 (a)), where the transmitter and receiver are flying along the same trajectory at an equal velocity, and the translational invariant mode (see Figure 1.4 (b)), where the transmitter and receiver are flying along parallel trajectories with the same velocity. An azimuth-invariant configuration can be seen as an equivalent monostatic con-

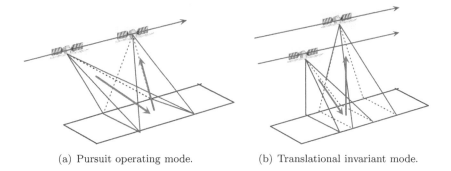

(a) Pursuit operating mode.   (b) Translational invariant mode.

FIGURE 1.4: Two typical azimuth-invariant bistatic SAR configurations.

figuration. In contrast, an azimuth-variant configuration cannot be seen as an equivalent monostatic configuration because the geometry between transmitter and receiver will change with azimuth time. A space-surface bistatic SAR with spaceborne transmitter and airborne receiver is an example of azimuth-variant configuration [52].

1.1.2.2   MIMO Multi-Antenna SAR

In multistatic MIMO SAR systems, two or more waveforms are transmitted by one or more antennas. MIMO using multiple antennas at both sides of the wireless link (see Figure 1.5) was a technique used in wireless communications to increase data throughout and link performance without additional bandwidth or transmit power [152]. It is known that, wireless communication concerns mainly the transmitted information across a wireless link and it is not necessary to estimate the channel parameters. In contrast, for radar, while it is possible to perform communication parasitically, the estimation of channel information is of primary interest. In MIMO communication systems, if the transmitter has the channel knowledge, the informed transmitter can then adapt its strategy to improve performance. Hence, similar to MIMO communication, the idea of MIMO radar has also drawn considerable attention in recent years [244]. The essence of the MIMO radar concept is to employ multiple antennas for transmitting several orthogonal waveforms and multiple antennas for receiving the echoes reflected by the target. The enabling concept for MIMO radar, e.g., the transmission of multiple orthogonal waveforms from different antennas, is usually referred to as *waveform diversity* [316, 444].

The emerging MIMO radar literature can be classified into two main types. The first type employs arrays of closely spaced transmitting/receiving antennas to cohere a beam towards a certain direction in space [245]. In this case, the target is usually assumed to be in the far-field. The waveform diversity boils down to increasing the virtual aperture of the receive array due to the fact that multiple independent waveforms are received by the same receiving

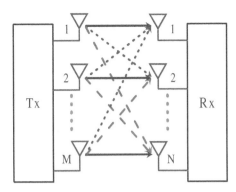

FIGURE 1.5: Illustration of a MIMO system with M transmit antennas and N receive antennas.

array. The second type uses widely separated transmitting/receiving antennas to capture the spatial diversity of the target's radar cross section (RCS) [170]. This type assumes an extended target model and, therefore, takes advantages of the properties of the associated spatially distributed signal model. In this case, the waveform diversity is similar to the multipath diversity concept used in wireless communications over fading channels. According to this concept, signals transmitted over multiple fading links/channels can be decoded reliably at the receiver due to the fact that it is unlikely that all links/channels undergo unfavorable fading conditions simultaneously [381].

Differently, in this book we consider mainly MIMO SAR for remote sensing. Note that the difference between general MIMO radar and MIMO SAR is that aperture synthesis is employed in the latter. Conventional SAR provides target recognition, resolution and imaging capability by the virtue of coherent processing gain. Consequently, SAR imaging performance will be degraded if the coherent assumptions are violated or if channel fading occurs within the synthetic aperture illumination. Evidence for such is increased sidelobes and tails behavior in the point spread function. One possible way to mitigate this effect is to maximize the received energy from the target, i.e., to maximize the coherent processing gain. Even so, SAR imaging theory and techniques at current state cannot effectively image the targets with fading, varying and unstable signature. Nevertheless, these targets represent an important class of SAR remote sensing applications. MIMO system and associated exploitation methodology provides a potential to address the shortcomings of current SARs for some specific applications [404].

Given that MIMO SAR is in its infancy, there is no clear definition of what it is. It is generally assumed that independent signals are transmitted through different antennas placed in moving platforms, and that these signals, after propagating through the environment, are received by multiple antennas [404]. In the MIMO SAR receiver, a matched filter-bank is used to extract the orthogonal waveform components. Generally speaking, when compared to conventional SARs, MIMO SAR has two advantages. The first advantage lies in spatial diversity gain and flexible transmit-receive beampattern design. The second advantage is resolution improvement [28]. When MIMO SAR transmits M orthogonal waveforms, the virtual array of the radar system is a filled array with an aperture length up to M times that of the receiving array. This configuration can be exploited to achieve an M-fold improvement in the spatial imaging resolution over conventional SAR systems.

1.2   Multi-Antenna SAR Potentials and Challenges

1.2.1   Benefits of Multi-Antenna SAR

Compared to a single-antenna SAR, multi-antenna SAR has many benefits such as improved system gain, increased degrees-of-freedom and increased system flexibility.

1.2.1.1   Improved System Gain

According to the radar range equation, if there are N receiving antennas, the total antenna area will significantly increase, and consequently, the system gain can be improved by a factor of N. If there are also M transmitting antennas, the system gain will increase by a factor of $M \times N$. This improvement can be used to improve the system signal-to-noise ratio (SNR). It can also be used to improve the azimuth resolution. If the antenna area for a single antenna and a linear array are equal, smaller elements can be employed in the array. Since the azimuth resolution is determined by the transmitting antenna size, a smaller antenna area offers finer azimuth resolution.

1.2.1.2   Increased Degrees-of-Freedom

Another benefit of multi-antenna SAR is the increased degrees-of-freedom. It allows ground moving target indication (GMTI) for target tracking, compensating for moving targets in a SAR image and the ability to reject interferences while forming a SAR image.

Moreover, the increased degrees-of-freedom can change the equivalent PRF, which is a key parameter in keeping range and azimuth ambiguities from affecting a SAR image. Range ambiguities occur when simultaneous returns from two pulses in flight occur. To avoid aliasing problems in the azimuth domain, the PRF must exceed the frequency which ensures that a stationary target's two-way distance never changes by more than half a wavelength between samples. If one transmitter illuminates a wide swath and N smaller antennas aligned in the along track direction record simultaneously the scattered signal from the illuminated footprint, the effective PRF is N

times larger than the operating PRF. Since each antenna has a broader receiving beam compared to the whole antenna, a longer synthetic aperture is possible. The azimuth resolution of the array can then be decreased to half the subaperture length and since combinations of the subaperture signals can be used to form narrow beams, the operating PRF can be reduced without introducing azimuth ambiguities.

1.2.2   Application Potentials

High operation flexibility and reconfigurability can be obtained for multi-antenna SAR by utilizing the equivalent phase centers, thus enabling several satisfactory remote sensing application potentials.

1.2.2.1   High-Resolution Wide-Swath Remote Sensing

Efficient remote sensing techniques should provide high-resolution imagery over a wide area of surveillance, but it is a contradiction limited by the minimum antenna area constraint. This arises because the illuminated area must be restricted, so as to avoid ambiguous returns. In this respect, a high operating PRF is desired to suppress azimuth ambiguities. But the PRF is limited by range ambiguity requirement. Therefore, conventional SAR allows only for a concession between azimuth resolution and swath width.

One representative solution is the DPCA technique [77]. The basic idea is to divide the along-track antenna into multiple apertures. However, the relation between platform speed and along-track subantenna offset have to be adjusted to obtain a uniform spatial sampling signal. Another representative solution is the ScanSAR [288] which achieves a wider swath at an expense of degraded azimuth resolution and fewer image looks. These approaches can be extended to multi-antenna SAR. The basic idea is to form multiple transmitting and receiving apertures which are steered toward different subswaths [410]. These subapertures can be disjointed or overlapped in space. Multiple pairs of virtual transmitting and receiving beams can then be formed simultaneously in the direction of a desired subswath. Equivalently, a large swath can be synthesized.

1.2.2.2   Ground Moving Targets Indication

GMTI is of great important for surveillance and reconnaissance, which is a twofold problem: one is the detection of moving targets within severe ground clutter, and the other is the estimation of their parameters such as velocity and location. Moving targets with a slant speed will generate a different phase shift which could be detected by the along-track interferometry (ATI) SAR. However, if the clutter contribution is not negligible, the estimation of the target radial velocity may lead to erroneous results. ATI SAR is thus a clutter-limited GMTI detector.

Another representative technique is the DPCA GMTI SAR. However, sta-

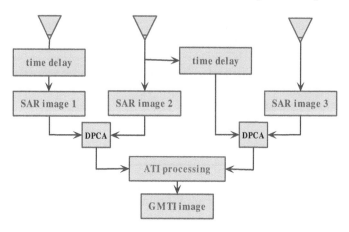

FIGURE 1.6: Illustration of a multi-antenna SAR-based GMTI processing.

tionary targets in the imaging scene may contribute as phase noise and consequently complicate the process of extracting moving targets. That is, DPCA GMTI SAR is a noise-limited GMTI detector [63]. Moreover, in a two-antenna DPCA GMTI SAR there will be no additional degrees-of-freedom that can be used to estimate the moving target's position information.

Multi-antenna SAR-based GMTI is interesting because the virtual array enables a large effective aperture to be obtained. The larger virtual array increases the distance between the clutter and the target within a given Doppler bin, making clutter suppression easier. Moreover, the minimum detectable velocity can be improved by the larger virtual aperture size. Figure 1.6 shows the geometry of a multi-antenna SAR for GMTI applications. As the multiple antennas are displaced in azimuth direction, it is effective to define "two-way" phase center as the midpoint between the transmitting and receiving phase centers. This provides a potential to clutter suppression, like the DPCA technique. The clutter cancellation is performed by subtracting the samples of the returns received by two-way phase center located at the same spatial position. Next, GMTI can be obtained by interferometry processing.

1.2.2.3   Three-Dimensional Imaging

The last but not the least of multi-antenna SAR potentials is three-dimensional imaging. Conventional SAR obtains its two-dimensional images by projecting distributed three-dimension targets onto two-dimensional planes. Consequently, it usually suffers from geometric distortions such as foreshortening and layover. The strong shadowing effects caused by buildings, hills, and valleys may result in information loss of the explored area. In order to overcome these disadvantages, three-dimensional SAR imaging has become an urgent need.

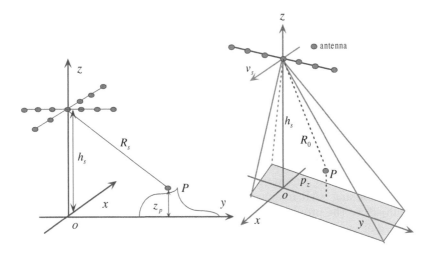

FIGURE 1.7: Two typical multi-antenna SAR antenna arrangements for three-dimensional imaging.

Figure 1.7 shows two typical multi-antenna arrangements for three-dimension imaging. These arrangements can also potentially be used for simultaneous three-dimensional SAR imaging and GMTI applications. Moreover, multi-dimensional spatial filter can be used to tackle possible range and/or Doppler ambiguities; at the same time, bandwidth synthesis provided by the degrees-of-freedom in multi-antenna SAR systems can be employed to overcome the contradiction between high resolution, wide swath and high SNR performance.

1.2.3 Technical Challenges

Multi-antenna SAR shows much promise for remote sensing applications, but there are several technical challenges to be overcome.

1.2.3.1 Waveform Diversity Design

Unlike conventional SAR systems, in MIMO multi-antenna SAR each antenna should transmit a unique waveform, orthogonal to the waveforms transmitted by other antennas. To obtain a high range resolution, the autocorrelation of the transmitted waveforms should be close to an impulse function. Moreover, since SAR is usually placed inside airplane or satellite, a high average transmit power is required; hence, the waveforms should have a large time-bandwidth product. Certainly the waveforms should also have good ambiguity characteristics such as high range resolution, Doppler tolerance, low adjacent-band interference and good matched filtering sidelobe performance. A typical orthogonal waveform is the Barker code; however, Barker code-based waveforms often have only a single carrier frequency [366]. Consequently they are not suitable for MIMO multi-antenna SAR high-resolution imaging due to its limited bandwidth.

1.2.3.2 Spatial, Time and Phase Synchronization

Distributed multi-antenna SAR, operates with separate transmitter and receiver that are mounted on different platforms, will play a great role in future radar applications [218]. Distributed multi-antenna SAR configuration can bring many benefits in comparison with monostatic systems, such as exploitation of additional information contained in the bistatic reflectivity of targets [40, 105], improved flexibility [260], reduced vulnerability [416], forward looking SAR imaging [54]. However, distributed multi-antenna SAR is subject to the problems and special requirements that are either not encountered or encountered in a less serious form for monostatic multi-antenna SAR. The biggest technological challenge lies in synchronization of the independent transmitter and receiver: time synchronization, the receiver must precisely know when the transmitter fires (in the order of nanoseconds); spatial synchronization, the receiving and transmitting antennas must simultaneously illuminate the same spot on the ground; phase synchronization, the receiver and the transmitter must be coherent over extremely long periods of time. The most difficult synchronization problem is the phase synchronization. To focus distributed multi-antenna SAR data, phase information of the transmitted pulse has to be preserved. In a monostatic SAR, the colocated transmitter and receiver use the same stable local oscillator, the phase can only be decorrelated over a very short period of time (about 1×10^{-3} sec.). In contrast, for a distributed multi-antenna SAR system, the transmitter and receiver fly on different platforms and use independent master oscillators, resulting no phase noise cancellation. This superimposed phase noise corrupts the received signal over the whole synthetic aperture time. Moreover, any phase noise (instability) in the master oscillator is magnified by frequency multiplication. As a consequence, the low phase noise requirements imposed on the oscillators of distributed multi-antenna SAR are much higher than that for monostatic SAR. In the case of indirect phase synchronization using identical local oscillators in the transmitter and receiver, phase stability is required over the coherent integration time. Even the toleration of low frequency or quadratic phase synchronization errors can be relaxed to 90°, the requirement of phase stability is only achievable for ultra-high-quality oscillators [433, 434]. Moreover, aggravating circumstances accompany airborne platforms because of different platform motions, thus the performance of phase stability will be further degraded.

1.2.3.3 High-Precision Imaging Algorithm

The last but not the least technical challenge is high-precision imaging algorithm. For multi-antenna SARs, especially azimuth-variant operating modes, it is a mandatory to know both the transmitter-to-receiver distance and target-to-receiver distance to properly focus its raw data, but they depend on the target height. In this case, conventional image formation algorithms are not suitable to accurately focus the collected data. Multi-antenna SAR data may be focused in time-domain and frequency-domain or hybrid-domain. Since an high computational cost is required for time-domain image formation algorithm, frequency-domain or hybrid-domain image formation algorithms are thus of great interest. However, the processing efficiency depends on the two-dimensional spectrum model of the radar returns. Moreover, since multi-antenna SARs are in their infancy, there lack efficient image formation algorithms and much further research work are required.

1.3   Organization of the Book

In this book, we discuss mainly the signal processing aspects of various multi-antenna SARs from a top-level system description. The focus is placed on elevation multi-antenna SARs, azimuth multi-antenna SARs, and two-dimensional multi-antenna SARs. The corresponding system scheme, signal models, time/phase/spatial synchronization methods, and high-precision imaging algorithms are detailed. Their potential applications in high-resolution wide-swath remote sensing, ground moving target imaging, three-dimensional imaging, surveillance and reconnaissance are investigated.

Figure 1.8 shows the chapter organization of the book. The remaining chapters are organized as follows.

Chapter 2 gives an introduction to the multi-antenna SAR basics, such as convolution and correlation, matched filtering and pulse compression, radar ambiguity function, synthetic aperture principle, SAR imaging performance parameters, SAR echo modelling, basic SAR image formation algorithms, etc.

Chapters 3 and 4 are devoted to SIMO multi-antenna SAR in high-resolution and wide-swath remote sensing. Chapter 3 introduces the azimuth multi-antenna SAR. The constraints on high-resolution and wide-swath are derived, followed by the DPCA technique and conceptual design systems. The biggest problem of the azimuth multi-antenna SAR is the possible non-uniformly sampled azimuth Doppler signal. Several potential solutions to this problem are investigated. Furthermore, an azimuth multi-antenna SAR-based GMTI approach is proposed. Chapter 4 introduces the elevation multi-antenna SAR in high-resolution and wide-swath remote sensing. The ill-condition problem of the sensing matrix is analyzed, along with extensive simulation examples. The blind range and nadir interferences are also discussed. A chirp scaling (CS)-based image formation algorithm is presented to focus the elevation multi-antenna SAR data.

Chapters 5 to 7 deal with MIMO multi-antenna SAR in high-resolution

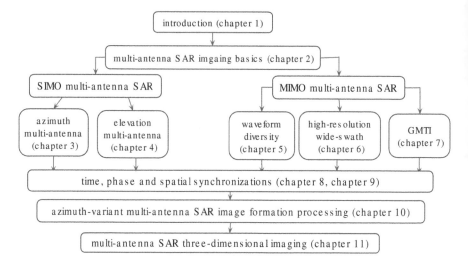

FIGURE 1.8: Chapter organization of the book.

wide-swath remote sensing and GMTI applications. Chapter 5 discusses several practical waveform diversity design methods for MIMO SARs. A practical orthogonal frequency diversion multiplexing (OFDM) chirp waveform with a large time-bandwidth product is proposed. Chapter 6 presents the MIMO SAR for high-resolution wide-swath remote sensing. A novel space-time coding MIMO-OFDM SAR is proposed in this chapter for high-resolution remote sensing. Chapter 7 discusses the role of MIMO SAR in GMTI applications. An efficient two-antenna MIMO SAR is proposed for imaging ground moving targets.

Chapters 8 and 9 deal with the synchronization compensation processing for distributed multi-antenna SAR systems. Chapter 8 discusses several time and phase synchronization approaches, such as the direct-path signal-based approach, the global positioning systems (GPS)-based approach, and the communication link-based approach. Chapter 9 discusses the spatial synchronization, i.e., antenna directing synchronization. The antenna baseline measurement in distributed multi-antenna SAR systems is also introduced.

Chapter 10 discusses the image formation algorithms for focusing multi-antenna SAR data. Taking a specific azimuth-variant bistatic SAR as an example, the corresponding azimuth-variant Doppler characteristics, two-dimensional point spread function, and image formation algorithms are provided.

Chapter 11 introduces the multi-antenna SAR for three-dimensional imaging. The downward-looking, forward-looking and side-looking multi-antenna SAR for three-dimensional imaging are discussed. Furthermore, we introduce the frequency diverse array (FDA), which provides promising potentials for future three-dimensional imaging (if a moving platform is employed, otherwise, two-dimensional imaging is possible for stationary platform).

2

Background Material

The presentation of multi-antenna SAR systems and signal processing relies on some basic concepts, particularly from signal analysis and processing. For the benefit of the reader, we shall motivate these concepts. In this way, readers will be introduced to the necessary concepts in a gradual and motivated manner. In this chapter, we collect several basic concepts of general interest. These concepts complement well the material in future chapters and will be called upon at different stages of the discussion.

This chapter is organized as follows. Section 2.1 introduces the convolution theorem and correlation function, Section 2.2 introduces the sampling and interpolation technique. Next, the linearly frequency modulated (LFM) signal and pulse compression are introduced in Section 2.3, followed by the radar ambiguity function in Section 2.4. Section 2.5 introduces the principle of synthetic aperture, along with the main imaging performance parameters. Finally, Section 2.6 derives the general SAR point spread function, followed by the basic image formation algorithms in Section 2.7.

2.1 Convolution and Correlation

2.1.1 Convolution Integral

A linear system can be represented as a box with input $\mathbf{x}$, output $\mathbf{y}$ and a system operator $\mathbf{H}$ that defines the relationship between $\mathbf{x}$ and $\mathbf{y}$. Note that both $\mathbf{x}$ and $\mathbf{y}$ can be a set of components. A system is linear if and only if

$$\mathbf{H}(a\mathbf{x}_1 + b\mathbf{x}_2) = a\mathbf{H}\mathbf{x}_1 + b\mathbf{H}\mathbf{x}_2 \tag{2.1}$$

where a and b are constants, $\mathbf{x}_1$ and $\mathbf{x}_2$ are the system's input signals. In addition, a linear system having the following input-output relation (see Figure 2.1)

$$\mathbf{H}\mathbf{x}(t) = \mathbf{y}(t) \tag{2.2}$$

is time-invariant if and only if

$$\mathbf{H}\mathbf{x}(t - \tau) = \mathbf{y}(t - \tau) \tag{2.3}$$

for any $\mathbf{x}(t)$ and any τ.

FIGURE 2.1: A time-invariant linear system.

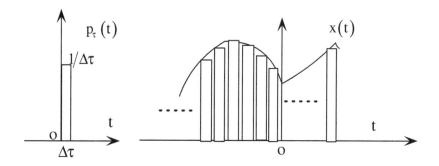

FIGURE 2.2: Illustration of a function represented in terms of pulses.

It is known that a function $x(t)$ can be represented as [308] (see Figure 2.2)

$$x(t) = \lim_{\Delta\tau} \sum_{n=0}^{\infty} x(n \cdot \Delta\tau)p_\tau(t - n \cdot \Delta\tau)$$

$$= \lim_{\Delta\tau} \sum_{n=0}^{\infty} [x(n \cdot \Delta\tau)/\Delta\tau]p_\tau(t - n \cdot \Delta\tau)\Delta\tau \tag{2.4}$$

where $p_\tau(t)$ is a pulse with amplitude $1/\Delta\tau$ and duration $\Delta\tau$. As $\Delta\tau \to 0$, the $p_\tau(t - n \cdot \Delta\tau)$ can be approximated as

$$[x(n \cdot \Delta\tau)/\Delta\tau]p_\tau(t - n \cdot \Delta\tau) \to x(n \cdot \Delta\tau)\delta(t - n \cdot \Delta\tau) \tag{2.5}$$

We then have

$$x(t) = \lim_{\Delta\tau} \sum_{n=0}^{\infty} x(n \cdot \Delta\tau)\delta(t - n \cdot \Delta\tau)\Delta\tau \tag{2.6}$$

According to the properties of a linear system, we have the following relations

$$\delta(t) \Rightarrow h(t) \tag{2.7}$$

$$\delta(t - n \cdot \Delta\tau) \Rightarrow h(t - n \cdot \Delta\tau) \tag{2.8}$$

$$[x(n \cdot \Delta\tau)\Delta\tau]\delta(t - n \cdot \Delta\tau) \Rightarrow [x(n \cdot \Delta\tau)\Delta\tau]h(t - n \cdot \Delta\tau) \tag{2.9}$$

$$\lim_{\Delta\tau} \sum_{n=0}^{\infty} [x(n \cdot \Delta\tau)\Delta\tau]\delta(t - n \cdot \Delta\tau) \Rightarrow \lim_{\Delta\tau} \sum_{n=0}^{\infty} [x(n \cdot \Delta\tau)\Delta\tau]h(t - n \cdot \Delta\tau) \tag{2.10}$$

where the items on the left side of the arrow are the input signals and the right side elements are the output signals.

As $n \cdot \Delta\tau \to 0$, for any input signal $x(t)$, the output signal of the linear system can be expressed as

$$
\begin{aligned}
y(t) &= \lim_{\Delta\tau} \sum_{n=0}^{\infty} [x(n \cdot \Delta\tau)\Delta\tau]h(t - n \cdot \Delta\tau) \\
&= \int_{-\infty}^{\infty} x(\tau)h(h - \tau)d\tau \\
&= x(t) * h(t)
\end{aligned}
\tag{2.11}
$$

where the asterisk ($*$) refers to convolution integral. This is just the convolution, which states the relationship between the input signal $x(t)$, the system function $h(t)$ and the output signal $y(t)$. It is an odd-looking definition but it serves as a fundamental concept and is widely used in linear and time-invariant systems. We should note that the variable of integration is τ. As far as the integration process is concerned, the t-variable is temporarily regarded as a constant.

The convolution integral can also be extended to a two-dimensional convolution integral

$$
\begin{aligned}
y(t_1, t_2) &= x(t_1, t_2) * h(t_1, t_2) \\
&= \int_{-\infty}^{\infty} \int_{-\infty}^{\infty} x(t_1, t_2)h(t_1 - \tau_1, t_2 - \tau_2)d\tau_1 d\tau_2
\end{aligned}
\tag{2.12}
$$

The convolution integral has the following key properties:

1. **Commutative Property**

   Letting $\lambda = t - \tau$ and substituting it to Eq. (2.11), we get

   $$
   \begin{aligned}
   y(t) &= \int_{-\infty}^{\infty} x(\tau)h(h - \tau)d\tau \\
   &= \int_{-\infty}^{\infty} x(t - \lambda)h(\lambda)d\lambda \\
   &= h(t) * x(t)
   \end{aligned}
   \tag{2.13}
   $$

   Where

   $$
   \begin{aligned}
   y(t) &= x(t) * h(t) \\
   &= h(t) * x(t)
   \end{aligned}
   \tag{2.14}
   $$

2. **Distributive Property**

   If

   $$
   y(t) = x_1(t) * h(t) + x_2(t) * h(t)
   \tag{2.15}
   $$

   then

   $$
   y(t) = [x_1(t) + x_2(t)] * h(t)
   \tag{2.16}
   $$

3. **Associate Property**
   If

   $$y(t) = x(t) * h_1(t) * h_2(t) \qquad (2.17)$$

   then

   $$\begin{aligned}
   y(t) &= x(t) * [h_1(t) * h_2(t)] \\
        &= x(t) * [h_2(t) * h_1(t)]
   \end{aligned} \qquad (2.18)$$

4. **Time Shift Property**
   If

   $$y(t) = x(t) * h(t) \qquad (2.19)$$

   then

   $$\begin{aligned}
   y(t - \tau) &= x(t - \tau) * h(t) \\
               &= x(t) * h(t - \tau)
   \end{aligned} \qquad (2.20)$$

2.1.2 Convolution Theorem

Convolution may be very difficult to calculate directly, but the convolution theorem gives us a way to simplify many calculations using Fourier transforms and multiplication. There are two ways of expressing the convolution theorem:

Theorem 1 *The Fourier transform of a convolution is the product of the Fourier transforms*

$$F[x(t) * h(t)] = F[x(t)] \cdot F[h(t)] \qquad (2.21)$$

where $F[\cdot]$ denotes a forward Fourier transform.

Proof Applying the Fourier transform to the convolution integral

$$\begin{aligned}
F[x(t) * h(t)] &= F\left[\int_{-\infty}^{\infty} x(\tau)h(t - \tau)d\tau\right] \\
&= \int_{-\infty}^{\infty} \int_{-\infty}^{\infty} x(\tau)h(t - \tau)d\tau e^{-j2\pi ft} dt \\
&= \int_{-\infty}^{\infty} x(\tau)\left[h(t - \tau)e^{-j2\pi ft} dt\right] d\tau \\
&= F[h(t)] \int_{-\infty}^{\infty} x(\tau)d\tau \\
&= F[h(t)] \cdot F[x(t)]
\end{aligned} \qquad (2.22)$$

Theorem 2 *The Fourier transform of a product is the convolution of the Fourier transforms*

$$F[x(t) \cdot h(t)] = F[x(t)] * F[h(t)] \qquad (2.23)$$

Proof We take an inverse Fourier transform $F^{-1}[\cdot]$ on each side of the equation

$$F^{-1}\left[F\left[x(t)\right] * F\left[h(t)\right]\right] = \int_{-\infty}^{\infty} \left[\int_{-\infty}^{\infty} X(f_0)H(f-f_0)\mathrm{d}f_0\right] e^{j2\pi ft}\mathrm{d}f$$

$$= \int_{-\infty}^{\infty} X(f_0)\left[\int_{-\infty}^{\infty} H(f')e^{j2\pi f'}\mathrm{d}f'\right] e^{j2\pi f_0 t}\mathrm{d}f_0$$

$$= x(t)\cdot h(t) \tag{2.24}$$

where $X(f)$ and $H(f)$ denote the Fourier transform representations of $x(t)$ and $h(t)$, respectively.

2.1.3 Correlation Function

The correlation function is closely related to the convolution integral, and it also turns out to be useful in SAR imaging. The cross-correlation function between $x(t)$ and $y(t)$ is defined as [311]

$$R_{xy}(\tau) = \int_{-\infty}^{\infty} x(t)y^*(t-\tau)\mathrm{d}t = \int_{-\infty}^{\infty} x(t+\tau)y^*(t)\mathrm{d}t \tag{2.25a}$$

$$R_{yx}(\tau) = \int_{-\infty}^{\infty} x^*(t-\tau)y(t)\mathrm{d}t = \int_{-\infty}^{\infty} x^*(t)y(t+\tau)\mathrm{d}t \tag{2.25b}$$

where $(\cdot)^*$ denotes a conjugate operator. When $x(t) = y(t)$, it becomes the auto-correlation $R_x(\tau)$. From the definition it follows readily that

$$R_{xy}(-\tau) = R_{yx}^*(\tau) \tag{2.26a}$$

$$R_x(-\tau) = R_x^*(\tau) \tag{2.26b}$$

If $x(t)$ is real, then $R_x(\tau)$ is real and $R_x(-\tau) = R_x(\tau)$.

The power spectrum of a signal $x(t)$ is the Fourier transform of its auto-correlation

$$S_x(f) = \int_{-\infty}^{\infty} R_x(\tau)e^{-j2\pi f\tau}\mathrm{d}\tau$$

$$= X(f)\cdot X^*(f) \tag{2.27}$$

$$= |X(f)|^2$$

The inverse Fourier transform yields

$$R_x(\tau) = \int_{-\infty}^{\infty} S_x(f)e^{j2\pi f\tau}\mathrm{d}f \tag{2.28}$$

Since $R_x(-\tau) = R_x^*(\tau)$, the power spectrum of a signal $x(t)$, real or complex, is a real function of f.

The cross-power spectrum of two signals $x(t)$ and $y(t)$ is the Fourier transform of their cross-correlation

$$S_{xy}(f) = \int_{-\infty}^{\infty} R_{xy}(\tau) e^{-j2\pi f\tau} d\tau \qquad (2.29)$$

The inverse Fourier transform yields

$$R_{xy}(\tau) = \int_{-\infty}^{\infty} S_{xy}(f) e^{j2\pi f\tau} df \qquad (2.30)$$

The function $S_{xy}(f)$ is, in general, complex even when $x(t)$ is real. Furthermore

$$S_{xy}(f) = S_{yx}^*(f) \qquad (2.31)$$

The correlation function can also be extended to two-dimensional form

$$R_{xy}(\tau_1, \tau_2) = \int_{-\infty}^{\infty} \int_{-\infty}^{\infty} x(t_1, t_2) y(t_1 + \tau_1, t_2 + \tau_2) dt_1 dt_2 \qquad (2.32a)$$

$$S_{xy}(f_1, f_2) = F\{R_{xy}(\tau_1, \tau_2)\} = X(f_1, f_2) \cdot Y^*(f_1, f_2) \qquad (2.32b)$$

2.1.4   Relations Between Convolution and Correlation

Comparing Eqs. (2.11) and (2.25), we get the following relationship

$$R_{xy}(t) = x(t) * y(t) \qquad (2.33)$$

However, it should be noted that

- The two sides of Eq. (2.33) are equal only in the sense of numerical value. They have rather different physical means. Convolution can be seen as a special filter. The convolution output signal thus has the same physical unit to the input signal $x(t)$. Differently, the correlation output signal has a different physical unit from the input signal $x(t)$. As an example, suppose $x(t)$ is a voltage signal, the convolution output signal is also a voltage signal; however, the correlation output signal is a power signal.

- Another difference is that convolution integral has a commutative property which is not valid for correlation function.

2.2   Sampling Theorem and Interpolation

In SAR systems, the signals are often transmitted over analog transmission facilities and are also processed digitally. To make the analog signals suitable for digital processing, we make use of two operations: sampling and quantization. The sampling operation is used to convert a continuous-time signal to a discrete sequence. The quantizing operation converts a continuous-amplitude signal to a discrete-amplitude signal. The sampling theorem states that any signal that is frequency band-limited to f_m can be reconstructed from samples taken at a uniform time interval of $T_s \leq 1/(2f_m)$. The time interval $T_s = 1/(2f_m)$ is named the *Nyquist interval*, and the corresponding sampling rate is called the *Nyquist rate* In this section, we will discuss the techniques for sampling a continuous-amplitude, continuous-time signal and reconstructing it from the discrete samples.

2.2.1 Sampling

Consider a real-valued signal $x(t)$ with its Fourier transform as $X(f)$, which is frequency bandlimited to f_m. Suppose the sampling function is

$$p_T(t) = \sum_{k=-\infty}^{\infty} \delta(t - nT_0) \tag{2.34}$$

The sampled signal can then be expressed as

$$x_s(t) = x(t)p_T(t)$$
$$= x(t) \sum_{k=-\infty}^{\infty} \delta(t - nT_0) \tag{2.35}$$

Since $p_T(t)$ is a periodic function, we can use a Fourier series to represent it as

$$p_T(t) = \sum_{k=-\infty}^{\infty} c_k e^{jk2\pi f_s t} \tag{2.36}$$

where $f_s = 1/T_0$, and

$$c_k = \frac{1}{T_0} \int_{-T_0/2}^{T_0/2} p_T(t)e^{-jk2\pi f_s t}, \quad k = 0, \pm 1, \pm 2, \ldots$$
$$= \frac{1}{T_0} \tag{2.37}$$

Substituting it to Eq. (2.35), we have

$$x_s(t) = \frac{1}{T_0} \sum_{k=-\infty}^{\infty} x(t)e^{jk2\pi f_s t} \tag{2.38}$$

Taking the Fourier transform of $x_s(t)$ yields

$$X_s(f) = \frac{1}{T_0} \sum_{k=-\infty}^{\infty} X(f - kf_s) \tag{2.39}$$

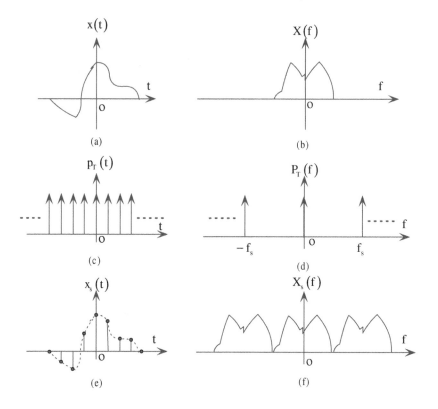

FIGURE 2.3: Graphical representation of the sampling theorem.

This equation states that after multiplication of $x(t)$ by the sampling function $p_T(t)$, the sampled signal spectra $X_s(f)$ consists of $X(f)$, plus replica located at $f = \pm k f_s$. Furthermore, the amplitude of $X_s(f)$ is attenuated by a factor of $1/T_0$.

Figure 2.3 illustrates the graphical representation of the sampling theorem. Figures 2.3(a) and 2.3(b) show the original signal $x(t)$ and its analog frequency spectrum. The sampling function $p_T(t)$ and its spectra are shown in Figures. 2.3(c) and 2.3(d). The sampled signal $x_s(t)$ and its spectra are shown in Figures. 2.3(e) and 2.3(f). It can be noticed from Figure 2.3(f) that to prevent overlap between the neighboring spectra, the sampling frequency f_s must satisfy $f_s \geq 2f_m$.

2.2.2   Interpolation

To reconstruct the original signal $x(t)$ from the sampled signal $x_s(t)$, we can filter out the spectrum $X(f)$ from the spectra $X_s(f)$ with a low-pass filter.

The ideal low-pass filter should be

$$H(f) = \begin{cases} T_0, & -f_c \leq f \leq f_c \\ 0, & \text{otherwise} \end{cases} \tag{2.40}$$

where f_c ($f_m \leq f_c \leq f_s - f_m$) is the cut-off frequency of the filter. The corresponding system function is

$$h(t) = 2f_c T_0 \text{sinc}(2f_c t) \tag{2.41}$$

where

$$\text{sinc}(t) = \frac{\sin(\pi t)}{\pi t} \tag{2.42}$$

Passing $x_s(t)$ through this low-pass filter, we can reconstruct the original signal $x(t)$

$$
\begin{aligned}
x(t) &= x_s(t) * h(t) \\
&= \int_{-\infty}^{\infty} \sum_{k=-\infty}^{\infty} x(kT_0)\delta(\tau - kT_0)h(t-\tau)\mathrm{d}\tau \\
&= \sum_{k=-\infty}^{\infty} \int_{-\infty}^{\infty} x(kT_0)\delta(\tau - kT_0)h(t-\tau)\mathrm{d}\tau \\
&= 2f_c T_0 \sum_{k=-\infty}^{\infty} x(kT_0)\text{sinc}\left[2f_c(t-kT_0)\right]
\end{aligned}
\tag{2.43}
$$

This equation states that $x(t)$ can be reconstructed from its sampled signal $x(kT_0)$ and the interpolation function $\text{sinc}\left[2f_c(t-kT_0)\right]$. In practical SAR systems, the interpolation function is often simplified by two approximations. First, the interpolation filter is chosen with a finite number of sidelobes. For radar signal processing, an 8-tap sinc function is normally used to generate a new interpolated sample [393]. The second approximation involves choosing a finite number of interpolation intervals. Many other interpolation functions can also be employed. The details can be found in [273].

2.2.3 Aliasing Effects

When $f_s > 2f_m$, we have shown that $x(t)$ can be reconstructed in the time domain from samples of $x(t)$ according to Eq. (2.43). But when $f_s < 2f_m$, the sampled signal spectra will be overlapped, as shown in Figure 2.4. This phenomenon often prevents us from reconstructing $x(t)$ from $x_s(t)$ with the required accuracy.

Since significant signal variation usually implies that high frequency components are present in the signal, we would expect that if the signal has significant variation then the sampling frequency f_s must be high enough to provide

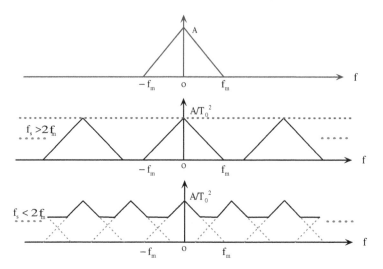

FIGURE 2.4: Comparative graphical representation of the aliasing effects.

an accurate approximation of the signal $x(t)$. Figure 2.5 shows the effect of different sampling frequencies when sampling a cosine signal $x(t) = \cos(100t)$. If the sampling frequency is not high enough to sample the signal correctly then a phenomenon called aliasing effects occurs.

The term aliasing refers to the distortion that occurs when a continuous time signal has frequencies larger than half of the sampling rate. The process of aliasing describes the phenomenon in which components of the signal at high frequencies are mistaken for components at lower frequencies. If there are aliasing effects, the sampled apparent spectra f_{apparent} will be [73]

$$f_{\text{apparent}} = f_{\text{continuous}} - \left(\text{round} \left[\frac{f_{\text{continuous}}}{f_s} \right] \right) f_s \qquad (2.44)$$

where $f_{\text{continuous}}$ and $\text{round}[x]$ denote the actual frequency of the continuous-time signal and the nearest integer of x.

There are several practical considerations about the aliasing effects

- If $x(t)$ has negligible power beyond some frequency f_0, then aliasing can be avoided by sampling at the time intervals

$$T_0 \leq \frac{\pi}{f_0} \qquad (2.45)$$

- If we are interested in the frequency spectra only over a restricted range, say for $|f| \leq f_0$, then we can pass the observed signal through a low-pass filter to remove all frequencies higher than f_0.

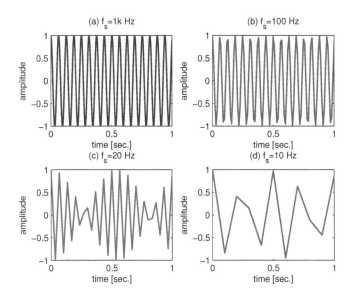

FIGURE 2.5: Effect of different different sampling frequencies when sampling a cosine signal.

- The aliasing effect can be avoided altogether by sampling at random time points. It has been proved that if we sample $x(t)$ at the time intervals which are sampled from a Poisson process, then we may recover the full spectral properties of $x(t)$ from the sampled signal.

2.3 Linearly Frequency Modulated Signal and Matched Filtering

2.3.1 Principle of Stationary Phase

Before discussing the LFM signal and matched filtering, we introduce first the principle of stationary phase. The main idea of stationary phase relies on the cancellation of sinusoids with a rapidly-varying phase. If many sinusoids have the same phase and they are added together, they will add constructively. If, however, these same sinusoids have phases which change rapidly as the frequency changes, they will add incoherently, varying between constructive and destructive addition at different times.

Consider a signal with real envelope $a(t)$ and phase $\phi(t)$

$$x(t) = a(t) \exp\left[j\phi(t)\right] \tag{2.46}$$

Its Fourier transform is

$$
\begin{aligned}
X(\omega) &= \int_{-\infty}^{\infty} x(t)\exp(-j\omega t)dt \\
&= \int_{-\infty}^{\infty} a(t)\exp\left[j\phi(t) - j\omega t\right]dt
\end{aligned}
\tag{2.47}
$$

The phase term in the integral, $\phi(t) - \omega t$ is "stationary" when

$$
\frac{d}{dt}\left[\phi(t) - \omega t\right] \approx 0
\tag{2.48}
$$

Solutions to this equation yield the dominant frequencies

$$
\omega = \phi'(t_k)
\tag{2.49}
$$

with t_k being the stationary phase points, where $\phi'(t)$ is the first derivative of $\phi(t)$.

Expanding the phase term $\phi(t) - \omega t$ in a Taylor series about the stationary phase points t_k, we get

$$
\begin{aligned}
\phi(t) - \omega t =&\phi(t_k) - \omega t_k + [\phi'(t_k) - \omega](t - t_k) \\
&+ \frac{\phi''(t_k)}{2!}(t - t_k)^2 - \dots \\
\approx&\phi(t_k) - \omega t_k + \frac{\phi''(t_k)}{2}(t - t_k)^2 - \dots
\end{aligned}
\tag{2.50}
$$

where $\phi''(t)$ is the second derivative of $\phi(t)$. Neglecting the terms of order higher than $(t - t_k)^2$ yields

$$
\phi(t) - \omega t \approx \phi(t_k) - \omega t_k + \frac{\phi''(t_k)}{2}(t - t_k)^2
\tag{2.51}
$$

Substituting it to Eq. (2.47), we can get

$$
\begin{aligned}
X(\omega) =&a(t_k)\exp\left[j(\phi(t_k) - \omega t_k)\right] \\
&\times \int_{t_k-\delta}^{t_k+\delta} \exp\left[j\frac{\phi''(t_k)}{2}(t - t_k)^2\right]dt
\end{aligned}
\tag{2.52}
$$

with δ being a small constant approximate to zero.

Letting

$$
t - t_k = \mu
\tag{2.53}
$$

and

$$
\frac{\phi''(t_k)}{2}\mu^2 = \frac{\pi\gamma^2}{2}
\tag{2.54}
$$

Then [162, 309]

$$
X(\omega) \doteq \sqrt{\frac{2\pi}{|\phi''(t_k)|}}|a(t_k)|\exp\left\{-j\omega t_k + j\phi(t_k) + j\cdot\mathrm{sgn}\left[\phi''(t_k)\right]\frac{\pi}{4}\right\}
\tag{2.55}
$$

where $\mathrm{sgn}[\cdot]$ is the signum function.

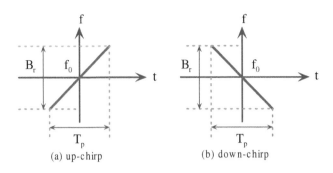

(a) up-chirp   (b) down-chirp

FIGURE 2.6: Typical up-chirp and down-chirp LFM pulse signals.

2.3.2 LFM Signal

LFM signals are commonly used in SAR systems. The frequency of a LFM signal is swept linearly across the pulse width, either upward (up-chirp) or downward (down-chirp), as shown in Figure 2.6. The pulse duration is T_p, and the bandwidth is B_r.

A LFM signal is defined by

$$x(t) = \text{rect}\left(\frac{t}{T_p}\right) \exp\left[j2\pi\left(f_c t + \frac{1}{2}k_r t^2\right)\right] \tag{2.56}$$

where k_r is the chirp rate, and $\text{rect}(\cdot)$ is a window function

$$\text{rect}\left(\frac{t}{T_p}\right) = \begin{cases} 1, & -T_p/2 \leq t \leq T_p/2 \\ 0, & \text{otherwise} \end{cases} \tag{2.57}$$

The instantaneous frequency is

$$\begin{aligned} f(t) &= \frac{1}{2\pi} \cdot \frac{d}{dt}\left[2\pi\left(f_c t + \frac{1}{2}k_r t^2\right)\right] \\ &= f_c + k_r t, \quad -T_p/2 \leq t \leq T_p/2 \end{aligned} \tag{2.58}$$

It is a linear function with variable t.

It can be easily shown that the stationary phase point of Eq. (2.56) is

$$t_k = \frac{f - f_c}{k_r} \tag{2.59}$$

From Eq. (2.55) we can obtain the LFM signal spectrum

$$\begin{aligned} X(f) &\approx \frac{1}{\sqrt{|k_r|}} \exp\left[j\frac{\pi}{4}\text{sgn}(k_r)\right] \\ &\times \exp\left[-j\pi\frac{(f - f_c)^2}{k_r}\right], \quad |f - f_c| < \frac{B_r}{2} \end{aligned} \tag{2.60}$$

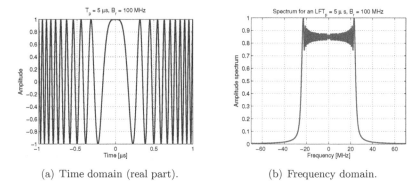

(a) Time domain (real part).   (b) Frequency domain.

FIGURE 2.7: Typical LFM signal time domain and frequency domain.

It can be noticed that the spectrum is also approximated as a LFM signal.

Figure 2.7 shows an example plot of the LFM signal in time domain and frequency domain. The square-like spectrum is widely known as the Fresnel spectrum. Thus, LFM signal can offer both wide bandwidth and long pulse duration. This is particularly valuable for SAR high-resolution imaging.

2.3.3   Matched Filtering

A matched filter is obtained by correlating a known signal, or template, with an unknown signal to detect the presence of the template in the unknown signal. This is equivalent to convolving the unknown signal with a conjugated and time-reversed version of the template.

The matched filter is the optimal linear filter for maximizing the signal-to-noise ratio (SNR) in the presence of additive stochastic noise. The output SNR at $t = t_0$ is given by the instantaneous signal power at $t = t_0$ divided by the average noise power [96] [1]

$$\text{SNR}_0 \big|_{t=t_0} \triangleq \frac{|s(t_0)|^2}{E\{|n_0(t)|^2\}} \tag{2.61}$$

Let $x(t)$ represent the observed signal consists of the desired signal $s_i(t)$ and additive noise $n_i(t)$

$$x(t) = s_i(t) + n_i(t) \tag{2.62}$$

Suppose the system transfer function of the matched filtering is $h(t)$, then the matched filter output signal is

$$s_0(t) = \int_{-\infty}^{\infty} S_i(f)H(f)\mathrm{d}f \tag{2.63}$$

---

[1]The numerator is not an average value of the signal power as in the case of the denominator. This makes sense since the incoming signal is assumed to be deterministic, as opposed to the noise which is modeled as a stationary stochastic process.

where $S_i(f)$ and $H(f)$ denote the Fourier transform representations of $s_i(t)$ and $h(t)$, respectively. If we assume $n_i(t)$ to be white noise with a power spectra density of σ_n^2, we get

$$E\{|n_0(t)|^2\} = \int_{-\infty}^{\infty} |H(f)|^2 \sigma_n^2 \mathrm{d}f \qquad (2.64)$$

Substituting Eqs. (2.63) and (2.64) into Eq. (2.61), we get

$$\mathrm{SNR}_0 = \frac{\left| \int_{-\infty}^{\infty} S(f) H(f) e^{j2\pi f t_0} \mathrm{d}f \right|^2}{\sigma_n^2 \int_{-\infty}^{\infty} |H(f)|^2 \mathrm{d}f} \qquad (2.65)$$

The unknown receiver $H(f)$ should be chosen so as to maximize the output SNR at the observation time $t = t_0$. According to the Schwarz inequality, it follows that

$$\left| \int_{-\infty}^{\infty} S(f) H(f) e^{j2\pi f t_0} \mathrm{d}f \right|^2 \leq \int_{-\infty}^{\infty} |S(f)|^2 \mathrm{d}f \cdot \int_{-\infty}^{\infty} |H(f)|^2 \mathrm{d}f \qquad (2.66)$$

Thus, the maximum value of the output SNR is given by

$$\mathrm{SNR}_{0_{\max}} = \frac{\int_{-\infty}^{\infty} |S(f)|^2 \mathrm{d}f}{\sigma_n^2} \qquad (2.67)$$

This maximum value is achieved if and only if

$$H(f) = \left[S(f) e^{j2\pi f t_0} \right]^* = S^*(f) e^{-j2\pi f t_0} \qquad (2.68)$$

Or, in the time domain we obtain

$$h(t) = s^*(t_0 - t) \qquad (2.69)$$

If $s(t)$ is real, the matched filter solution reduces to

$$h(t) = s(t_0 - t) \qquad (2.70)$$

which is a time-reversed and shifted version of $s(t)$.

Thus, under additive white noise, the matched filter $h(t)$ depends only on the receiver input signal $s(t)$, or the receiver is matched to its input signal $s(t)$. The matched filter output is equivalent to convolving the unknown signal with $h(t)$

$$y(t) = x(t) * h(t) = \int_{-\infty}^{\infty} x(\tau) h(t - \tau) \mathrm{d}\tau \qquad (2.71)$$

2.3.4   Pulse Compression

Pulse compression allows us to achieve the average transmitted power of a relatively long pulse, while obtaining the range resolution corresponding to a short pulse. SAR systems use mainly the linear pulse compression technique, which is accomplished by adding linear frequency modulation to a long pulse at transmission, and by using a matched filter in the receiver to compress the received signal [22, 143, 195].

Consider the transmitted LFM signal expressed in Eq. (2.56). It follows that the demodulated baseband signal is

$$x_r(t) = \text{rect}\left(\frac{t - t_0}{T_p}\right) \exp\left[j\pi\left(k_r(t - t_0)^2\right)\right] \tag{2.72}$$

and the corresponding matched filter is

$$h(t) = \text{rect}\left(\frac{t}{T_p}\right) \exp\left[-j\pi\left(k_r t^2\right)\right] \tag{2.73}$$

From Eq. (2.71), it follows that

$$
\begin{aligned}
y(t) =& x_r(t) * h(t) \\
=& \int_{-\infty}^{\infty} \left\{ \text{rect}\left(\frac{u - t_0}{T_p}\right) \exp\left[j\pi\left(k_r(u - t_0)^2\right)\right] \right. \\
& \left. \times \text{rect}\left(\frac{t - u}{T_p}\right) \exp\left[-j\pi\left(k_r(t - u)^2\right)\right] \right\} du \\
\approx& T_p \text{sinc}\left[k_r T_p(t - t_0)\right]
\end{aligned}
\tag{2.74}
$$

To determine the pulse compression realized by the matched filter output compared to its input, we can examine the input to output effective pulse duration ratio $T_{\text{in}}/T_{\text{out}}$. It is determined by the chirp rate k_r and bandwidth B_r [316]

$$\frac{T_{\text{in}}}{T_{\text{out}}} \geq T_p B_r \tag{2.75}$$

It can be further derived that, after the LFM pulse compression, the $3dB$ resolution [2] (see Figure 2.8) is determined by

$$\rho_{3\text{dB}} \approx \frac{c_0}{2|k_r|T_p} = \frac{c_0}{2B_r} \tag{2.76}$$

where c_0 is the speed of light.

---

[2] Formally, the definition of resolution requires two tests of equal strength, and is given as the minimum spacing for which the targets in the image are discernable as being two separate objects; In practice, however, radar resolution is taken to be the width of a single impulse response, referred to as the impulse response width. It may be specified either as the 3-dB width, which is the width of the pulse at its half-power level, or as the equivalent rectangle width, which is the width of a rectangular representation of the pulse that has the same peak power and the same integrated power. This book uses the 3-dB width definition to measure resolution.

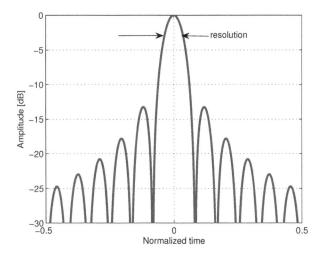

FIGURE 2.8: The pulse compression result of a LFM signal.

Thus, a large time compression ratio can be obtained at the matched filter output for the signal with a large time-bandwidth product T_pB_r. In this case, the matched filter output will appear as a narrow pulse whose peak corresponds to the input delay. As an example, Figure 2.9 shows the pulse compression results with the following simulation parameters: $T_p = 10\mu s$, $B_r = 50MHz$, and $f_c = 1GHz$. The targets can be easily estimated from the peaks.

2.4 Radar Ambiguity Function

Woodward[3] pointed out that a matched filter will preserve all information contained in the received signal. To see what happens if the targets are moving, we have to correlate the emitted pulse with a copy shifted in range and Doppler. This leads to the definition of the ambiguity function of $s(t)$ [452]

$$\chi(\tau, f_d) = \int s(t)s^*(t-\tau)\exp(j2\pi f_d t)\mathrm{d}t \qquad (2.77)$$

---

[3]It is sometimes claimed that Wigner derived the ambiguity function in 1932 within the field of quantum mechanics and that Ville made a similar contribution to signal theory in 1948. This view is anachronistic, because neither Wigner nor Ville was thinking of radar [115].

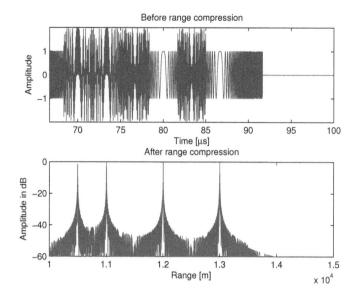

FIGURE 2.9: An example pulse compression for LFM radar system.

Although the basic question of what to transmit remains substantially unanswered, Woodward has taught us to look for new possibilities when designing radar waveforms. It is only necessary to analyze the imposed restrictions through the ambiguity function.

The ambiguity function expressed in Eq. (2.77) is usually restricted to narrowband signals. The wideband ambiguity function of $s(t)$ is defined as [346, 441]

$$\chi_{\text{wb}}(\tau, f_d) = \sqrt{|\alpha|} \int s(t)s^*[\alpha(t - \tau)]\mathrm{d}t \qquad (2.78)$$

where α is a time scale factor of the received signal relative to the transmitted signal given by

$$\alpha = \frac{c_0 - v_r}{c_0 + v_r} \qquad (2.79)$$

for a target moving with constant radial velocity v_r. When the wave speed in the medium is sufficiently faster than the target speed, as is common with radar, this wideband ambiguity function is closely approximated by the narrowband ambiguity function given above, which can be computed efficiently by making use of the FFT algorithm.

The ambiguity function is determined by the properties of the pulse and the matched filter, and not any particular target scenario. An ambiguity function of interest is a two-dimensional Dirac delta function

$$\chi(\tau, f_d) = \delta(\tau)\delta(f_d) \qquad (2.80)$$

However, an ambiguity function of this kind would have no ambiguities at all, and both the zero-delay and zero-Doppler cuts would be impulses. Any Doppler shift would make the target disappear. This is not desirable because the target will disappear from the radar picture if it has unknown velocity. On the other hand, if Doppler processing is independently performed, knowledge of the precise Doppler frequency allows ranging without interference from any other targets that are not also moving at exactly the same velocity. Moreover, this type of ambiguity function is not physically realizable.

In this book, we define the radar ambiguity function as the modulus squared of Eq. (2.77)

$$|\chi(\tau, f_d)|^2 = \left| \int s(t)s^*(t - \tau) \exp(j2\pi f_d t) dt \right|^2 \qquad (2.81)$$

This ambiguity evaluated at $(\tau, f_d) = (0, 0)$ is equal to the matched filter output that is matched perfectly to the signal reflected from the target of interest, and the ambiguity diagram is centered at the same point.

Since a more concise way of representing the ambiguity function consists of examining the one-dimensional zero-delay $\chi(0, f_d)$ and zero-Doppler $\chi(\tau, 0)$ "cuts", we discuss them respectively in the following subsections.

2.4.1 Range Ambiguity Function

Suppose there are two targets with the same radial velocity but with different radial ranges, which means that their reflected signals only have a relative time delay τ. To differentiate the two targets, we consider the following mean variance

$$
\begin{aligned}
\varepsilon_\tau^2 &= \int_{-\infty}^{\infty} |s(t) - s(t + \tau)|^2 \, dt \\
&= \int_{-\infty}^{\infty} [s(t) - s(t + \tau)] [s(t) - s(t + \tau)]^* \, dt \qquad (2.82) \\
&= 2 \int_{-\infty}^{\infty} |s(t)|^2 \, dt - 2\mathrm{Re} \left[\int_{-\infty}^{\infty} s(t)s^*(t + \tau) dt \right]
\end{aligned}
$$

where Re denotes the real-part of the signal. Since the two reflected signals can be seen as having the equal energy E

$$\int_{-\infty}^{\infty} |s(t)|^2 \, dt = \int_{-\infty}^{\infty} |s(t + \tau)|^2 \, dt = E \qquad (2.83)$$

Eq. (2.82) can be evaluated by

$$\chi_r(\tau) = \int_{-\infty}^{\infty} s(t)s^*(t + \tau) dt \qquad (2.84)$$

This equation is often named as the range ambiguity function.

The time-delay resolution can then be evaluated by

$$\rho_\tau = \frac{\int_{-\infty}^{\infty} |\chi_r(\tau)|^2 \, d\tau}{|\chi_r(0)|^2} \tag{2.85}$$

and the equivalent correlation bandwidth can be represented by

$$W_e = \frac{\left| \int_{-\infty}^{\infty} |S(f)|^2 \, df \right|^2}{\int_{-\infty}^{\infty} |S(f)|^4 \, df} \tag{2.86}$$

where $S(f)$ is the Fourier transform of $s(t)$. Correspondingly the range resolution is

$$\rho_r = \frac{c_0}{2W_e} \tag{2.87}$$

2.4.2   Velocity Ambiguity Function

Suppose there are two targets with the same radial range but with different radial velocities, which means that their reflected signals have only a relative Doppler frequency shift f_d. To differentiate the two targets, we consider the following mean variance

$$
\begin{aligned}
\varepsilon_{f_d}^2 &= \int_{-\infty}^{\infty} |s(t) - s(t)\exp(-j2\pi f_d t)|^2 \, dt \\
&= \int_{-\infty}^{\infty} [s(t) - s(t)\exp(-j2\pi f_d t)]\,[s(t) - s(t)\exp(-j2\pi f_d t)]^* \, dt \quad (2.88) \\
&= 2\int_{-\infty}^{\infty} |s(t)|^2 \, dt - 2\mathrm{Re}\left[\int_{-\infty}^{\infty} s(t)s^*(t)\exp(j2\pi f_d t) dt\right]
\end{aligned}
$$

In a manner like Eq. (2.82), we define the velocity ambiguity function

$$\chi_D(f_d) = \int_{-\infty}^{\infty} s(t)s^*(t)\exp(j2\pi f_d t) dt \tag{2.89}$$

The Doppler resolution can be evaluated by

$$\rho_{f_d} = \frac{\int_{-\infty}^{\infty} |\chi_D(f_d)|^2 \, df_d}{\chi_D^2(0)} \tag{2.90}$$

From the Parseval's theorem [194]

$$\int_{-\infty}^{\infty} |x(t)|^2 \, dt = \int_{-\infty}^{\infty} |X(f)|^2 \, df \tag{2.91}$$

Eq. (2.90) can be represented by

$$\rho_{f_d} = \frac{\int_{-\infty}^{\infty} |s(t)|^4 \, df_d}{\left| \int_{-\infty}^{\infty} |s(t)|^2 \, dt \right|^2} \tag{2.92}$$

and the equivalent correlation time is

$$T_e = \frac{\left|\int_{-\infty}^{\infty} |s(t)|^2 \, dt\right|^2}{\int_{-\infty}^{\infty} |s(t)|^4 \, dt} \tag{2.93}$$

Correspondingly the velocity resolution is

$$\rho_v = \frac{c_0}{2 f_c T_e} \tag{2.94}$$

with f_c being the carrier frequency.

2.4.3 Properties of the Ambiguity Function

Radar ambiguity function has the following properties:

(1) The maximum value for the ambiguity function occurs at $(\tau, f_d) = (0, 0)$

$$|\chi(\tau, f_d)|^2 \le |\chi(0, 0)|^2 \tag{2.95}$$

(2) The ambiguity function is symmetric about the origin

$$|\chi(\tau, f_d)|^2 = |\chi(-\tau, -f_d)|^2 \tag{2.96}$$

(3) The total volume under the ambiguity function is constant

$$\int_{-\infty}^{\infty} \int_{-\infty}^{\infty} |\chi(\tau, f_d)|^2 \, d\tau df_d = |\chi(0, 0)|^2 \tag{2.97}$$

(4) If the function $S(f)$ is the Fourier transform of the signal $s(t)$, then

$$S(f)S^*(f) = \int_{-\infty}^{\infty} \chi(\tau, 0) \exp(j2\pi f_d \tau) d\tau \tag{2.98}$$

(5) If the ambiguity function of $s(t)$ is $\chi(\tau, f_d)$, i.e., $s(t) \to \chi(\tau, f_d)$, then

$$s(t)e^{jbt^2} \to e^{-jb\tau^2} \chi\left(\tau, f_d - \frac{b\tau}{\pi}\right) \tag{2.99a}$$

$$s(at) \to \frac{1}{|a|} \chi\left(a\tau, \frac{f_d}{a}\right) \tag{2.99b}$$

where a and b are constants.

2.4.4   Example: LFM Ambiguity Function

For simplicity and without loss of generality, we consider only the LFM complex envelope signal defined by

$$s(t) = \text{rect}\left[\frac{t}{T_p}\right] \exp\left(j\pi k_r t^2\right) \tag{2.100}$$

From Eq. (2.77), we have

$$\chi(\tau, f_d) = \int_{-\infty}^{\infty} \text{rect}\left[\frac{t}{T_p}\right] \text{rect}\left[\frac{t-\tau}{T_p}\right] e^{j\pi k_r t^2} e^{j\pi k_r (t-\tau)^2} e^{j2\pi f_d t} dt \tag{2.101}$$

When $0 \leq \tau \leq T_p$, it follows that

$$\chi(\tau, f_d) = e^{-j\pi k_r \tau^2} \int_{-T_p/2}^{T_p/2-\tau} e^{j2\pi(k_r \tau + f_d)t} dt$$

$$= e^{j\pi f_d(T_p-\tau)} \frac{\sin\left[\pi f_d(T_p - \tau)\right]}{\pi f_d(T_p - \tau)} (T_p - \tau) \tag{2.102}$$

Similar analysis for the case when $-T_p \leq \tau \leq 0$ can be carried out. The same results can be obtained by using the symmetry property of the ambiguity function. It follows that the $\chi(\tau, f_d)$ can be written in a uniform form

$$|\chi(\tau, f_d)|^2 = \begin{cases} \left|\frac{\sin[\pi(f_d-k_r\tau)(T_p-|\tau|)]}{\pi(f_d-k_r\tau)(T_p-|\tau|)}(T_p - |\tau|)\right|^2, & |\tau| < T_p \\ 0, & \text{otherwise} \end{cases} \tag{2.103}$$

Suppose $T_p = 10\mu s$ and $k_r = 1 \times 1000$, Figure 2.10 shows the three-dimensional and contour plots of the LFM ambiguity function. It follows that the LFM ambiguity function cut along the time delay axis is narrower than that of the unmodulated pulse by a factor [269]

$$\xi = \frac{T_p}{1/B_r} = B_r T_p \tag{2.104}$$

This factor is referred to as compression ratio or compression gain, which is equal to the time-bandwidth product $(B_r T_p)$.

2.5   Basic Principle of Synthetic Aperture

A technique closely related to SAR uses a phased array of real antenna elements spatially distributed over either one or two dimensions across the range dimension. All elements of the array receive simultaneously in real time, and

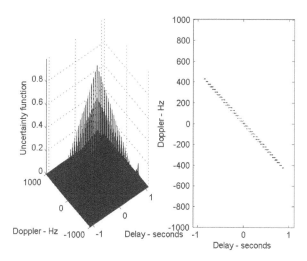

FIGURE 2.10: The three-dimensional and contour plots of an LFM ambiguity function.

the signals passing through them can be individually subjected to controlled phase shifts. One result can be obtained by processing the signals reflected from a specific small scene area, focusing on that area to determine its contribution to the total received signal. The coherently detected set of signals received over the entire array aperture can be replicated in several data-processing channels and processed differently in each channel.

In comparison, a SAR usually employs a single physical antenna element to gather signals at different positions and different times. When the radar is carried on an aircraft or a satellite, those positions are functions of a single variable, distance along the radar's path. When the stored radar signals are read out later and combined with specific phase shifts, the results are the same as if the recorded data had been gathered by an equally long and shaped phased array. What is thus synthesized is a set of signals equivalent to what could have been received simultaneously by such an actual large-aperture phased array. The SAR simulates (rather than synthesizes) that long phased array.

2.5.1 Synthetic Aperture Radar Imaging

A basic SAR gathers its data as it (or its target) moves at some speed by an aircraft or satellite with a side-looking antenna, which transmits a stream of radar pulses and records the backscattered signal corresponding to each pulse. Figure 2.11 shows a side-looking strip-map SAR geometry with three radar positions, A, B, C, together with a ground point target located at $(0, P, 0)$. At

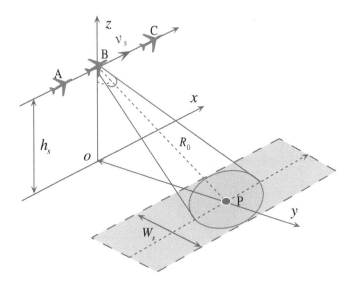

FIGURE 2.11: Geometry of a side-looking strip-map SAR.

radar position A, the radar beam begins to illuminate the target. At position B the center of the radar beam is on the target. At position C the radar beam ends the illumination on the target. Position B is chosen to be at $(0, 0, h_s)$, with beamwidth $\theta_b = \lambda/L$ (λ and L are the carrier wavelength and antenna length in azimuth, respectively). The distance between positions A and C is the synthetic aperture length, L_s

$$L_s = R_0 \theta_b = R_0 \frac{\lambda}{L} \tag{2.105}$$

where R_0 is the closest distance between the target and radar.

The radar antenna is pointed to the side of the aircraft towards the ground and orthogonal to the flight track. As the aircraft moves, a swath is mapped out on the ground by the antenna footprint. The radar transmits pulses at the pulse repetition frequency, PRF, and for each pulse the backscatter return from the ground is sampled in range at the analog-to-digital (A/D) sampling frequency. A simplified version of Figure 2.11 is shown in Figure 2.12, where the plane covered by points A, B, C and the point target on the ground forms a new coordinate system.

When the radar moves at a speed v_s, at azimuth slow time τ the instantaneous slant range between radar platform and the point target is

$$R(\tau) = \sqrt{R_0^2 + (v_s \tau)^2}$$
$$\approx R_0 + \frac{(v_s t)^2}{2R_0} \tag{2.106}$$

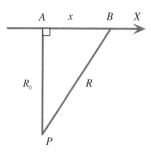

FIGURE 2.12: Geometry of a simplified side-looking strip-map SAR.

Suppose the transmitted signal is a LFM waveform (as given in Eq. (2.56)). The radar returns can be represented by

$$
\begin{aligned}
s_{r_1}(t,\tau) =& \operatorname{rect}\left[\frac{t - \frac{2R(\tau)}{c_0}}{T_p}\right] \\
& \times \exp\left[j2\pi f_0\left(t - \frac{2R(\tau)}{c_0}\right) + j\pi k_r\left(t - \frac{2R(\tau)}{c_0}\right)^2\right]
\end{aligned}
\tag{2.107}
$$

where t is the fast time in range dimension.

According to the "stop-and-go" assumption, pulse compressing about variable t yields

$$
s_{r_2}(t,\tau) = \operatorname{rect}\left[\frac{t - \frac{2R(\tau)}{c_0}}{T_p}\right]\operatorname{sinc}\left[k_r T_p\left(t - \frac{2R(\tau)}{c_0}\right)\right]\exp\left(-j2\pi f_0\frac{2R(\tau)}{c_0}\right)
\tag{2.108}
$$

Its phase term can be expressed as

$$
\varphi(\tau) \approx -\frac{4\pi R_0}{\lambda} - \frac{2\pi(v_s\tau)^2}{\lambda R_0}
\tag{2.109}
$$

Obviously, Eq. (2.108) is still a LFM signal about variable τ. The Doppler chirp rate and Doppler bandwidth are expressed, respectively, by

$$
k_a = -\frac{2v_s^2}{\lambda R_0}
\tag{2.110a}
$$

$$
B_a = |k_a|T_s = \frac{2v_s}{L_a}
\tag{2.110b}
$$

Pulse compressing about variable τ yields

$$
s_0(t,\tau) \doteq \operatorname{sinc}\left[k_r T_p\left(t - \frac{2R(\tau)}{c_0}\right)\right]\operatorname{sinc}\left[\frac{2v_s L_s}{\lambda R_0}\tau\right]
\tag{2.111}
$$

Then, the azimuth resolution is only determined by the antenna length, L_a

$$\rho_a = \frac{0.886\lambda R_0}{2L_s} = \frac{0.886\lambda R_0}{2 \times 0.886\frac{\lambda}{L_a}R_0} = \frac{L_a}{2} \qquad (2.112)$$

2.5.2   Remote Sensing Swath Width

SAR usually employs a pulse signal. The primary advantage of pulse signal over continuous-wave signal is that pulsing allows the transmitter and receiver to share the same antenna. However, if the period between successive pulses is too short, the return from a distant target may arrive after the transmitter has emitted another pulse. This would make it impossible to tell whether the observed pulse is the return of the pulse just transmitted or that of the preceding pulse. In this case, the derived range information will be ambiguous.

To avoid range ambiguities resulting from the preceding and succeeding pulse returns arriving at the antenna simultaneously, the slant range R should be satisfactory with

$$\frac{c_0}{2}\left(\frac{i}{\text{PRF}} + T_p + \Delta T_{\text{tr}}\right) < R < \frac{c_0}{2}\left(\frac{i+1}{\text{PRF}} - T_p\right) \qquad (2.113)$$

where i is an integer and ΔT_{tr} is the switching time between transmission and reception of a pulse. Considering Figure 2.13, where R_e is the Earth radius, the maximum swath width on ground is

$$W_s = R_e \text{acos}\left[\frac{2R_e^2 + h_s^2 + 2h_sR_e - \frac{c_0^2}{4}\left(\frac{i+1}{\text{PRF}} - T_p\right)^2}{2R_e(h_s + R_e)}\right]$$
$$- R_e \text{acos}\left[\frac{2R_e^2 + h_s^2 + 2h_sR_e - \frac{c_0^2}{4}\left(\frac{i}{\text{PRF}} + T_p - \Delta T_{tr}\right)^2}{2R_e(h_s + R_e)}\right] \qquad (2.114)$$

2.5.3   System Sensitivity

According to the radar equation [349], the power P_r returning to the receive antenna is given by

$$P_r = \frac{P_tG\sigma_0}{(4\pi)^2R^4} \qquad (2.115)$$

where P_t is the transmit power, G is the antenna gain, and σ_0 is the radar cross section (RCS) of the target. The corresponding SNR is

$$\text{SNR} = \frac{P_tG^2\lambda^2\sigma_0}{(4\pi)^3R^4(K_0T_0B_nF_n)} \qquad (2.116)$$

where K_0 is the Boltzmann constant, T_0 is the system noise temperature, B_n is the receiver noise bandwidth and F_n is the receiver noise figure.

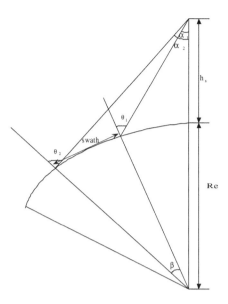

FIGURE 2.13: Geometry of remote sensing swath.

The subsequent SAR processing will integrate multiple radar pulses of bandwidth B_n and pulse duration T_p, thereby improving the SNR by a factor of $G_{\mathrm{pr}} \cdot G_{\mathrm{pa}}$, where

$$G_{\mathrm{pr}} = B_r \cdot T_p \qquad (2.117a)$$
$$G_{\mathrm{pa}} = \mathrm{PRF} \cdot T_s \qquad (2.117b)$$

are the number of independent data samples in range and azimuth, respectively. This implicitly assumes a match between the transmitted pulse bandwidth and the receiver filter, a sufficient sampling frequency for unambiguous signal representation and a constant azimuth antenna pattern during the synthetic aperture time T_s. The final SAR image SNR can then be determined by

$$\mathrm{SNR} = \frac{P_{\mathrm{av}} G^2 \lambda^2 \sigma_0}{(4\pi)^3 L_a R^3 v_s (K_0 T_0 B_n F_n) L_{\mathrm{loss}}} \qquad (2.118)$$

where $P_{\mathrm{av}} = P_t T_p \cdot \mathrm{PRF}$ is the average transmit power and L_{loss} is the loss factor.

A quantity directly related to SAR imaging performance is the noise equivalent sigma zero (NESZ), which is the mean radar cross section (RCS) necessary to produce an SNR of unity. The NESZ can be interpreted as the smallest target cross section that is detectable by the radar system against thermal noise. Letting $\mathrm{SNR} = 1$, from Eq. (2.118) we have

$$\mathrm{NESZ} = \frac{(4\pi)^3 L_a R^3 v_s (K_0 T_0 F_n) L_{loss}}{P_{\mathrm{av}} G^2 \lambda^3} \qquad (2.119)$$

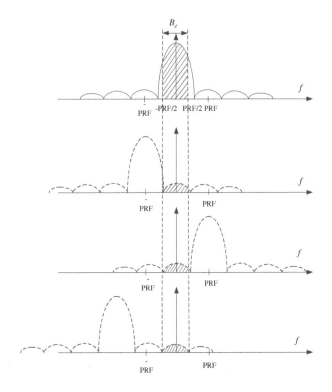

FIGURE 2.14: Illustration of azimuth ambiguities.

2.5.4   Ambiguity-to-Signal Ratio

The presence of range and azimuth (or Doppler) ambiguities in SAR imaging is an important problem. When the PRF is set too high, the radar return from two successive pulses will overlap at the receiver and there will be range ambiguities. This type of ambiguity is referred to as range ambiguity. When the PRF is set too low, on the other hand, there will be returns from the targets that have the same Doppler history as the signal due to aliasing. These targets appear as if illuminated by a portion of the physical antenna beam away form the antenna boresight in azimuth. These returns are referred to as azimuth ambiguities.

2.5.4.1   Azimuth Ambiguity-to-Signal Ratio

Azimuth ambiguities arise from finite sampling of the Doppler spectra. Since the spectra repeats at PRF intervals, the signal components outside this frequency interval will fold back into the main part of the spectra, as shown in Figure 2.14.

Bayma and McInnes [21] described an approach to evaluate the azimuth ambiguity-to-signal ratio (AASR) performance for a spaceborne SAR. In this approach, the far-field antenna gain pattern is expressed as a function of Doppler frequency f and time delay τ. The radar return at a particular Doppler frequency f_0 and time delay τ_0 is then expressed as

$$R(f_0, \tau_0) = \sum_{m,n=-\infty}^{\infty} G(f_0 + m \cdot \text{PRF}, \tau_0 + \frac{n}{\text{PRF}})$$

$$\times \sigma_A(f_0 + m\text{PRF}, \tau_0 + \frac{n}{\text{PRF}}) \tag{2.120}$$

where m, n are integers, PRF is the pulse repetition frequency, $G(f, \tau)$ is the antenna gain pattern, $\sigma_A(f, \tau)$ is the radar reflectivity.

However, it is difficult to evaluate in practice due to two reasons: First, the antenna gain patterns are generally given as a function of the elevation and azimuth angle off the antenna boresight and not as a function of f and τ. The second reason is that the dependence of σ_A on range, angle of incidence, etc., are not explicitly indicated. To transform this portion of Eq. (2.120) into a form more readily suited for calculation, we can express the radar reflectivity obtained over a portion of the azimuth spectra bandwidth as

$$\sigma(i, R) = \frac{\sum(\sigma_i)}{(4\pi R^2)^2} \tag{2.121}$$

where any dependence on the antenna (such as transmitter power, losses, etc.) are not included. $\sum(\sigma_i)$ denotes the radar reflectivity as a function of the angle of incidence i, and $(4\pi R^2)^2$ represents the usual radar return dependence on slant range R. $\sum(\sigma_i)$ can be expressed as

$$\sum(i) \approx \sigma_0(\sigma_i) [R\delta_\theta \delta_R] \tag{2.122}$$

where $\sigma_0(i)$ is the RCS per unit area at incidence angle i, δ_θ is the angle subtended by targets that are separated by Doppler frequency bandwidth, and δ_R is the ground range resolution. In general, δ_θ depends only on the bandwidth and does not depend on the actual value of Doppler frequencies.

Many of the factors appear in Eq. (2.120) cancel out when one evaluates the AASR because they appear both in the signal and in the ambiguity calculation. Thus, the AASR can be written as [241]

$$\text{AASR} = \frac{\sum_{m,n=-\infty}^{\infty} \int_{-B_a/2}^{B_a/2} G\left[e_{f,\tau}(m, n), a_{f,\tau}(m, n)\right] \left[\frac{\sum(\sigma_i)}{R^4}\right] df}{\int_{-B_a/2}^{B_a/2} G\left[e_{f,\tau}(0, 0), a_{f,\tau}(0, 0)\right] \left[\frac{\sum(\sigma_i)}{R^4}\right]} \tag{2.123}$$

where $m \neq 0, n \neq 0$ with

$$e_{f,\tau}(m, n) = e\left(f + m \cdot \text{PRF}, \tau + \frac{n}{\text{PRF}}\right) \tag{2.124a}$$

$$a_{f,\tau}(m, n) = a\left(f + m \cdot \text{PRF}, \tau + \frac{n}{\text{PRF}}\right) \tag{2.124b}$$

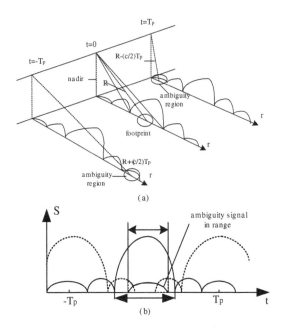

FIGURE 2.15: Illustration of range ambiguities.

where e and a denote the elevation and azimuth angles off the antenna bore-sight, respectively. The remaining task is to relate the Doppler frequency and slant range to the elevation and azimuth angles off the antenna boresight. This problem has been addressed by Curlander [74]. He developed an iterative algorithm that determines the location of the target on the ellipsoidal Earth's surface that satisfies the Doppler frequency and slant range conditions.

2.5.4.2   Range Ambiguity-to-Signal Ratio

Range ambiguities arise from that the preceding and succeeding pulses arrive back at the antenna simultaneously with the desired returns. Consider the illustration of range ambiguities shown in Figure 2.15, for a specific target with a slant range of R_0, the ambiguous signals arrive from the ranges of

$$R_i = R_0 + \frac{c_0}{2} \cdot \frac{i}{\text{PRF}}, \quad i = \pm 1, \pm 2, \ldots, N_h \quad (2.125)$$

where i, the pulse number ($i = 0$ for the desired pulse), is positive for preceding interfering pulses and negative for succeeding ones.

The range ambiguity is evaluated by the range ambiguity-to-signal ratio (RASR), which is determined by summing all signal components within the data record window arising from the preceding and succeeding pulse returns, and taking the ratio of this sum to the integrated signal return from the

FIGURE 2.16: Image formation from the convolution of signal with the PSF.

desired pulse. The RASR of a basic SAR system can be expressed as

$$\text{RASR} = \frac{\sum_i \int\limits_{R_{\text{min}}}^{R_{\text{max}}} \frac{G_i^2 \sigma_i}{R_i^3 \sin(\varphi_i)}}{\int\limits_{R_{\text{min}}}^{R_{\text{max}}} \frac{G_0^2 \sigma_0}{R_0^3 \sin(\varphi_0)}}, \quad i \neq 0 \tag{2.126}$$

where G_i is the cross-track antenna pattern at the i-th time interval of the data recording window at a given angle of φ_i, σ_i is the normalized backscatter coefficient at the i-th time interval. The G_0, σ_0, R_0 and φ_0 are the corresponding parameters of the desired unambiguous return.

2.6 Point Spread Function

The point spread function (PSF) describes the response of an imaging system to a point source or point object. A more general term for the PSF is a system's impulse response, the PSF is the impulse response of a focused radar system. In functional term it is the spatial domain version of the modulation transform function, as shown in Figure 2.16.

The precise nature of the radar PSF is dependent on several platform and radar parameters. For side-looking SAR undergoing constant cross-track acceleration a_y, the variation in phase with time is

$$\phi(t) = (\beta_0 + \delta\beta)t^2 \tag{2.127}$$

where

$$\beta_0 = \frac{2\pi v_x^2}{R_0 \lambda} \tag{2.128a}$$

$$\delta\beta = \frac{2\pi a_y}{\lambda} \tag{2.128b}$$

The signal received by the radar for a point target is

$$s(t) = \exp(-j\phi(t)) \tag{2.129}$$

Suppose the reference signal is

$$h(t) = \exp(j\beta_0 t^2) \tag{2.130}$$

The matched filter output is given by

$$
\begin{aligned}
y(t) &= \frac{1}{T_p} \int_{-T_p/2}^{T_p/2} s(t+\tau)h(\tau)d\tau \\
&= \frac{1}{T_p} \int_{-T_p/2}^{T_p/2} \exp\left[-j(\beta_0+\delta\beta)(t+\tau)^2 + j\beta_0\tau^2\right] d\tau \\
&= \frac{1}{T_p} \exp\left[-j(\beta_0+\delta\beta)t^2\right] \int_{-T_p/2}^{T_p/2} \exp\left[-j(\beta_0+\delta\beta)2t\tau - j\delta\beta\tau^2\right] d\tau
\end{aligned}
\tag{2.131}
$$

For zero cross-track acceleration $\delta\beta = 0$ and this simplifies to the familiar sinc function

$$y(t) = \text{sinc}(\beta_0 T_p t)\exp(-j\beta_0 t^2) \tag{2.132}$$

When the cross-track acceleration is non-zero the integral in Eq. (2.131) can be Fourier transformed to

$$
\begin{aligned}
Y(\omega) &= \frac{1}{T_p} \int_{-T_p/2}^{T_p/2} \exp(-j\delta\beta\tau^2) \int_{-T_p/2}^{T_p/2} \exp\left[-j(\beta_0+\delta\beta)2t\tau\right]\exp(j\omega t)dtd\tau \\
&= \int_{-T_p/2}^{T_p/2} \exp(-j\delta\beta\tau^2)\text{sinc}\left([(\beta_0+\delta\beta)2\tau - \omega]\frac{T_p}{2}\right) d\tau
\end{aligned}
\tag{2.133}
$$

If

$$\frac{\pi}{(\beta_0+\delta\beta)T_p} \ll \frac{T_p}{2} \tag{2.134}$$

which holds well for the parameters used here, the sinc function may be approximated by an expression proportional to the Dirac delta function. We then have

$$H(\omega) \propto \exp\left[\frac{-j\delta\beta\omega^2}{4(\beta_0+\delta\beta)^2}\right] \tag{2.135}$$

The inverse Fourier transform of this function cannot be evaluated in terms of simple functions. However, the analytical form when using integration limits $\omega = \pm B_a/2$ is given as [228]

$$
y(t) \propto \frac{\sqrt{\pi}(-1)^{1/4}(\beta_0 + \delta\beta)}{B_a\sqrt{\delta\beta}} \exp\left[\frac{j(\beta_0 + \delta\beta)^2 t^2}{\delta\beta}\right]
$$

$$
\times \left\{ \text{erfi}\left[\frac{(-1)^{3/4}\left[4(\beta_0 + \delta\beta)^2 t - \delta\beta B_a\right]}{4\sqrt{\delta\beta}(\beta_0 + \delta\beta)}\right] \right.
$$

$$
\left. - \text{erfi}\left[\frac{(-1)^{3/4}\left[4(\beta_0 + \delta\beta)^2 t + \delta\beta B_a\right]}{4\sqrt{\delta\beta}(\beta_0 + \delta\beta)}\right] \right\} \tag{2.136}
$$

where $\text{erfi}(x) = \text{erf}(jx)/j$ and

$$
\text{erf}(x) = \frac{2}{\sqrt{\pi}} \int_0^x \exp(-t^2)\,\mathrm{d}t \tag{2.137}
$$

The previous analysis is expressed in one dimension only. For the two-dimensional SAR it is a very good approximation to factorise the PSF and its Fourier transform into range and azimuth parts

$$
Y(\omega) \equiv Y(\omega_r, \omega_a) = Y_r(\omega_r)Y_a(\omega_a) \tag{2.138}
$$

where $Y_r(\omega_r)$ and $Y_a(\omega_a)$ are the spatial Fourier transforms of the range and azimuth factors, respectively. This approximation becomes invalid when "range walk" occurs. We also assume that the PSF is translation invariant which is very good approximation only for the range direction because the characteristic drift time of the range response is very long, and so we may determine the range position using the above analysis. The azimuth response is not translationally invariant because of the effects of antenna swing, which makes this approach to determining azimuth position somewhat questionable.

2.7 Basic Image Formation Algorithm

Processing of SAR data requires a two-dimensional space-variant correlation of the received echo data with the point scatterer response of the SAR data acquisition system. Both the shape and phase history of the correlation kernel vary systematically in the across-track (range) direction. A full two-dimensional time domain correlator can handle the space-variance, but is computationally inefficient. In order to take advantage of fast frequency domain correlation techniques, several transform-domain algorithms that impose different approximations on the correlation kernel have been developed in the past years [132, 188, 332].

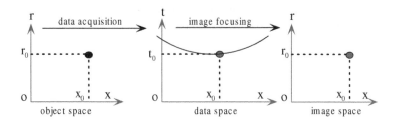

FIGURE 2.17: Coordinate systems for SAR imaging processor.

2.7.1   Two-Dimensional Spectrum Model

SAR image formation is a two-step process: The data acquisition performs the transformation from the object space to the data space and "smears out" the energy of a single point scatterer $\delta(x - x_0, r - r_0)$ to a two-dimensional function $h(x - x_0, t : r0)$, the point scatterer response (PSR). The processor tries to focus the PSR back to a single point, as shown in Figure 2.17.

Let

$$a(t) \exp(j\pi k_r t^2 + j2\pi f_0 t) \tag{2.139}$$

be the pulse transmitted by the radar where $a(t)$ is the signal envelope. Then the echo from a scatterer at a distance $R(\tau)$ with τ being the azimuth time is

$$
\begin{aligned}
s(t, \tau) = &\sigma w(\tau) a\left(t - \frac{2R(\tau)}{c_0}\right) \\
&\times \exp\left\{j\pi k_r \left[t - \frac{2R(\tau)}{c_0}\right]^2\right\} \exp\left\{-j\frac{4\pi}{\lambda} R(\tau)\right\}
\end{aligned}
\tag{2.140}
$$

with $w(\tau)$ being the antenna azimuth pattern.

Considering the range model shown in Figure 2.18, the instantaneous range $R(\tau)$ can be represented by

$$
\begin{aligned}
R(\tau) &= \sqrt{R_0^2 + v_s^2 \tau^2 - 2R_0 v_s \tau \cos(\theta)} \\
&\approx R_0 - v_s \tau \cos(\theta) + \frac{v_s^2 \sin^2(\theta)}{2R_0} \tau^2
\end{aligned}
\tag{2.141}
$$

Let

$$f_{dc} = \frac{2v_s}{\lambda} \cos(\theta) \tag{2.142a}$$

$$f_{dr} = -\frac{2v_s^2 \sin^2(\theta)}{\lambda R_0} \tag{2.142b}$$

we then have

$$R(\tau) = R_0 - \frac{\lambda}{2} f_{dc} \tau - \frac{\lambda}{4} f_{dc} \tau^2 \tag{2.143}$$

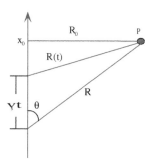

FIGURE 2.18: Geometry of a strip-map SAR range model.

Applying Fourier transform with respect to t, according to the stationary phase principle we get

$$s(f_r, \tau) = \int s(t, \tau) \exp(-j2\pi f_r t)\mathrm{d}t$$

$$= w(\tau)a\left(-\frac{f_r}{k_r}\right)\exp\left\{j\pi\frac{f_r^2}{k_r}\right\}\exp\left\{-j\frac{4\pi}{c_0}(f_r + f_0)R(\tau)\right\} \qquad (2.144)$$

Fourier transforming with respect to τ yields

$$s(f_r, f_a) = \int s(f_r, \tau)\exp(-j2\pi f_a \tau)\mathrm{d}\tau$$

$$= a\left(-\frac{f_r}{k_r}\right)\exp\left(j\pi\frac{f_r^2}{k_r}\right) \qquad (2.145)$$

$$\times \int w(\tau)\exp\left[-j\frac{4\pi}{c_0}(f_r + f_0)R(\tau) - j2\pi f_a \tau\right]\mathrm{d}\tau$$

Defining

$$\Phi_1(f_r, \tau) = -\frac{4\pi}{c_0}(f_r + f_0)R(\tau) - 2\pi f_a \tau \qquad (2.146)$$

we then have

$$\frac{\partial}{\partial \tau}\Phi_1(f_r, \tau) = \frac{4\pi}{c_0}(f_r + f_0)\frac{v_s(x_0 - v_s\tau)}{\sqrt{R_0^2 + (x_0 - v_s\tau)^2}} - 2\pi f_a = 0 \qquad (2.147)$$

Letting $x^* = x_0 - v_s\tau^*$ and substituting it into Eq. (2.147), we get

$$x^* = \frac{R_0 f_a}{v_s\sqrt{\left(\frac{f_r + f_0}{c_0/2}\right)^2 - \left(\frac{f_a}{v_s}\right)^2}} \qquad (2.148\text{a})$$

$$R(f_a) = R_0\frac{\frac{f_r + f_0}{c_0/2}}{\left(\frac{f_r + f_0}{c_0/2}\right)^2 - \left(\frac{f_a}{v_s}\right)^2} \qquad (2.148\text{b})$$

According to the stationary phase principle, we have

$$\Phi_1(f_r, f_a) = -\frac{4\pi}{c_0}(f_r + f_0)\frac{R_0\frac{f_r+f_0}{c_0/2}}{\sqrt{\left(\frac{f_r+f_0}{c_0/2}\right)^2 - \left(\frac{f_a}{v_s}\right)^2}}$$

$$- 2\pi f_a\left(\frac{x_0}{v_s} - \frac{R_0 f_a}{v_s^2\sqrt{\left(\frac{f_r+f_0}{c_0/2}\right)^2 - \left(\frac{f_a}{v_s}\right)^2}}\right) \tag{2.149}$$

$$= - 2\pi R_0\sqrt{\left(\frac{f_r+f_0}{c_0/2}\right)^2 - \left(\frac{f_a}{v_s}\right)^2} - 2\pi\frac{x_0}{v_s}f_a$$

Substituting Eq. (2.149) to Eq. (2.145) yields

$$s(f_r, f_a) = w\left(\frac{x_0}{v_s} - \frac{\lambda R_0 f_a}{2v_s^2\sqrt{1 - \left(\frac{\lambda f_a}{2v_s}\right)^2}}\right)a\left(-\frac{f_r}{k_r}\right)\exp\left(j\pi\frac{f_r^2}{k_r}\right)$$

$$\times \exp\left(-j2\pi R_0\sqrt{\left(\frac{f_r+f_0}{c_0/2}\right)^2 - \left(\frac{f_a}{v_s}\right)^2} - j2\pi\frac{x_0}{v_s}f_a\right) \tag{2.150}$$

The phase term in the second line can be further extended in the Taylor series with respect to f_r

$$\Phi_2(f_r, f_a) = \exp\left(-j2\pi R_0\sqrt{\left(\frac{f_r+f_0}{c_0/2}\right)^2 - \left(\frac{f_a}{v_s}\right)^2} - j2\pi\frac{x_0}{v_s}f_a\right) \tag{2.151}$$

$$= \phi_0(f_a) + \phi_1(f_a)f_r + \phi_2(f_a)f_r^2 + \phi_3(f_a)f_r^3 + \cdots$$

The first term is useful for azimuth focusing, the second term is range mitigation, the third term is quadratic range chirp signal, the fourth and more are high-frequency components.

2.7.2   Range-Doppler (RD) Algorithm

The basic RD algorithm works in the range-Doppler domain and consists of the following steps (see Figure 2.19):

1. **Range FFT** Fourier transforming the collected radar returns, $s(t, \tau)$ with respect to fast time t yields the $s(f_r, \tau)$ shown in Eq. (2.144).

2. **Range Compressing** Using the range reference function

$$H_r(f_r) = \text{rect}\left(-\frac{f_r}{k_r}\right)\cdot\exp\left(-j\pi\frac{f_r^2}{k_r}\right) \tag{2.152}$$

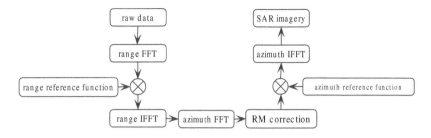

FIGURE 2.19: Block diagram of the range-Doppler algorithm.

we perform the range pulse compressing

$$s_c(f_r, \tau) = s(f_r, \tau)H_r(f_r, \tau)$$

$$= w(\tau)\text{rect}\left(\frac{f_r}{k_r}\right)\exp\left[-j\frac{4\pi}{c_0}f_r R(\tau)\right]\exp\left[-j\frac{4\pi}{\lambda}R(\tau)\right]$$

$$(2.153)$$

3. **Range IFFT** Inverse Fourier transforming $s_c(f_r, \tau)$ with respect to f_r yields

$$s_c(t, \tau) = w(\tau)\text{sinc}\left[\pi B_r\left(t - \frac{2R(\tau)}{c_0}\right)\right]\exp\left[-j\frac{4\pi R(\tau)}{\lambda}\right] \quad (2.154)$$

4. **Azimuth FFT** Fourier transforming $s_c(t, \tau)$ with respect to slow time τ

$$s_c(t, f_a) = w\left(\frac{f_a - f_{dc}}{f_{dr}}\right)\exp\left[-j\pi\frac{(f_a - f_{dc})^2}{f_{dr}}\right]\exp\left(-j\frac{4\pi R_0}{\lambda}\right)$$

$$\times \text{sinc}\left[\pi B_r\left[t - \frac{2}{c_0}\left(R_0 - \frac{\lambda}{2}f_{dc}\frac{f_a - f_{dc}}{f_{dr}} - \frac{\lambda}{4}\frac{(f_a - f_{dc})^2}{f_{dr}}\right)\right]\right]$$

$$(2.155)$$

5. **Range Mitigation Correction** Range mitigation correction is performed by a time shift

$$\Delta t(t, f_a) = \frac{2}{c_0}\left(-\frac{\lambda}{2}f_{dc}\frac{f_a - f_{dc}}{f_{dr}} - \frac{\lambda}{4}\frac{(f_a - f_{dc})^2}{f_{dr}}\right) \quad (2.156)$$

This operation implies a time domain interpolation which can cause artifacts in the image.

6. **Azimuth Compression** After range mitigation correction, we then have

$$s_c(t, f_a) = w\left(\frac{f_a - f_{dc}}{f_{dr}}\right)\exp\left[-j\pi\frac{(f_a - f_{dc})^2}{f_{dr}}\right]$$

$$\times \exp\left(-j\frac{4\pi R_0}{\lambda}\right)\times \text{sinc}\left[\pi B_r\left(t - \frac{2}{c_0}R_0\right)\right]$$

$$(2.157)$$

An one-dimension filter function

$$H_a(f_a) = \exp\left[j\pi\frac{(f_a - f_{\rm dc})^2}{f_{\rm dr}}\right] \tag{2.158}$$

is applied for the azimuth compression

7. **Azimuth IFFT** The focused imagery can then be obtained by an IFFT with respect to f_a

$$s_i(t,\tau) = {\rm sinc}\left[\pi B_r\left(t - \frac{2R_0}{c_0}\right)\right] \cdot {\rm sinc}\left[\pi B_a\left(\tau - \tau_0\right)\right] \tag{2.159}$$

A further improvement of the basic Range-Doppler algorithm is possible by considering the signal model given in Eq. (2.154). As the transmitted radar signal usually has a large time-bandwidth product, the Eq. (2.154) can be approximated by

$$s_c(t,\tau) = \delta\left(t - \frac{2R(\tau)}{c_0}\right)\exp\left(-j\frac{4\pi}{\lambda}R(\tau)\right) \tag{2.160}$$

In this way, the Range-Doppler algorithm with second range compression (SRC) can be developed. The details can be found in [17].

2.7.3   Chirp-Scaling (CS) Algorithm

The range-Doppler algorithm is an efficient imaging technique and, in principle, solves the problems of azimuth focusing and range cell mitigation correction (RCMC). However, the range-Doppler algorithm has the following disadvantages: 1) the secondary range compression cannot easily incorporate azimuth frequency dependence, and 2) RCMC requires an interpolator. To achieve accurate results, the kernel of the interpolator must have range-varying coefficients and should span many samples. In practice, this leads to a loss than ideal accuracy/efficiency tradeoff. Most processors use 4-point to 8-point kernels, which favors the efficiency side of the tradeoff: precision SAR processing loses out in the bargain.

SAR data can also be focused by wavenumber domain algorithms, which have the advantage of dealing directly with the natural polar coordinate system arising from wave propagation, and achieve most signal processing operations in two-dimensional frequency domain with a relatively simple phase multiply. The dominant disadvantage of wavenumber domain algorithms is that an interpolator is needed to match the range-dependent RCMC parameter variations, either for the Stolt change of variables in the two-dimensional frequency domain [41] or for the residual RCMC in the range-Doppler domain [321]. Since range is not available as an independent parameter in the two-dimensional frequency domain, when there are several signals from different range locations simultaneously present, as is generally the case, then their transforms overlap. A fixed value of the range parameter corrects range cell migration for one specific range position, but signals from all other ranges have only an approximate correction. Thus, it is not possible to correct fully for range cell migrations using linear filters in the two-dimensional transform domain when no other RCMC stages are used.

The chirp scaling algorithm is designed around *curvature equalization* [320], so that by the time the signal is transformed to the two-dimensional frequency domain, all of the range cell migration trajectories have been adjusted to have congruent loci, equivalent to the trajectory of scatterer at the selected reference range, R_{ref}. As all of the resulting range migration trajectories are single valued in the two-dimensional frequency domain, then RCMC may be completed by a phase multiply that is known and single valued at each and every point. Curvature equalization is done simply by application of a phase multiply in the range signal/Doppler domain. If the radar uses a large time-bandwidth signal modulation having a dominant linear frequency modulation (FM) characteristics, then there exists a unique solution. The required function, which depends on Doppler frequency, has quadric phase modulation in range, whose FM rate is very small compared to the FM rate of the range pulse modulation. The equalizing phase function causes a range/Doppler-dependent change in range scale over signal space. Secondary range compression may be implemented by using a range compression filter whose rate is Doppler frequency dependent.

Before discussing the detailed chirp scaling algorithm, we introduce briefly the scaling principle applied to a large time-bandwidth LFM signal [310]. Consider a one-dimensional signal with chirp rate k_r and phase center at $t = t_1$ as

$$\exp\left\{-j\pi k_r(t-t_1)^2\right\} \tag{2.161}$$

The phase structure of this signal may be slightly reshaped by multiplying with another linearly modulated signal having an FM rate C_s that is a small fraction of that of the original signal. The multiplication factor is

$$\Phi_1 = \exp\left\{-j\pi C_s K_s t^2\right\} \tag{2.162}$$

The result, after multiplication of the two modulated signal, is

$$\exp\left\{-j\pi K_s\left[(1+C_s)t^2 - 2t_1 t + t_1^2\right]\right\} \tag{2.163}$$

which is a chirp signal resembling the original, but with a new phase center

$$t_1' = \frac{1}{1+C_s}t_1 \tag{2.164}$$

and new chirp rate

$$k_r' = k_r(1+C_s) \tag{2.165}$$

The position of the new phase center is proportional to t_1, and thus is scaled linearly. The new chirp rate is known, and in the chirp scaling algorithm, is matched, to lead to range signal focus, during range compression. A consequence of the chirp scaling multiply is that the phase of the signal at the new phase center is

$$-\pi k_r \frac{C_s}{1 + C_s} t_1^2 \tag{2.166}$$

which, having a known value, may be eliminated by a final phase multiply.

The chirp scaling algorithm uses operations in both the range-Doppler domain and the two-dimensional frequency domain. Rather than completing range compression first, the chirp scaling algorithm starts and finishes with azimuth transforms; range operations are embedded in the middle. In the following, we discuss the key steps in the algorithm:

1. **Azimuth FFT**

   Having available a block of raw data, the first step is an FFT in the azimuth dimension, which carries the signal data to the range signal/Doppler domain. At this stage, it is helpful to distinguish between the "range (image)/Doppler" domain and the "range signal/Doppler" domain. Since the signal is characterized by a large time-bandwidth product, there is a degree-of-freedom available in the signal domain that is not available in the compressed range (image) dimension. The chirp scaling algorithm exploits the LFM signal structure that exists only prior to range compression. Considering the baseband signal impulse response expressed in Eq. (2.140), the corresponding range-Doppler transform may be written as

$$S_1(t, f_a) = A_1 \cdot w_r \left(t - \frac{2R(f_a; R_0)}{c_0} \right) \cdot W_a \left(-\frac{\lambda R_0 f_a}{2v_s^2 \sqrt{1 - \left(\frac{\lambda f_a}{2v_s} \right)^2}} \right)$$

$$\times \exp \left\{ -j\pi K_s(f_a; R_0) \cdot \left[t - \frac{2R(f_a; R_0)}{c_0} \right] \right\}$$

$$\times \exp \left\{ -j\frac{4\pi}{\lambda} \sqrt{1 - \left(\frac{\lambda f_a}{2v_s} \right)^2} \right\} \tag{2.167}$$

   where A_1 is a complex constant and $W_a(\cdot)$ is the Fourier representation of $w_a(\tau)$. The range migration $R(f_a; R_0)$, the range curvature factor $C_s(f_a)$ and the effective FM chirp rate $K_s(f_a; R_0)$ can be expressed, respectively, by

$$R(f_a; R_0) = R_0 \left[1 + C_s(f_a) \right] \tag{2.168}$$

$$C_s(f_a) = \frac{1}{\sqrt{1 - \left(\frac{\lambda f_a}{2v_s} \right)^2}} - 1 \tag{2.169}$$

$$K_s(f_a; R_0) = \cfrac{k_r}{1 + k_r \cfrac{2\lambda}{c_0^2} \cfrac{\left(\frac{\lambda f_a}{2v_s}\right)^2}{\left[1 - \left(\frac{\lambda f_a}{2v_s}\right)^2\right]^{3/2}}} \qquad (2.170)$$

2. **Chirp Scaling Phase Multiply**

While in the range signal/Doppler domain, the data array is multiplied by a function whose phase is chosen so that the range migration phase term of each and every scatterer is equalized to that of the reference range, R_{ref}. In principle, the choice of the reference locus is not critical, and even may be outside the swath being imaged. The time locus of the reference range in the range signal/Doppler domain is

$$t_{ref}(f_a) = \frac{2R_{ref}}{c_0}[1 + C_s(f_a)] \qquad (2.171)$$

The required chirp scaling multiplier is [332]

$$H_1(t, f_a; R_{ref}) = \exp\left\{-j\pi K_s(f_a; R_{ref}) \cdot C_s(f_a) \cdot [t - t_{ref}(f_a)]^2\right\} \qquad (2.172)$$

It is exactly a quadratic function of t only if: 1) the radar pulse is LFM signal, and 2) the azimuth chirp rate is constant with range. In practice, the azimuth chirp rate is range dependent, and the range chirp may not be perfectly linear. These situation may be handled directly in the algorithm by inclusion of a matching nonlinear component in the chirp scaling phase multiply [83, 320, 448].

All parameters of the phase multiplier function are single valued and known over the range signal/Doppler domain. Multiplying $S_1(t, f_a)$ by $H_1(t, f_a; R_{ref})$, we get

$$S_1(t, f_a) = A_1 \cdot w_r\left(t - \frac{2R(f_a; R_0)}{c_0}\right) \cdot W_a\left(-\frac{\lambda R_0 f_a}{2v_s^2 \sqrt{1 - \left(\frac{\lambda f_a}{2v_s}\right)^2}}\right)$$

$$\times \exp\left\{-j\pi K_s(f_a; R_{ref}) \cdot [1 + C_s(f_a)]\left[t - \frac{2}{c_0}(R_0 + R_{ref}C_s(f_a))^2\right]\right\}$$

$$\times \exp\left\{-j\frac{4\pi}{\lambda}\sqrt{1 - \left(\frac{\lambda f_a}{2v_s}\right)^2} - j\pi K_s(f_a; R_{ref})\frac{C_s(f_a) \cdot (t - t_{ref})^2}{1 + C_s(f_a)}\right\} \qquad (2.173)$$

This multiplication causes a very small range and Doppler-dependent deformation of each range chirp phase structure so that the phase centers of all signals follow the same reference curvature trajectory

$$t_{ref}(f_a) = \frac{2}{c_0}[r + R_{ref}C_s(f_a)] \qquad (2.174)$$

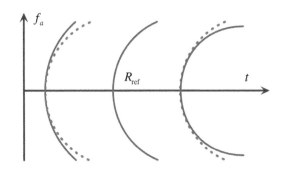

FIGURE 2.20: Equalized range curvatures resulting from the chirp scaling algorithm.

The spatial loci of the resulting range curvature are illustrated in Figure 2.20, shown as they would appear after range focusing.

3. **Range FFT**

The effect of the chirp scaling phase multiply is evident after application of the range FFT. The result is the two-dimensional frequency domain expression

$$S_2(f_r, f_a) = A_2 \cdot W_r \left(-\frac{f_r}{K_s(f_a; R_{\text{ref}}) \cdot [1 + C_s(f_a)]} \right)$$

$$\times W_a \left(-\frac{\lambda R_0 f_a}{2v_s^2 \sqrt{1 - \left(\frac{\lambda f_a}{2v_s} \right)^2}} \right)$$

$$\times \exp \left\{ -j \frac{4\pi R_0}{\lambda} \sqrt{1 - \left(\frac{\lambda f_a}{2v_s} \right)^2} - j\Theta_\Delta(f_a; R_0) \right\}$$

$$\times \exp \left\{ j\pi \frac{f_r^2}{K_s(f_a; R_{\text{ref}}) \cdot [1 + C_s(f_a)]} \right\}$$

$$\times \exp \left\{ -j \frac{4\pi}{c_0} f_r \left[R_0 + R_{\text{ref}} \cdot C_s(f_a) \right] \right\}$$

$$(2.175)$$

where $W_r(\cdot)$ is the Fourier transform of $w_r(\cdot)$, A_2 is a complex constant determined by the A_1 and range FFT, and the $\Theta_\Delta(f_a; R_0)$ is

$$\Theta_\Delta(f_a; R_0) = \frac{4\pi}{c_0} K_s(f_a; R_{\text{ref}}) \cdot [1 + C_s(f_a)] C_s(f_a) \cdot (R_0 - R_{\text{ref}})^2$$

$$(2.176)$$

Each phase term of Eq. (2.175) has a direct interpretation. The first phase function is the azimuth modulation which is constant with respect to the range (frequency) variable. It includes a parametric dependence on range r, which is matched in the range/Doppler phase multiplier H_3 as discussed below. The second phase term, quadratic in range frequency f_r, is the effective range chirp modulation, whose initial chirp rate has been modified by the range curvature, and modified even more by the Doppler-dependent chirp rate of the chirp scaling applied during the previous step. The chirp-scaled frequency modulation (FM) rate is single valued and known over the two-dimensional frequency domain, and may be exactly compensated. The third phase term in Eq. (2.175), linear in range frequency f_r, carries the correct range position r of each scatterer as well as its range curvature. Thanks to the previous chirp scaling multiply, however, the relative range curvature is the same value for all ranges; hence, it is single valued over the two-dimensional frequency domain.

4. **RCMC, Range Compression and SRC Multiply**

   RCMC and range focus, including SRC (second range compression) with its Doppler-dependent variation and compensation for the chirp scaling change in FM rate, may be done with a phase multiplication by a function whose value is known at each point on the two-dimensional space. The phase multiplication is

$$H_2(f_r, f_a) = \exp\left\{ -j\pi \frac{f_r^2}{K_s(f_a; R_{\text{ref}}) \cdot [1 + C_s(f_a)]} \right\}$$
$$\times \exp\left\{ -j\frac{4\pi}{c_0} f_r R_{\text{ref}} C_s(f_a) \right\}$$

$$(2.177)$$

The first factor uses for the range compression including the SRC. The RCMC is done by the second factor. The signal, after RCMC, range compression and SRC multiply, can then be expressed as

$$S_3(f_r, f_a) = A_2 \cdot W_r\left(-\frac{f_r}{K_s(f_a; R_{\text{ref}}) \cdot [1 + C_s(f_a)]} \right)$$
$$\times W_a\left(-\frac{\lambda R_0 f_a}{2v_s^2 \sqrt{1 - \left(\frac{\lambda f_a}{2v_s}\right)^2}} \right)$$
$$\times \exp\left\{ -j\frac{4\pi R_0}{\lambda}\sqrt{1 - \left(\frac{\lambda f_a}{2v_s}\right)^2} - j\Theta_\Delta(f_a; R_0) \right\}$$

$$(2.178)$$

5. **Range IFFT**

With near perfect phase compensation of all range modulation, the range IFFT collapses to the focused range envelope at the correct range position, leaving only azimuth phase terms. The output signal in range-Doppler domain is

$$
S_4(f_r, f_a) = A_3 \cdot \text{sinc}\left[B_r \left(t - \frac{2R_0}{c_0} \right) \right] \cdot W_a \left(-\frac{\lambda R_0 f_a}{2v_s^2 \sqrt{1 - \left(\frac{\lambda f_a}{2v_s} \right)^2}} \right)
$$

$$
\times \exp\left\{ -j\frac{4\pi R_0}{\lambda}\sqrt{1 - \left(\frac{\lambda f_a}{2v_s} \right)^2} - j\Theta_\Delta(f_a; R_0) \right\}
$$

$$(2.179)$$

where A_3 is a complex constant.

6. **Azimuth Filtering and Phase Residual Compensation**
   The first phase term of Eq. (2.179) is the normal Doppler modulation. This term must be matched to focus each signal in azimuth. It is a single-valued two-dimensional quantity known over the domain. The phase term, $\Theta_\Delta(f_a; R_0)$, has a known value over the range/Doppler domain. It is a residual generated by the original chirp scaling phase multiply. The required phase multiplier is

$$
H_3(f_r, f_a) = \exp\left\{ j\frac{4\pi R_0}{\lambda}\sqrt{1 - \left(\frac{\lambda f_a}{2v_s} \right)^2} - j\frac{2\pi}{\lambda}c_0 t + j\Theta_\Delta(f_a; R_0) \right\}
$$

$$(2.180)$$

Multiplying $S_4(f_r, f_a)$ by $H_3(f_r, f_a)$ yields

$$
S_5(t, f_a) = A_3 \cdot W_a \left(-\frac{\lambda R_0 f_a}{2v_s^2 \sqrt{1 - \left(\frac{\lambda f_a}{2v_s} \right)^2}} \right)
$$

$$
\times \text{sinc}\left[B_r \left(t - \frac{2R_0}{c_0} \right) \right] \cdot \exp\left\{ -j\frac{2\pi}{\lambda}c_0 t \right\}
$$

$$(2.181)$$

7. **Azimuth IFFT**
   The algorithm is completed by an azimuth IFFT

$$
|S_5(f_r, f_a)| = A_4 \text{sinc}\left[B_r \left(t - \frac{2R_0}{c_0} \right) \right] \cdot \text{sinc}\left[B_a \left(\tau - \frac{x_0}{v_s} \right) \right]
$$

$$(2.182)$$

where A_4 is a complex constant, and x_0 is the scatterer azimuth position.

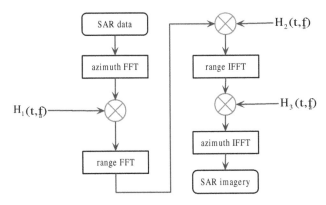

FIGURE 2.21: Flow diagram of the chirp scaling algorithm.

TABLE 2.1: Simulation parameters.

system parameter	value (unit)	system parameter	value (unit)
flying speed	150 (m/s)	range resolution	2.5 (m)
operational range	15 (km)	azimuth resolution	2.5 (m)
carrier frequency	1.25 (GHz)	sampling frequency	72 (MHz)
signal bandwidth	60 (MHz)	synthetic aperture time	4.8 (s)

The logical flow diagram of the chirp scaling algorithm is shown in Figure 2.21. It can be noticed that only multiplications and Fourier transforms are required. There are no interpolations needed for the RCMC. Note that the azimuth reference frequency used in Eq. (2.175) is chosen to be zero for simplicity. It is valid when non-zero azimuth reference frequency is used. As is true for most SAR processor, for squinted data, the Doppler centroid frequency of the azimuth antenna pattern must be used to unwrap the frequencies in the compensations used to rectify signal and image domain. This requires knowledge of the integer and the fractional part of the Doppler ambiguity.

2.7.4 Numerical Simulation Examples

To evaluate the performance of the derived imaging algorithm, the four example stripmap SAR data from five point targets, located at $(0,0)$, $(10,10)$, $(20,0)$, $(50,30)$ and $(-20,50)$, are simulated using the parameters listed in Table 2.1. Figure 2.22 shows the corresponding amplitude of the radar returns. Figure 2.23 is an image domain contour plot of a typical compressed test signal for the five point targets. It is noticed that the targets are well focused by the chirp scaling algorithm.

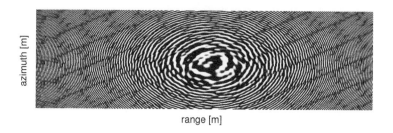

FIGURE 2.22: The amplitude of the radar returns.

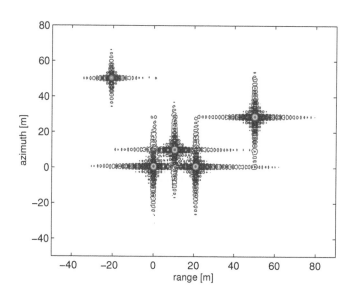

FIGURE 2.23: Imaging result of the chirp scaling algorithm.

3

Azimuth Multi-Antenna SAR

In SAR remote sensing, a broad footprint implies that fewer missions are required to map the region of interest. More importantly, for general applications, this will also imply that a given region of the earth can be repeatedly mapped with a greater frequency. The width of the ground area covered on a single pass is usually referred to as the *swath width* of the system. In a typical spaceborne SAR, the swath width is of order 100 km, coupled with a revisit time to a given area of approximately three days. Although this may be adequate for the surveillance of certain features, such as geologic formations and crops, for many other more dynamic features, such as ocean, ice and man-made targets, there could be a requirement for much shorter revisit time. In such cases, an increase in the swath coverage is desired. However, this is not a straight-forward issue and involves consideration of fundamental SAR operating constraints as well as technological limitations.

In this chapter, we will summarize SAR high-resolution and wide-swath imaging problems and discuss azimuth multi-antenna-based solutions. This chapter is organized as follows. Section 3.1 introduces the constraints on azimuth resolution and swath width. Two different approaches to implementing the displaced phase center antenna (DPCA) are introduced Sections 3.2 and 3.3, respectively. Next, the azimuth nonuniform sampling problem is discussed in Section 3.4, followed by azimuth multi-antenna SAR system performance analysis in Section 3.5. Section 3.6 designs azimuth multi-antenna SAR conceptual systems. Finally, the use of azimuth multi-antenna SAR in ground moving target indication and imaging (GMTI) is discussed in Section 3.7.

3.1 Constraints on Resolution and Swath

An efficient SAR should provide high-resolution over a wide area of surveillance, although there is a contradiction between azimuth resolution and swath width [77, 241]. A good azimuth resolution requires a short antenna to illuminate a long synthetic aperture which results in a wide Doppler bandwidth, but this calls for a high pulse repeated frequency (PRF) to sample the Doppler spectrum. To better understand this trade-off, we consider the geometry given in Figure 3.1, where a flat earth model is assumed.

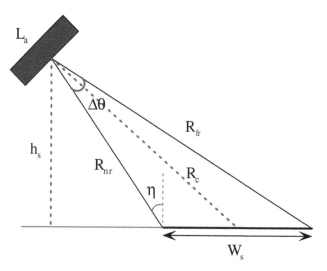

FIGURE 3.1: Equalized range curvatures resulting from the chirp scaling algorithm.

The swath width on the ground (W_s) corresponds to the ground region illuminated within the elevation beamwidth of the antenna [135]

$$W_s = \frac{W_r}{\sin(\eta)} = \frac{\Delta\theta \cdot R_c}{\cos(\eta)} = \frac{\lambda \cdot R_c}{L_a \cdot \sin(\eta)} \tag{3.1}$$

where W_r is the swath width in slant range, η is the local incidence angle for the center of the swath, $\Delta\theta_{el}$ is the antenna 3-dB beamwidth in elevation, R_c is the slant range distance from the radar sensor to the center of the swath and L_a is the antenna height.

In order to avoid range ambiguities, the difference in the reception time between the farthest point in the swath (farthest-range R_{fr}) and the nearest point (nearest-range R_{nr}) should be smaller than the pulse repetition period

$$\frac{2R_{fr}}{c_0} - \frac{2R_{nr}}{c_0} < \frac{1}{\text{PRF}} \tag{3.2}$$

Correspondingly the swath in slant range is bounded by

$$W_r < \frac{c_0}{2\text{PRF}} \tag{3.3}$$

On the other hand, in order to avoid azimuth frequency spectra aliasing and taking into account the *Nyquist sampling theorem*, a lower bound on the PRF should be

$$\text{PRF} > \frac{2v_s}{L_a} = \frac{v_s}{\rho_a} \tag{3.4}$$

where L_a and ρ_a denote the antenna length and the best possible azimuth resolution, respectively. This relation states that the PRF should be larger than the range of Doppler frequencies within the bounds of the area illuminated by the physical antenna in azimuth, which is the Doppler bandwidth for the antenna length. Note that the azimuth ambiguities need only be evaluated over the processing bandwidth required to achieve the needed azimuth resolution, not over the entire range of frequencies in PRF spans.

The combination of the two previous requirements to avoid ambiguities yields the classical SAR trade-off

$$\frac{W_r}{\rho_a} < \frac{c_0}{2v_s}, \quad \text{or} \quad \frac{W_s}{\rho_a} < \frac{c_0}{2v_s \sin(\eta)} \tag{3.5}$$

which requires that the swath width in slant-range W_r decreases as the azimuth resolution ρ_a improves (i.e., becomes smaller). This equation can be reformulated into the minimum antenna area constrain [76]

$$A_a = H_a L_a \geq \frac{4v_s \lambda R_c \tan(\eta)}{c_0} \tag{3.6}$$

This requirement shows that the illuminated area of the ground must be restricted so that the radar does not receive ambiguous returns in range dimension and/or azimuth dimension. In this respect, a high operating PRF is desired to suppress azimuth ambiguities. But the magnitude of the operating PRF is limited by the range ambiguity requirement.

It is derived in [135] that Eq. (3.6) is only suitable to a special case, which is when the radar designer seeks to achieve both the best possible resolution and the widest possible swath at the same time. The fundamental constraint is actually given in Eq. (3.5), which places a limit on the ratio of the swath width versus azimuth resolution that really depends only on the platform speed v_s. If we want to achieve a swath width smaller than the best possible and an azimuth resolution that is worse than the best possible resolution [190], it should be satisfactory with the following relationship

$$\frac{W_r}{\rho_a} \leq \frac{\max\{W_s\}}{\min\{\rho_a\}} = 2\frac{\lambda R_c \tan(\eta)}{A_{\text{antenna}}} \tag{3.7}$$

or, alternatively,

$$A_a \leq \frac{2\lambda R_c \tan(\eta)}{W_r} \rho_a \tag{3.8}$$

It states that there is actually an upper bound for the antenna area, depending on the size of the swath and the desired azimuth resolution. However, it is straightforward to show that this upper bound on antenna area is always greater than or equal to the lower bound expressed in Eq. (3.6) for all realizable cases.

Therefore, in conventional SAR the unambiguous swath width and the achievable azimuth resolution impose contradicting requirements on the system design, and consequently allows only for a concession between azimuth resolution and swath width. These considerations can be combined into an inequality which expresses the range of slant range values R that are appropriate to each of the unambiguous swath intervals

$$\left(\frac{n_a}{\text{PRF}} + T_p\right) \frac{c_0}{2} \leq R \leq \left(\frac{n_a + 1}{\text{PRF}} - T_p\right) \frac{c_0}{2} \tag{3.9}$$

where n_a is an integer that effectively labels each swath interval, and T_p is the pulse duration. Note that the possible presence of *guard bands*, which would lead to a small narrowing of the unambiguous swath intervals, is ignored in this equation.

Another consideration is to avoid nadir returns, which tend to swamp any other radar returns arriving at the same time. Although the relative strength of nadir returns may be reduced by arranging the radar antenna pattern to have small gain in the nadir direction, the antenna pattern synthesis requirements will be considerably relaxed if the PRF is set such that nadir returns arrive back at the antenna at the same time that pulse transmissions occur, i.e., the nadir returns are placed in the blind-range intervals. The condition for this situation to occur is simply that the operating PRF should be a multiple of the nadir round-trip time

$$\text{PRF} = \frac{k c_0}{2 h_s} \tag{3.10}$$

where k is an integer (called the 'rank' of the PRF).

3.2   Displaced Phase Center Antenna Technique

The DPCA technique was proposed to improve the performance of moving target indication (MTI) radars mounted on moving platforms [293]. Prior to the invention of the DPCA technique, airborne MTI radars suffered from the butterfly effect, a phenomenon associated with clutter leakage that causes wing-shaped patterns to appear on the position indicator.

General Electric (GE) is generally credited with inventing the DPCA technique to largely eliminate the butterfly effect [95]. In 1952, GE Electronics Laboratory in Syracuse, New York, was working to improve MTI on a ground-based radar called 'anti-clutter technique'. The study noted that by adding and subtracting the output of the azimuth difference port to the output of the azimuth sum port of a monopulse antenna, two beams—a leading beam and a trailing beam could be formed. At the proper separation angle and width the antenna rotating, signals from the trailing beam could be received at exactly the same pointing angle as the previously received signals from the leading

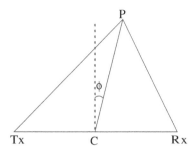

FIGURE 3.2: Geometry of the phase center approximation.

beam. Subtracting the two signals can largely cancel the clutter. GE engineer F.R. Dickey noted the similarity of this problem caused by rotation to the butterfly effect caused by translation in airborne MTI. This observation led him to propose using a monopulse antenna for motion compensation of the airborne MTI radar platform, a technique later named DPCA.

To introduce the DPCA technique, we consider the geometry of the phase center approximation shown in Figure 3.2. Let Tx be the position of the transmitter center at a given azimuth time and Rx the position of the receiver at the instant of the reception of the echo from an arbitrary scatterer. The basis of the DPCA is to replace the true bistatic situation by a fictitious monostatic one, which assumes that transmission and reception occurs from the virtual position C=(Tx+Rx)/2.

Let R_c = CP and L = Tx-Rx, expanding Tx + Rx − 2Cx in series of L ($L \ll$ Cx) yields [23]

$$\text{Tx} + \text{Rx} - 2\text{Cx} = \frac{L^2}{4R_c} \cos^2(\theta) + \frac{L^4}{64R_c^3} \cos^2(\theta) \left[4 - 5\cos^2(\theta)\right] + \ldots \quad (3.11)$$

where θ is the bearing of the target P. The phase center approximation holds when $L^2/4R_c \ll \lambda$, which can be interpreted as a far field condition. More generally, the phase center approximation also holds in near field at the condition that the received signal is advanced by $L^2/(4R_c c_0)$ and the transmission sector satisfies

$$\frac{L^2}{4\lambda R_c} \left[1 - \cos(\theta_e)\right] \ll 1 \quad (3.12)$$

where θ_e is half of the transmission beamwidth. This condition ensures that the round trip travel path is the same for all scatterer within 3dB transmission beamwidth, so that their interference pattern is also the same.

In addition, this derivation has assumed that the L/R_c is small compared to the beamwidth of both the transmitter and the receiver, so that the corresponding changes in transmission and reception directivity gains between the bistatic case and the equivalent monostatic one are negligible. Fortunately, this condition is almost always valid for usual SAR systems.

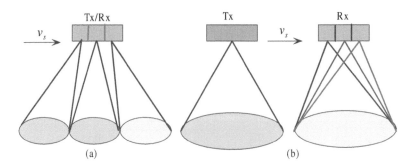

FIGURE 3.3: Geometry of two different DPCA techniques: (a) single-phase center DPCA, (b) multiple-phase center DPCA.

The basic idea of the DPCA SAR is to cancel the along-track displacement two successive pulses. In the followings, two different implementation approaches (see Figure 3.3) are described separately.

3.3   Single-Phase Center Multibeam SAR

The single-phase center multibeam (SPCM) SAR transmits pulses by a single broad azimuth beam and receives the returns by multiple narrow azimuth beams that span the mainlobe width of the transmitting beam, as shown in Figure 3.4. A distinct channel is associated with each of the receiving beams, and hence, the data are split according to the azimuth angular position or, equivalently, the instantaneous Doppler frequency center in the azimuth. As a result, while given knowledge of the relative squint angles of each beam (hence the Doppler center frequency for each beam) and assuming suitable isolation between the beams, each channel can be sampled at a Nyquist rate appropriate to the bandwidth covered by each narrow beams instead of that covered by the full beamwidth.

This arrangement enables correct sampling of the azimuth spectrum with a PRF fitting the total antenna azimuth length, which is N-times smaller than the PRF necessary for the antenna azimuth length, L_a. The area of each beam antenna is then restricted by

$$A_{\mathrm{as}} \geq \frac{4v_s \lambda R_c \tan(\eta)}{c_0} \cdot \frac{1}{N} \tag{3.13}$$

Clearly, the minimum area of each beam antenna is N-times smaller than the respective area of a monostatic SAR. Correspondingly, the relationship

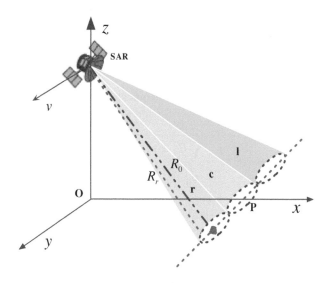

FIGURE 3.4: Geometry of the SPCM SAR imaging.

expressed Eq. (3.5) is changed to

$$\frac{W_s}{\rho_a} \leq \frac{N c_0}{2 v_s \sin(\eta)} \tag{3.14}$$

It can be noticed that the relation depends on not only the platform velocity v_s and the incidence angle η but also the number of the antenna beams.

The DPCA technique [24, 262] can be used to gain additional samples along the synthetic aperture which enables an efficient suppression of the azimuth ambiguities, i.e., the multiple beams in azimuth allow for the division of a broad Doppler spectrum into multiple narrow-band subspectra with different Doppler centroids. A coherent combination of the subspectra can yield a broad Doppler spectra for high azimuth resolution. Thus this technique is especially attractive for high-resolution SAR that uses a long antenna for unambiguous wide-swath remote sensing.

The data in each receiving channel will be aliased to the relative zero-frequency. This problem can be resolved by utlizing the prior knowledge of the relevant Doppler center frequencies to regain the full azimuth Doppler bandwidth [419], which will be detailed in the following discussions.

3.3.1 Azimuth Multichannel Signal Processing

Consider Fig. 3.4 which shows a DPCA SAR with three beams in azimuth, where the radar moves along the y-axis. For a given coordinate of the SAR, the imaged area is simultaneously illuminated by the three beams. It can be regarded as a conventional SAR (the central beam), operating with a PRF, which is one third of that required to adequately sample its beamwidth, together with two additional beams on either side of the central one. As described previously, the basic idea is that the additional samples obtained by the outer beams will fill in the gaps in a target's azimuth phase history which occur due to operating at a low PRF.

For the left beam, its instantaneous slant range in terms of the slow time variable τ is given by [73]

$$
\begin{aligned}
R_l(\tau) &= \sqrt{R_0^2 + v_s^2\tau^2 - 2R_0 v_s \tau \sin(\theta_s)} \\
&\approx R_0 - v_s\tau \sin(\theta_s) + \frac{v_s^2\tau^2 \cos^2(\theta_s)}{2R_0}
\end{aligned}
\tag{3.15}
$$

where R_0 is the nearest slant range of the target for the left beam, and θ_s is the squint angle. The Doppler centroid frequency and azimuth bandwidth can be derived from Eq. (3.15) as

$$
f_{dl} = \frac{2}{\lambda} v_s \sin\theta_s
\tag{3.16a}
$$

$$
B_{dl} = \frac{2v_s \cos\theta_s}{L_{as}}
\tag{3.16b}
$$

where L_{as} denotes the length of each beam antenna.

Similarly, for the central beam and right beam, we have

$$
f_{dc} = 0
\tag{3.17a}
$$

$$
B_{dc} = \frac{2v_s}{L_{as}}
\tag{3.17b}
$$

and

$$
f_{dr} = -\frac{2}{\lambda} v_s \sin\theta_s
\tag{3.18a}
$$

$$
B_{dr} = \frac{2v_s \cos\theta_s}{L_{as}}
\tag{3.18b}
$$

Generally, for small θ_s, we have the following approximations

$$
B_{dl} \approx B_{dc} \approx B_{dl} = B_{ds}
\tag{3.19a}
$$

$$
f_{dl} - f_{dc} = f_{dc} - f_{dr} = B_{ds}
\tag{3.19b}
$$

The ambiguous Doppler spectrum should be unambiguously recovered for subsequent signal processing. Some reconstruction algorithms have been proposed in literature, e.g., [146, 214]. A block diagram for the reconstruction in the case of three channels is shown in Figure 3.5. This algorithm is based on considering the multichannel SAR data acquisition as a linear system with multiple channels, each described by a linear filter. From the sampling theorem [70], we know that the sampled signal spectrum $X_s(f)$ is the sum of the

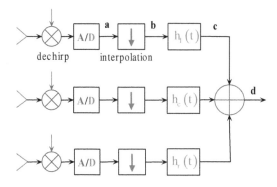

FIGURE 3.5: Multichannel signal reconstruction algorithm in case of three channels.

unsampled signal spectrum, $X_0(f)$. It repeats every f_s Hz with f_s being the sampling frequency. That is

$$X_s(f) = f_s \sum_{n=-\infty}^{\infty} X_0(f - nf_s) \qquad (3.20)$$

If $f_s \geq 2B_{ds}$, the replicated spectra will not be overlapped, and the original spectrum can be regenerated by chopping $X_s(f)$ off above $f_s/2$. Thus, $X_0(f)$ can be reproduced from $X_s(f)$ through an ideal low-pass filter that has a cutoff frequency of $f_s/2$. The subsequent data processing involves interpolating the data from each channel. We then have

$$y_s(n) = \begin{cases} x_s(\frac{n}{N_0}), & n = 0, \pm N_0, \pm 2N_0, \cdots \\ 0, & \text{otherwise} \end{cases} \qquad (3.21)$$

where N_0 is the interpolation scaling factor. The corresponding spectra is

$$Y_s(f) = \sum_{n=-\infty}^{\infty} y_s(n)e^{-j2\pi fn} = X_s(N_0 f) \qquad (3.22)$$

Then, the linear filters are derived as

$$H_k(f) = \begin{cases} N_0, & \left|f - \frac{f_{ck}}{N_0}\right| \leq \frac{B_{ds}}{N_0}, k \in (l, c, r) \\ 0, & \text{otherwise} \end{cases} \qquad (3.23)$$

where $k \in (l, c, r)$ is shown in Figure 3.5, and f_{ck} is the corresponding Doppler frequency centroid.

Finally, the filtered signals can be coherently combined into a wideband Doppler signal, as shown in Figure 3.6. More details can be found in Figure 3.7.

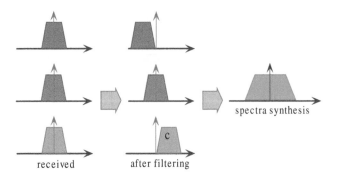

FIGURE 3.6: Azimuth Doppler spectra synthesis for multichannel subsampling in case of three channels.

The ambiguity suppression allows for an improved resolution and an enlarged swath coverage. Notice that, for optimum performance the relation between sensor velocity and the along-track offset of the three subchannels must result in equally spaced effective phase center, so that a uniform spatially sampling of the received signal can be obtained [214].

3.3.2   System Performance Analysis

To keep the analysis independent of specific SAR configurations, a general mathematical mode is established in the following discussions.

For a given range and azimuth antenna pattern, the PRF must be optimally designed such that the total ambiguity noise contribution is small enough relative to the signal, i.e., -18 to -20 dB. A low PRF will increase the azimuth ambiguity level due to increased aliasing of the azimuth spectra. On the other hand, a high PRF value will reduce the interpulse period and result in overlapped pulses. To resolve these problems, the transmit interference restriction on the PRF should be satisfactory with

$$\frac{n'}{\frac{2R_{\min}}{c_0} - T_p} < \text{PRF} < \frac{n' + 1}{\frac{2R_{\max}}{c_0}} \tag{3.24}$$

where $R_{\min}$ is the nearest slant range, $R_{\max}$ is the farthest slant range, and n' is a given integer. At the same time, to avoid the nadir interference restriction, the PRF should be

$$\frac{m'}{\frac{2R_{\min}}{c_0} - 2T_p - \frac{2h_s}{c_0}} < \text{PRF} < \frac{m'}{\frac{2R_{\max}}{c} - \frac{2h_s}{c_0}} \tag{3.25}$$

where m' is a given integer.

Alternatively, given a PRF or a range of PRFs, the antenna dimension and/or beamforming weight vector (to lower the sidelobe energy) must

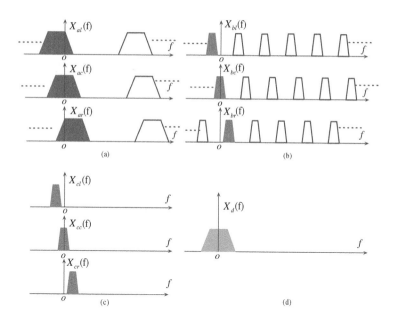

FIGURE 3.7: The detailed azimuth Doppler spectra synthesis.

be specifically designed such that the required ambiguity-to-signal ratio (ASR) specification can be achieved. Since the range ambiguity-to-signal ratio (RASR) of the SPCM SAR can be analyzed similar to the general SAR, in the following we analyze only the azimuth ambiguity-to-signal ratio (AASR) performance.

Consider the three beams illustrated in Figure 3.4, the k-th azimuth antenna beampattern can be represented by

$$G_k(\theta) = \text{sinc}^2 \left[\frac{\pi L_{as} \cos(i \cdot \theta_s)}{\lambda} \sin(\theta - i \cdot \theta_s) \right] \qquad (3.26)$$

where $i \in (-1, 0, 1), k \in (l, c, r)$. Note that $i \in (-1, 0, 1)$ is determined from the positions of the three beams, i.e., $i = -1$ is for the left beam, $i = 0$ is for the central beam and $i = 1$ is for the right beam. Because the 3-dB beam-width is approximately determined by [241]

$$\theta \approx \frac{\lambda}{2v_s} f \qquad (3.27\text{a})$$

$$k_a = \frac{\lambda}{L_{as}} \qquad (3.27\text{b})$$

with k_a being a given constant. Equation (3.27) can then be further simplified into

$$G_k(f) \approx \text{sinc}^2 \left[\pi L_{as} \cos \left(i \cdot \frac{k_a c_0}{f L_{as}} \right) \left(\frac{f}{2v_s} - i \cdot \frac{k_a}{L_{as}} \right) \right] \qquad (3.28)$$

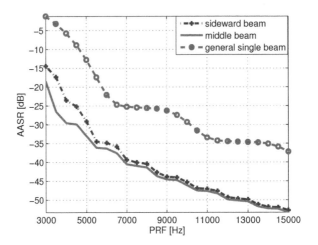

FIGURE 3.8: The AASR of an example azimuth multi-antenna SAR as a function of PRF.

Then, from Eq. (2.123) we can get

$$
\mathrm{AASR}_k(\mathrm{PRF}) = \left\{ \sum_{\substack{m=-\infty \\ m\neq 0}}^{\infty} \left[\int_{(i-0.5)B_{ds}}^{(i+0.5)B_{ds}} G_k^2(f + m \cdot \mathrm{PRF})\mathrm{d}f \right. \right.
$$

$$
\left. \left. + \sum_{j\neq k} \int_{(i-0.5)B_{ds}}^{(i+0.5)B_{ds}} G_k(f + m \cdot \mathrm{PRF})G_j(f + m \cdot \mathrm{PRF})\mathrm{d}f \right] \right\} \qquad (3.29)
$$

$$
\times \left\{ \int_{(i-0.5)B_{ds}}^{(i+0.5)B_{ds}} G_k^2(f)\mathrm{d}f + \sum_{j\neq k} \int_{(i-0.5)B_{ds}}^{(i+0.5)B_{ds}} G_k(f)G_j(f)\mathrm{d}f \right\}^{-1}
$$

with $j \in (l, c, r)$.

An example wide-swath SAR is considered with the following parameters [419]: $h_s = 60$ km, $v_s = 1000$ m/s, $\lambda = 0.03$ m, $L_{as} = 1.2$ m, the calculated AASR results are illustrated in Figure 3.8. Note that, the AASR is typically specified to be on the order of -20 dB; however, even at this value ambiguous signals may be observed in images that have very bright targets adjacent to dark targets, because SAR imagery can have an extremely wide dynamic range due to the compression processing gain. As such, a lower AASR, e.g., -30 dB, is desirable.

The set of PRFs is therefore determined by the acceptable maximum range and AASR requirement, as well as the nadir interference. Note that, there may be no acceptable PRFs at some incidence angles that can meet the minimum requirements. In this case, the designer has the option to relax the performance specifications for the imaging area or exclude these modes from the operations plan.

The reflected signal power available at the receiver is determined by [432]

$$P_r = \frac{P_t G}{4\pi R_c^2} \cdot \frac{\sigma_0}{4\pi R_c^2} \cdot \frac{\lambda^2 G}{4\pi} \tag{3.30}$$

where P_t is the transmit power, G is the antenna gain, and σ_0 is the target radar cross section (RCS) per unit area. In radar imaging applications, the total data samples are processed coherently to produce a single image resolution cell. The thermal noise samples can be taken as independent from sample to sample within each pulse, and from pulse to pulse. After coherent range and azimuth compression, the image signal-to-noise ratio (SNR) is determined by

$$\text{SNR}_{\text{image}} = \frac{P_t G}{4\pi R_c^2} \cdot \frac{N A_{as} \sigma_0}{4\pi R_c^2} \cdot \frac{1}{K_0 T_{\text{sys}} B_n F_n L_f} \cdot \frac{\tau_i}{\tau_0} \cdot \frac{\text{PRF} \cdot R_c \cdot \lambda}{v_s \cdot \rho_a} \cdot \xi \tag{3.31}$$

The ratio of the compressed signal during τ_0 and uncompressed signal during τ_i is equivalent to the SNR gain after range compression. $(\text{PRF} \cdot R_c \cdot \lambda)/(v_s \cdot \rho_a)$ is the integrated signals during the time of the aperture synthesis. Assuming that the receiver bandwidth and the transmit signal bandwidth are matched and $\xi = 0.5$, then $B_n \tau_0 \approx 1$ with B_n being the noise bandwidth. Additionally, F_n is the receiver noise figure, K_0 is the Boltzmann constant, L_f is the loss factor, and T_{sys} is the system noise temperature.

A quantity directly related to SAR processing is the noise equivalent sigma zero (NESZ) defined as the RCS for which the $\text{SNR}_{\text{image}}$ is equal to 1 ($\text{SNR}_{\text{image}} = 0$ dB). The NESZ can also be interpreted as the smallest target RCS that can be detected by the SAR system against thermal noise. From Eq. (3.31) we can get [218]

$$\text{NESZ} = \frac{8\pi R_c^3 v_s \lambda K T_{\text{sys}} F_n L_f}{P_{\text{avg}} N A_{as}^2 \rho_r} \tag{3.32}$$

where P_{avg} and ρ_r are the average transmit power and the size of the range resolution cell for one look, respectively.

To calculate the system performance we assume a range resolution $\rho_r = 0.2$ m, an overall loss factor $L_f = 3$ dB, and a receiver noise figure $F_n = 3$ dB. It is further assumed that the signal bandwidth is adjusted for varying incidence angles such that the ground-range resolution is constant across the whole swath. An example system design is provided in Table 3.1. We note that, for the incidence angles given in Table 3.1, the swath width is about 8 km and the NESZ is approximately −48 dB. These results show that a satisfactory performance can be achieved, however with only a small number of subapertures that have relative small antenna area.

TABLE 3.1: Performance parameters of an example azimuth multi-antenna SAR.

Parameters	Variables	Values
mean transmit power	P_{avg}	$1W$
number of sub-aperture	N	3
sub-aperture antenna length	L_{as}	$0.40m$
sub-aperture antenna width	H_a	$0.10m$
near-space vehicle velocity	v_s	$500m/s$
incidence angle	η	$20°$
swath width	W_s	$6.80km$
radiometric resolution	NESZ	$-48.58dB$
incidence angle	η	$25°$
swath width	W_s	$7.30km$
radiometric resolution	NESZ	$-48.11dB$
incidence angle	η	$35°$
swath width	W_s	$8.94km$
radiometric resolution	NESZ	$-46.73dB$

3.4   Multiple-Phase Center Multibeam SAR

The multiple-phase center multibeam (MPCM) SAR also employs multiple receiving beams in azimuth; however, its operating mode is quite different from that of the previous SPCM SAR. In this case, the system transmits a single broad beam and receives the radar returns in multiple beams which are displaced in the along-track direction. The motivation is that multiple independent sets of target returns can be obtained for each transmitted pulse if the distance between phase centers is suitably set. This method basically implies that we may broaden the azimuth beam from the diffraction-limited width, giving rise to improved resolution, without having to increase the system operating PRF.

3.4.1   System Scheme and Signal Model

Consider Figure 3.9, the slant range for the central receiving beam can be represented by

$$
\begin{aligned}
R(\tau) &= \sqrt{R_c^2 + v_s^2(\tau - \tau_0)^2 - 2R_0 v_s(\tau - \tau_0)\sin(\theta_s)} \\
&\approx R_c - v_s \sin(\theta_s)(\tau - \tau_0) + \frac{v_s^2 \cos^2(\theta_s)}{2R_c}(\tau - \tau_0)^2
\end{aligned}
\tag{3.33}
$$

where τ_0 is the azimuth reference time. The corresponding phase history can be expressed as

$$
\Phi_c(\tau) = \frac{4\pi R(\tau)}{\lambda}
\tag{3.34}
$$

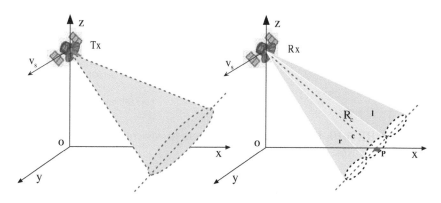

FIGURE 3.9: Illustration of the MPCM SAR systems.

Suppose the subaperture element separated by d_a, the phase histories of the left beam and right beam can be represented, respectively, by

$$\Phi_l(\tau) = \frac{2\pi R(\tau)}{\lambda} + \frac{2\pi R(\tau - \frac{d_a}{v_s})}{\lambda} \tag{3.35}$$

$$\Phi_r(\tau) = \frac{2\pi R(\tau)}{\lambda} + \frac{2\pi R(\tau + \frac{d_a}{v_s})}{\lambda} \tag{3.36}$$

Expanding (3.35) and (3.36) in Taylor series up to the second order yields

$$\Phi_l(\tau) = \frac{2R_c - 2v_s \sin(\theta_s)\left(\tau - \tau_0 - \frac{d_a}{2v_s}\right) + v_s^2\left(\frac{\cos^2(\theta_s)}{R_c}\right)\left(\tau - \tau_0 - \frac{d_a}{2v_S}\right)}{\lambda}$$
$$+ \frac{d_a^2 \cos^2(\theta)}{2R_c\lambda} \tag{3.37}$$

$$\Phi_r(\tau) = \frac{2R_c - 2v_s \sin(\theta_s)\left(\tau - \tau_0 + \frac{d_a}{2v_s}\right) + v_s^2\left(\frac{\cos^2(\theta_s)}{R_c}\right)\left(\tau - \tau_0 + \frac{d_a}{2v_S}\right)}{\lambda}$$
$$+ \frac{d_a^2 \cos^2(\theta)}{2R_c\lambda} \tag{3.38}$$

Since the last term $d_a^2 \cos^2(\theta)/2R_c\lambda$ is small and can be ignored, Eqs. (3.37) and (3.38) can be further simplified into

$$\Phi_l(\tau) = \frac{R(\tau - \frac{d_a}{2v_s})}{\lambda} \tag{3.39}$$

$$\Phi_r(\tau) = \frac{R(\tau + \frac{d_a}{2v_s})}{\lambda} \tag{3.40}$$

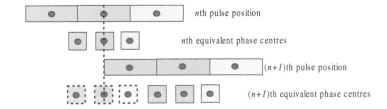

FIGURE 3.10: Equivalent phase centers in case of three receiving subapertures.

According to the DPCA principle, subsequent data processing simply involves in interleaving the multi-channel data properly in the azimuth direction so that the proper relative phasing is maintained. Note that the application of the DPCA technique assumes that the cubic (and higher order) terms of the azimuth phase history are ignorable. A general criterion for this condition to hold is that these terms cause less than $\pi/2$ over the synthetic aperture time.

For optimum performance. the along-track displacement of the subapertures $i = \{2, \ldots, N\}$ relative to the first receiver $(i = 1)$ should be chosen as [214]

$$x_i - x_1 \approx \frac{2v_s}{\text{PRF}} \left(\frac{i-1}{n} + k_i \right), \qquad k_i \in Z \qquad (3.41)$$

which will result in a uniform sampling of the received SAR signal. In a multistatic constellation the k_i may be different for each receiver, which enables a great flexibility in choosing the along-track distance between the platforms. In a single-platform system, all k_i will be zero. Since the subaperture distance and the platform velocity are fixed in this case, the required PRF is

$$\text{PRF} = \frac{2v_s}{N d_a} = \frac{2v_s}{L_a} \qquad (3.42)$$

where we assume that the overall antenna length in azimuth L_a with N subaperture elements, each length is d_a. In this case, a higher equivalent sampling frequency can be obtained due to the virtual phase centers, as shown in Figure 3.10.

3.4.2   Nonuniform Spatial Sampling

The PRF in a single-platform MPCM system has thus to be chosen such that the SAR platform moves just one half of the total antenna length between subsequent radar pulses. This optimum PRF yields a data array equivalent to that of a single-aperture system operating with $N \cdot \text{PRF}$. However, such a rigid selection of the PRF may be in conflict with the timing diagram for some incident angles. Moreover, it will exclude the opportunity to use an increased PRF for improved azimuth ambiguity suppression.

Any deviation from the relation given in Eq. (3.42) will result in a non-equally sampled data array along the synthetic aperture that is no longer equivalent to a monostatic case. To analyze the impact of nonuniform displaced phase center sampling, we consider the received radar returns

$$s_i(t, \tau) \approx \sigma_i \left[h_0(t) \otimes_t h_{1,i}(t, \tau) \right], \quad i = 1, 2, \dots, N \tag{3.43}$$

where σ_i is the corresponding RCS parameter and $\otimes_t$ is a convolution operator on the range fast time variable t. The $h_0(t)$ and $h_{1,i}(t, \tau)$ denote, respectively, the range reference function and azimuth reference function

$$h_0(t) = w_r(t) \cdot \exp\left(-j\pi k_r t^2\right) \tag{3.44}$$

$$h_{1,i}(t, \tau) = \exp\left\{ -j\frac{2\pi}{\lambda} \left[R_c(\tau) + R_c\left(\tau + i\frac{d_a}{v_s}\right) \right] \right\}$$
$$\times w_a(\tau) \cdot \delta\left[\tau - \frac{R_c(\tau) + R_c\left(\tau + i\frac{d_a}{v_s}\right)}{c_0} \right] \tag{3.45}$$

where $w_r(t)$ and $w_a(\tau)$ denote the antenna pattern in range dimension and azimuth dimension, respectively.

From the approximation relationship

$$R_c(\tau) + R_c\left(\tau + i\frac{d_a}{v_s}\right) \approx 2R_c\left(\tau + i\frac{d_a}{2v_s}\right) \tag{3.46}$$

we can get

$$s_i(t, \tau) \approx \sigma_i \left[h_0(t) \otimes_t h_{1,i}\left(t, \tau + i\frac{d_a}{2v_s}\right) \right] \tag{3.47}$$

where

$$h_{1,i}\left(t, \tau + i\frac{d_a}{2v_s}\right) = w_a\left(\tau + i\frac{d_a}{2v_s}\right) \exp\left\{ -j\frac{4\pi}{\lambda} R_c\left(\tau + i\frac{d_a}{2v_s}\right) \right\}$$
$$\times \delta\left[\tau - \frac{2R_c\left(\tau + i\frac{d_a}{2v_s}\right)}{c_0} \right] \tag{3.48}$$

Equivalently, the nonuniform PRF can be considered as an azimuth time shift

$$\tau_{\text{er}} = \frac{d_a}{2v_s} - \frac{1}{N\text{PRF}} \tag{3.49}$$

Substituting it into Eq. (3.48) yields

$$h_{1,i}\left(t, \tau + i\frac{d_a}{2v_s}\right) = w_a\left(\tau + i\frac{d_a}{2v_s}\right) \exp\left\{ -j\frac{4\pi}{\lambda} R_c\left(\tau + i\frac{1}{N \cdot \text{PRF}} + i\tau_{\text{er}}\right) \right\}$$
$$\times \delta\left[\tau - \frac{2R_c\left(\tau + i\frac{1}{N \cdot \text{PRF}} + i\tau_{\text{er}}\right)}{c_0} \right] \tag{3.50}$$

After range compression and range mitigation correction, we can get

$$
s_i\left(k\frac{1}{N\cdot\text{PRF}}\right) = w_a\left(k\frac{1}{N\cdot\text{PRF}} + i\tau_{\text{er}}\right)
$$
$$
\times \exp\left\{-j\left[2\pi f_d\left(\frac{1}{N\cdot\text{PRF}} + i\tau_{\text{er}}\right) + \pi k_a\left(k\frac{1}{N\cdot\text{PRF}} + i\tau_{\text{er}}\right)^2\right]\right\}
$$

$$(3.51)$$

where k is an integer, f_d is the Doppler centroid, and k_a is the Doppler rate.

It is noticed that the azimuth time shift $e(\tau)$ is periodic with the period of $1/\text{PRF}$, which can be expressed as

$$
e(\tau) = \tau_{\text{er}}\frac{\tau}{N\cdot\text{PRF}}, \quad |\tau| \le \frac{1}{2\text{PRF}} \tag{3.52}
$$

Expanding $e(\tau)$ in Fourier series yields

$$
e(\tau) = \tau_{\text{er}}\sum_{n=1}^{\infty} b_n \sin\left(2\pi n\text{PRF}\tau\right) \tag{3.53}
$$

where b_n is the Fourier coefficients. Substituting Eq. (3.53) to Eq. (3.51), we can get

$$
s_i(\tau) = w_a(\tau + e(\tau))\exp\left[-j\left(2\pi f_d\tau + \pi k_a\tau^2\right)\right]
$$
$$
\times \exp\left\{-j2\pi k_a\tau\cdot e(\tau) - j2\pi f_d\cdot e(\tau) - j\pi k_a\cdot e^2(\tau)\right\} \tag{3.54}
$$

It can be expanded into

$$
s_i(\tau) \approx w_a(\tau)\exp\left(-j\pi k_a\tau^2\right)
$$
$$
\times \left\{J_0(x) - j2\pi k_a\tau_{\text{er}}b_n J_0(x)\tau\sin\left(\frac{2\pi\tau}{N\cdot\text{PRF}}\right) - 2jJ_1(x)\tau\sin\left(\frac{2\pi\tau}{N\cdot\text{PRF}}\right)\right\} \tag{3.55}
$$

where $x = 2\pi f_d\tau_{\text{er}}b_n$. Note that the Doppler frequency component $\exp\left(-j2\pi f_d\tau\right)$ has been compensated.

Applying Fourier transform to Eq. (3.55) yields

$$
S_i(\omega) = S_a(\omega) + S_b(\omega) + S_c(\omega) + S_d(\omega) + S_e(\omega) \tag{3.56}
$$

where

$$
S_a(\omega) = \frac{1}{\sqrt{k_a}}J_0\left(2\pi f_d\tau_{\text{er}}b_n\right)W_a\left(\frac{-\omega}{2\pi k_a}\right)\exp\left(j\frac{\omega^2}{4\pi k_a} + j\frac{\pi}{4}\right) \tag{3.57}
$$

$$
S_b(\omega) = -\frac{1}{\sqrt{k_a}}\frac{2\pi\text{PRF} - \omega}{2}\tau_{\text{er}}b_n J_0\left(2\pi f_d\tau_{\text{er}}b_n\right)
$$
$$
\times W_a\left(\frac{2\pi\text{PRF} - \omega}{2\pi k_a}\right)\exp\left(j\frac{(2\pi\text{PRF} - \omega)^2}{4\pi k_a} + j\frac{\pi}{4}\right) \tag{3.58}
$$

$$S_c(\omega) = -\frac{1}{\sqrt{k_a}} \frac{2\pi\text{PRF} + \omega}{2} \tau_{\text{er}} b_n J_0 \left(2\pi f_d \tau_{\text{er}} b_n\right)$$

$$\times W_a \left(-\frac{2\pi\text{PRF} + \omega}{2\pi k_a}\right) \exp\left(j\frac{(2\pi\text{PRF} + \omega)^2}{4\pi k_a} + j\frac{\pi}{4}\right)$$

(3.59)

$$S_d(\omega) = -\frac{1}{\sqrt{k_a}} J_1 \left(2\pi f_d \tau_{\text{er}} b_n\right)$$

$$\times W_a \left(\frac{2\pi\text{PRF} - \omega}{2\pi k_a}\right) \exp\left(j\frac{(2\pi\text{PRF} - \omega)^2}{4\pi k_a} + j\frac{\pi}{4}\right)$$

(3.60)

$$S_e(\omega) = -\frac{1}{\sqrt{k_a}} J_1 \left(2\pi f_d \tau_{\text{er}} b_n\right)$$

$$\times W_a \left(-\frac{2\pi\text{PRF} + \omega}{2\pi k_a}\right) \exp\left(j\frac{(2\pi\text{PRF} + \omega)^2}{4\pi k_a} + j\frac{\pi}{4}\right)$$

(3.61)

$S_a(\omega)$ is the ideal signal spectra for subsequent image formation, the remaining terms are the undesired spectra caused by the nonuniform spatial sampling effects. Since the undesired components are similar to the ideal components, they cannot be processed by conventional SAR image formation algorithms without performance degradation. Efficient azimuth signal reconstruction algorithms are thus required for the MPCM SAR.

3.4.3   Azimuth Signal Reconstruction Algorithm

As noted previously, the azimuth signal may be periodic nonuniform with a period of 1/PRF. This information is particularly important for developing nonuniform signal reconstruction algorithms. Krieger et al. [214] proposed an innovative reconstruction algorithm, which allows for an unambiguous recovery of the Doppler spectrum even for a nonuniform sampling of the SAR signal. The only requirement is that the samples do not coincide. Such an algorithm has a great potential for MPCM SAR azimuth signal reconstruction processing.

The data acquisition in a MPCM SAR can be considered as a linear system with multiple receiving channels, each described by a linear filter $h_i(t)$ with transfer function $H_i(f)$. The algorithm proposed in [214] just makes a generalization of the sampling theorem according to which a bandlimited signal $u(t)$ is uniquely determined in terms of the samples $h_i(nT)$ of the responses $h_i(t)$ of n linear systems with input $u(t)$ sampled as $1/n$ of the Nyquist frequency [34]. A block diagram for the reconstruction from the subsampled signals is shown in Figure 3.11. The reconstruction consists essentially of n linear filters $P_i(f)$ that are individually applied to the subsampled signals of the receive channels and then superimposed. Each of the reconstruction filters $P_i(f)$ can again be regarded as a composition of n bandpass filters $P_{ij}(f)$, where $1 \leq j \leq n$.

The reconstruction filters can be derived from the following matrix $\mathbf{H}(f)$

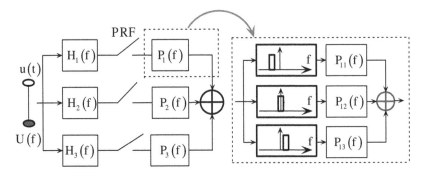

FIGURE 3.11: Reconstruction for multichannel subsampling in case of three channels.

[34]

$$\mathbf{H}(f) = \begin{bmatrix} H_1(f) & \cdots & H_n(f) \\ H_1(f + \mathrm{PRF}) & \cdots & H_n(f + \mathrm{PRF}) \\ \vdots & \ddots & \vdots \\ H_1(f + (n-1)\mathrm{PRF}) & \cdots & H_n(f + (n-1)\mathrm{PRF}) \end{bmatrix} \qquad (3.62)$$

It consists of the n transfer function $H_i(f)$, which have to be shifted by integer multiples of the PRF in the frequency domain. The reconstruction filters $P_{ij}(f)$ are then derived from an inversion of the matrix $\mathbf{H}(f)$ as

$$\mathbf{H}^{-1}(f) = \begin{bmatrix} P_{11}(f) & P_{12}(f + \mathrm{PRF}) & \cdots & P_{1n}(f + (n-1)\mathrm{PRF}) \\ P_{21}(f) & P_{22}(f + \mathrm{PRF}) & \cdots & P_{2n}(f + (n-1)\mathrm{PRF}) \\ \vdots & \vdots & \ddots & \vdots \\ P_{n1}(f) & P_{n2}(f + \mathrm{PRF}) & \cdots & P_{nn}(f + (n-1)\mathrm{PRF}) \end{bmatrix} \qquad (3.63)$$

The MPCM SAR can be regarded as a monostatic SAR that is followed by additional time and phase shifts for each receiving channel [214]. This is illustrated in Figure 3.12. As a simple example for an analytic derivation of the reconstruction filters we consider a monostatic MPCM SAR with three receive channels having the same azimuth antenna pattern. Since it is possible to incorporate the transfer function due to the antenna pattern into the monostatic SAR response, the remaining differences between the three channels are given by the following three impulse responses

$$h_1(\tau) = \delta\left(\tau - T'\right) \qquad (3.64)$$

$$h_2(\tau) = \delta\left(\tau\right) \qquad (3.65)$$

$$h_3(\tau) = \delta\left(\tau + T'\right) \qquad (3.66)$$

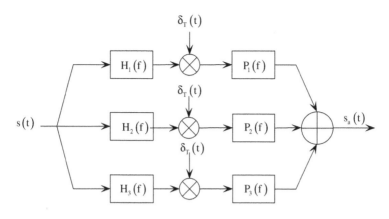

FIGURE 3.12: The equivalent reconstruction filter in case of three channels.

where $T' = d_a/(2v_s)$. Correspondingly, the sampled signals for the m-th pulse are

$$\left\{ s_m\left(\frac{n}{\text{PRF}} - \frac{d_a}{2v_s}\right), s_m\left(\frac{n}{\text{PRF}}\right), s_m\left(\frac{n}{\text{PRF}} + \frac{d_a}{2v_s}\right) \right\} \tag{3.67}$$

The matrix $\mathbf{H}(f)$ is then given by

$$\mathbf{H}(f) = \begin{bmatrix} 1 & \exp\left(-j2\pi fT'\right) & \exp\left(j2\pi fT'\right) \\ 1 & \exp\left(-j2\pi(f+\text{PRF})T'\right) & \exp\left(j2\pi(f+\text{PRF})T'\right) \\ 1 & \exp\left(-j2\pi(f+2\text{PRF})T'\right) & \exp\left(j2\pi(f+2\text{PRF})T'\right) \end{bmatrix} \tag{3.68}$$

The inverse matrix $\mathbf{H}^{-1}(f)$ is derived as

$$\mathbf{H}^{-1}(f) = \frac{1}{j\left(4\sin(2\pi\text{PRF}T') - 2\sin(4\pi\text{PRF}T')\right)}$$

$$\times \begin{bmatrix} 2j\sin(2\pi\text{PRF}T') & -2j\sin(4\pi\text{PRF}T') & 2j\sin(2\pi\text{PRF}T') \\ -K\exp\left\{j2\pi(f+\text{PRF})T'\right\} & G\exp\left\{j2\pi fT'\right\} & -K\exp\left\{j2\pi fT'\right\} \\ K^*\exp\left\{-j2\pi(f+\text{PRF})T'\right\} & -G^*\exp\left\{-j2\pi fT'\right\} & K^*\exp\left\{-j2\pi fT'\right\} \end{bmatrix} \tag{3.69}$$

where $K = \exp(j2\pi\text{PRF}T') - 1$ and $G = \exp(j4\pi\text{PRF}T') - 1$. The transfer functions of the reconstruction filters $P_1(f)$, $P_2(f)$ and $P_3(f)$ are expressed, respectively, by

$$P_1(f) = \begin{cases} \dfrac{2j\sin\left(2\pi\text{PRF}T'\right)}{j[4\sin(2\pi\text{PRF}T') - 2\sin(4\pi\text{PRF}T')]}, & \dfrac{-3\text{PRF}}{2} \le f \le \dfrac{-\text{PRF}}{2} \\[2ex] \dfrac{-2j\sin\left(4\pi\text{PRF}T'\right)}{j[4\sin(2\pi\text{PRF}T') - 2\sin(4\pi\text{PRF}T')]}, & \dfrac{-\text{PRF}}{2} \le f \le \dfrac{\text{PRF}}{2} \\[2ex] \dfrac{2j\sin\left(2\pi\text{PRF}T'\right)}{j[4\sin(2\pi\text{PRF}T') - 2\sin(4\pi\text{PRF}T')]}, & \dfrac{\text{PRF}}{2} \le f \le \dfrac{3\text{PRF}}{2} \end{cases} \tag{3.70}$$

$$P_2(f) = \begin{cases} \dfrac{-K\exp\left(j2\pi(f+\mathrm{PRF})T'\right)}{j[4\sin(2\pi\mathrm{PRF}T')-2\sin(4\pi\mathrm{PRF}T')]}, & \dfrac{-3\mathrm{PRF}}{2} \leq f \leq \dfrac{-\mathrm{PRF}}{2} \\[2mm] \dfrac{G\exp\left(j2\pi(f-\mathrm{PRF})T'\right)}{j[4\sin(2\pi\mathrm{PRF}T')-2\sin(4\pi\mathrm{PRF}T')]}, & \dfrac{-\mathrm{PRF}}{2} \leq f \leq \dfrac{\mathrm{PRF}}{2} \\[2mm] \dfrac{-K\exp\left(j2\pi(f-2\mathrm{PRF})T'\right)}{j[4\sin(2\pi\mathrm{PRF}T')-2\sin(4\pi\mathrm{PRF}T')]}, & \dfrac{\mathrm{PRF}}{2} \leq f \leq \dfrac{3\mathrm{PRF}}{2} \end{cases} \tag{3.71}$$

$$P_3(f) = \begin{cases} \dfrac{K^*\exp\left(-j2\pi(f+\mathrm{PRF})T'\right)}{j[4\sin(2\pi\mathrm{PRF}T')-2\sin(4\pi\mathrm{PRF}T')]}, & \dfrac{-3\mathrm{PRF}}{2} \leq f \leq \dfrac{-\mathrm{PRF}}{2} \\[2mm] \dfrac{-G^*\exp\left(-j2\pi(f-\mathrm{PRF})T'\right)}{j[4\sin(2\pi\mathrm{PRF}T')-2\sin(4\pi\mathrm{PRF}T')]}, & \dfrac{-\mathrm{PRF}}{2} \leq f \leq \dfrac{\mathrm{PRF}}{2} \\[2mm] \dfrac{K^*\exp\left(-j2\pi(f-2\mathrm{PRF})T'\right)}{j[4\sin(2\pi\mathrm{PRF}T')-2\sin(4\pi\mathrm{PRF}T')]}, & \dfrac{\mathrm{PRF}}{2} \leq f \leq \dfrac{3\mathrm{PRF}}{2} \end{cases} \tag{3.72}$$

The numerator of the reconstruction filters can be regarded as compensating the different time delays and phase shifts within each branch.

The reconstructed time-domain azimuth signal can be expressed as

$$
\begin{aligned}
s_a(\tau) = \sum_n s_m & \left\{ \left(\frac{n}{\mathrm{PRF}}\right) p_1\left(\tau - \frac{n}{\mathrm{PRF}}\right) \right. \\
& + s_m\left(\frac{n}{\mathrm{PRF}} - \frac{d_a}{2v_s}\right) p_2\left(\tau - \frac{n}{\mathrm{PRF}}\right) \\
& \left. + s_m\left(\frac{n}{\mathrm{PRF}} + \frac{d_a}{2v_s}\right) p_3\left(\tau - \frac{n}{\mathrm{PRF}}\right) \right\}
\end{aligned}
\tag{3.73}
$$

where $p_1(\tau)$, $p_2(\tau)$ and $p_3(\tau)$ denote the three time-domain transfer functions of the reconstruction filters, and n is an integer. To make it suitable for subsequent image formation processing, Eq. (3.73) can be reformed as

$$
\begin{aligned}
s_a\left(\frac{m}{3\mathrm{PRF}}\right) = \sum_n s_m & \left\{ \left(\frac{n}{\mathrm{PRF}}\right) p_1\left(\frac{m}{3\mathrm{PRF}} - \frac{n}{\mathrm{PRF}}\right) \right. \\
& + s_m\left(\frac{n}{\mathrm{PRF}} - \frac{d_a}{2v_s}\right) p_1\left(\frac{m}{3\mathrm{PRF}} - \frac{n}{\mathrm{PRF}}\right) \\
& \left. + s_m\left(\frac{n}{\mathrm{PRF}} + \frac{d_a}{2v_s}\right) p_1\left(\frac{m}{3\mathrm{PRF}} - \frac{n}{\mathrm{PRF}}\right) \right\} \\
= \sum_n s_m & \left\{ \left(\frac{3n}{3\mathrm{PRF}}\right) p_1\left(\frac{m}{3\mathrm{PRF}} - \frac{3n}{3\mathrm{PRF}}\right) \right. \\
& + s_m\left(\frac{3n}{3\mathrm{PRF}} - \frac{d_a}{2v_s}\right) p_2\left(\frac{m}{3\mathrm{PRF}} - \frac{3n}{3\mathrm{PRF}}\right) \\
& \left. + s_m\left(\frac{3n}{3\mathrm{PRF}} + \frac{d_a}{2v_s}\right) p_3\left(\frac{m}{3\mathrm{PRF}} - \frac{3n}{3\mathrm{PRF}}\right) \right\}
\end{aligned}
\tag{3.74}
$$

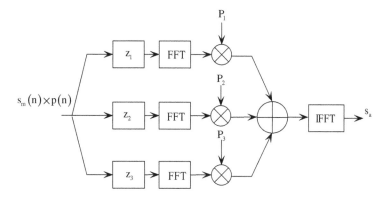

FIGURE 3.13: Reconstructing the azimuth signal with FFT algorithm.

Letting

$$
\begin{cases}
z_1(k) = \begin{cases} s_m\left(\frac{k}{3\mathrm{PRF}}\right), & k = 3n \\ 0, \; k \neq 3n \end{cases} \\[3mm]
z_2(k) = \begin{cases} s_m\left(\frac{k}{3\mathrm{PRF}} - \frac{d_a}{2v_s}\right), & k = 3n \\ 0, \; k \neq 3n \end{cases} \\[3mm]
z_3(k) = \begin{cases} s_m\left(\frac{k}{3\mathrm{PRF}} + \frac{d_a}{2v_s}\right), & k = 3n \\ 0, \; k \neq 3n \end{cases}
\end{cases}
\tag{3.75}
$$

we can rewrite Eq. (3.74) as

$$
\begin{aligned}
s_a\left(\frac{m}{3\mathrm{PRF}}\right) = & z_1\left(\frac{m}{3\mathrm{PRF}}\right) \otimes p_1\left(\frac{m}{3\mathrm{PRF}}\right) + z_2\left(\frac{m}{3\mathrm{PRF}}\right) \\
& \otimes p_2\left(\frac{m}{3\mathrm{PRF}}\right) + z_3\left(\frac{m}{3\mathrm{PRF}}\right) \otimes p_3\left(\frac{m}{3\mathrm{PRF}}\right)
\end{aligned}
\tag{3.76}
$$

It can be implemented through the fast Fourier transform (FFT) algorithm shown in Figure 3.13, where $\{P_1(k)\}$, $\{P_2(k)\}$ and $\{P_3(k)\}$ denote, respectively, the discrete FFT of the $\{p_1\left(\frac{m}{3\mathrm{PRF}}\right)\}$, $\{p_2\left(\frac{m}{3\mathrm{PRF}}\right)\}$ and $\{p_3\left(\frac{m}{3\mathrm{PRF}}\right)\}$.

3.4.4   System Performance Analysis

The reconstruction algorithm cannot cancel the energy outside the band $I_s = [-N_a\mathrm{PRF}/2, N_a\mathrm{PRF}/2]$ of the original signal spectrum. After focusing, this will give rise to azimuth ambiguities in the final SAR image.

As shown in Figure 3.14, the original spectrum is located around the center frequency $f_{0,n} = -n \cdot \mathrm{PRF}$, and consequently the ambiguous contributions are situated at frequencies $f_{a,n}$ that deviate more than $\pm N\mathrm{PRF}/2$ from $f_{0,n}$

$$
|f_{a,n} - f_{0,n}| > \frac{N_a\mathrm{PRF}}{2}
\tag{3.77}
$$

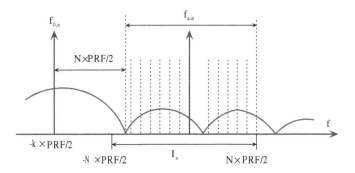

FIGURE 3.14: Illustration of the original sepctrum with ambiguous elements.

But, only contributions situated after sampling within the system band I_s are relevant

$$|f_{a,n}| < \frac{N_a \text{PRF}}{2} \tag{3.78}$$

Combining these two constraints yields the following expressions for the Doppler frequencies where the ambiguous parts are located.

$$\frac{N_a \text{PRF}}{2} - n\text{PRF} < f_{a,n} < \frac{N_a \text{PRF}}{2}, \quad n > 0 \tag{3.79}$$

$$-\frac{N_a \text{PRF}}{2} < f_{a,n} < -\frac{N_a \text{PRF}}{2} - n\text{PRF}, \quad n < 0 \tag{3.80}$$

Therefore, depending on n, spectral subbands containing frequencies according to the above equations are not cancelled by the reconstruction algorithm and have to be considered when determining the remaining ambiguous signals. As the problem is symmetrical, it is sufficient to concentrate only on $n > 0$. The complete ambiguous spectrum due to aliasing is obtained by summing all individual ambiguous contributions after reconstruction, $e_n(f)$, yielding $e_\Sigma(f)$ [145]

$$e_\Sigma(f) = 2 \sum_{n=1}^{\infty} e_n(f) \tag{3.81}$$

where the factor 2 accounts for both signs of n. This equation can be divided into the characteristics of the azimuth signal and the overall configuration. The azimuth signal spectrum is defined by size and tapering of the transmit antenna and the receiving aperture.

The mean squared amplitude of $e_\Sigma(f)$ represents the azimuth ambiguous power. Relating it to the signal power S_p results in an equivalent AASR for the MPCM SAR systems.

$$\text{AASR} = \frac{E\left\{|e_\Sigma(f)|^2\right\}}{S_p} \qquad (3.82)$$

It is calculated for the complete Doppler bandwidth given by $N_a\text{PRF}$. If the respective processed Doppler bandwidth B_a and the associated lowpass filtering are included, the AASR can be represented by

$$\text{AASR} = \frac{E\left\{\left|e_\Sigma(f) \cdot \text{rect}\left(\frac{f}{B_a}\right)\right|^2\right\}}{S_{p,B_a}} \qquad (3.83)$$

where S_{p,B_a} denotes the signal power in the Doppler bandwidth B_a. The numerator quantifies the ambiguous energy within the focused MPCM SAR image, accounting for the applied multichannel reconstruction algorithm.

3.4.5 Numerical Simulation Results

In this section, taking a dual-channel MPCM SAR as an example, we simulate the impact of platform speed deviation and the performance of the multichannel reconstruction algorithm with the system parameters listed in Table 3.2. Suppose the radar platform has an unideal speed deviation

$$\Delta v = a\tau, \quad -T_s/2 < \tau < T_s/2 \qquad (3.84)$$

where a is a constant and T_s is the synthetic aperture time. Then the deviation of the azimuth chirp rate is

$$\Delta k_a = \frac{4v_s a\tau}{\lambda R_c} + \frac{2(a\tau)^2}{\lambda R_c} \qquad (3.85)$$

which yields a quadratic phase error (QPE)

$$\text{QPE}_a = \pi\Delta k_a \left(\frac{T_s}{2}\right)^2 \qquad (3.86)$$

Figures 3.15 and 3.16 show, respectively, the QPE as a function of azimuth slow time and acceleration speed. It can be noticed that the QPE is very sensitive to the platform speed deviation. As shown in Figure 3.17, this kind of phase error will lead to a broadening of main-lobe, an increasing of integral side-lobe ratio (ISLR) and a phase error on the peak power point. Figure 3.18 shows the imaging results with the multichannel reconstruction algorithm. The detailed imaging performances are listed in Table 3.3. It can be noticed that the imaging performances have been significantly improved by the multichannel reconstruction algorithm.

TABLE 3.2: MPCM SAR system simulation parameters.

Parameter	Symbol	Value
platform speed	v_s	$180m/s$
carrier wavelength	λ	$0.24m$
number of subapertures	N_a	2
sub-aperture antenna length	L_a	$2m$
acceleration speed	a	$2m/s^2$
nearest slant range	R_c	$11.66km$
operating PRF	f_p	$57.6Hz$
azimuth Doppler bandwidth	B_d	$80Hz$

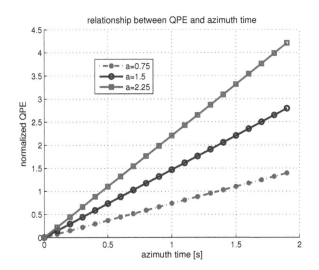

FIGURE 3.15: The QPE as a function of azimuth slow time.

TABLE 3.3: MPCM SAR imaging simulation results.

parameter	Fig. 3.17(a)	Fig. 3. 17(b)	Fig. 3.18
ISLR in azimuth [dB]	-4.83	0.12	-3.71
peak side-lobe ratio (PSLR) in azimuth [dB]	-5.08	-0.95	-4.86
normalized ambiguous signal power [dB]	-28.10	-21.96	-25.68

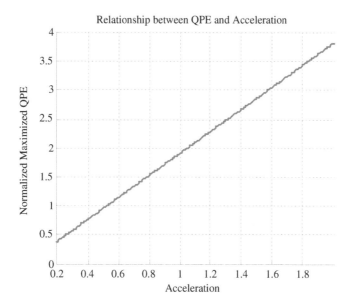

FIGURE 3.16: The QPE as a function of acceleration speed.

3.5 Azimuth Scanning Multibeam SAR

Wide-swath image can also be synthesized by scanning different azimuth sub-swaths and sequentially synthesizing the images for different beam positions [167, 288]. The physical beam of the antenna initially points to the near azimuth cell and dwells there sufficiently long to synthesize a radar image covering the entire area of the beam. The beam then is pointed to the next azimuth cell to synthesize a radar image there, and so on. The scanning and image synthesis terminates at the farthest beam cell when the beam arrives at the end of the area covered by the innermost cell, at which time the beam is pointed once again to the innermost angle and the process is repeated for the second set of cells.

In this section, we introduce an azimuth scanning multibeam SAR for wide-swath imaging. For simplicity and without loss of generality, we suppose there are only two azimuth beams and two subswaths. As shown in Figure 3.19, in azimuth time τ_1, the two beams work in slightly backward-looking squint mode and dwells there sufficiently long to synthesize a radar image covering the area of A and D (see Figure 3.20(left)). In azimuth time τ_2, the two beams work in slightly forward-looking squint mode and dwells there sufficiently long to synthesize a radar image covering the area of B and C (see

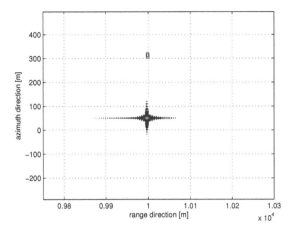

(a) The ideal imaging results without any derivations.

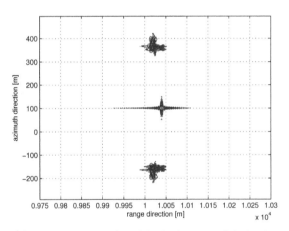

(b) The imaging results with platform speed derivation.

FIGURE 3.17: The impact of platform speed derivation on MPCM SAR imaging results.

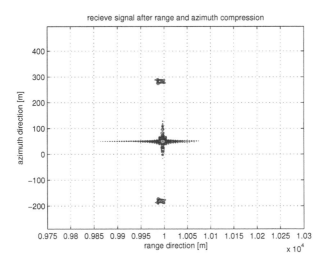

FIGURE 3.18: The imaging results with the multi-channel reconstruction algorithm.

Figure 3.20(right)). These two modes are operated alternatively along with the azimuth time.

3.5.1 Signal Model

Figure 3.21 shows the equivalent geometry model of the azimuth scanning multibeam SAR in case of two beams, where the solid line in the left denotes the first azimuth subaperture and the dashed line denotes the second azimuth subaperture. Suppose the equivalent transmit phase center is x_0, the two receive phase centers can then be represented respectively by $x_0 - 0.5d_a + \Delta x$ and $x_0 + 0.5d_a + \Delta x$, where d_a is the antenna length and Δx is the antenna moving distance during one pulse repetition period.

Consider a point target locating at (x, r), the range histories for the two antennas can be expressed, respectively, by

$$
\begin{aligned}
R_1(\tau) &= \sqrt{r^2 + (x - x_0)^2} + \sqrt{r^2 + (x - x_0 + 0.5d_a - \Delta x)^2} \\
&\approx 2\left[r + \frac{(x - x_0 - 0.5\Delta x + 0.25d_a)^2}{2r}\right] + \frac{(0.5\Delta x - 0.25d_a)^2}{r}
\end{aligned} \quad (3.87)
$$

$$
\begin{aligned}
R_2(\tau) &= \sqrt{r^2 + (x - x_0)^2} + \sqrt{r^2 + (x - x_0 - 0.5d_a - \Delta x)^2} \\
&\approx 2\left[r + \frac{(x - x_0 - 0.5\Delta x - 0.25d_a)^2}{2r}\right] + \frac{(0.5\Delta x + 0.25d_a)^2}{r}
\end{aligned} \quad (3.88)
$$

Since the second terms in Eqs. (3.87) and (3.88) are constants for a given

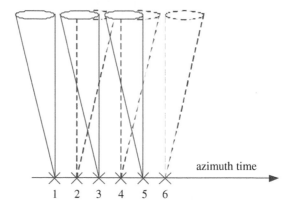

FIGURE 3.19: Illustration of the azimuth scanning multibeam SAR in the case of two beams.

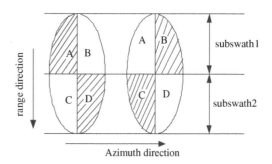

FIGURE 3.20: Illustration of the ground covered by the two beams at different azimuth time.

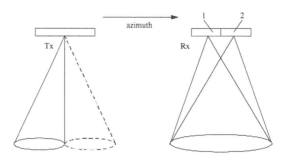

FIGURE 3.21: Equivalent geometry model of the azimuth scanning multibeam SAR.

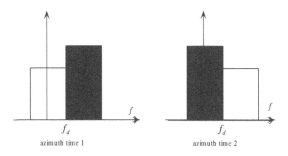

FIGURE 3.22: Spectra of the echoes received at two adjacent azimuth positions.

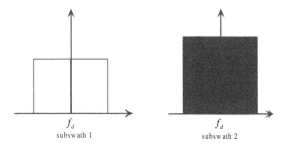

FIGURE 3.23: The reconstructed spectra for the two subswaths.

geometry, they can be ignored for subsequent image formation processing. We then have

$$R_1(\tau) = 2\left[r + \frac{(x - x_0 - 0.5\Delta x + 0.25d_a)^2}{2r}\right] \tag{3.89}$$

$$R_2(\tau) = 2\left[r + \frac{(x - x_0 - 0.5\Delta x - 0.25d_a)^2}{2r}\right] \tag{3.90}$$

The $R_1(\tau)$ and $R_2(\tau)$ can be seen as a single-antenna SAR at two azimuth positions with a distance of $0.5d_a$. Therefore, the spectra of the echoes received at two adjacent azimuth positions can be illustrated as Figure 3.22, where f_d is the Doppler centroid. They can be reconstructed as Figure 3.23.

3.5.2 System Performance Analysis

As noted previously in Section 2.5.4, in general single-antenna SAR the azimuth ambiguity results from the fact that the signal components outside the PRF interval fold back into the main part of the spectra. Consider the illustration of azimuth ambiguities of the azimuth scanning multibeam SAR shown in Figure 3.24, the returned signal spectrum of the first subswath is limited

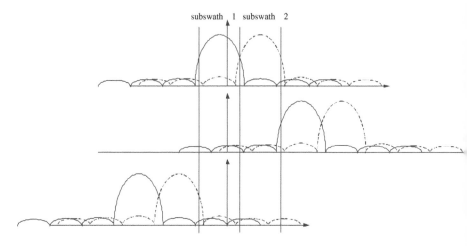

FIGURE 3.24: Illustration of azimuth scanning multibeam SAR azimuth ambiguities.

by

$$f_d - \frac{\text{PRF}}{2} < f_a < f_d \tag{3.91}$$

The possible ambiguous spectrum are

$$f_d - (i+1)\frac{\text{PRF}}{2} < f < f_d - i\frac{\text{PRF}}{2}, \quad i = \pm 2, \pm 4, \ldots \tag{3.92}$$

Similarly, for the second subswath, the possible ambiguous spectrum are

$$f_d - (i+1)\frac{\text{PRF}}{2} < f < f_d - i\frac{\text{PRF}}{2}, \quad i = \pm 1, \pm 3, \ldots \tag{3.93}$$

They are different from that of a general single-antenna SAR

$$f_d - (i+1)\frac{\text{PRF}}{2} < f < f_d - i\frac{\text{PRF}}{2}, \quad i = \pm 1, \pm 2, \pm 3, \ldots \tag{3.94}$$

Next, we consider the range ambiguities. Taking two subswaths as an example, the whole swath is limited by

$$\frac{ic_0}{2\text{PRF}} < R < \frac{(i+2)c_0}{2\text{PRF}} \tag{3.95}$$

Suppose there is a point target locating at R_0 ($\frac{i+2}{2\text{PRF}} < R_0 < \frac{n+3}{2\text{PRF}}$). In case of single-antenna SAR, this target may bring ambiguous signals on the targets locating at $R_0 - \frac{c_0}{\text{PRF}}$. Differently, in case of azimuth scanning multibeam SAR, the target will bring ambiguous signals on the targets locating at $R_0 - \frac{c_0}{\text{PRF}}$ and $R_0 - \frac{c_0}{2\text{PRF}}$; however, the overall AASR is equal to that of the single-antenna SAR.

3.6 Azimuth Multi-Antenna SAR in GMTI

GMTI is of great important for surveillance and reconnaissance, but it is not an easy job because separating the moving targets' returns from stationary clutter is a technical challenge [400]. GMTI is twofold [109]: one is the detection of moving targets within severe ground clutter, and the other is the estimation of their parameters such as velocity and location. It is well known that the moving target with a slant range velocity will generate a differential phase shift. This phase shift could be detected by interferometric combination of the signals from a two-channel along-track interferometry (ATI) SAR.

3.6.1 GMTI via Two-Antenna SAR

ATI SAR was initially proposed for detecting ground moving targets [39, 98, 458]; it uses two antennas to detect targets by providing essentially two identical views of the illuminated scene but at slightly different time.

Consider an ATI SAR (see Figure 3.25) consists of two antennas moving along the azimuth direction X and suppose that the two antennas are separated along the azimuth direction. For the sake of simplicity, we assume that there are two interfering point targets, one moving and one stationary. Suppose they are perfectly compressed in the two SAR images, the signals reflected from the moving target and received by the two antennas can be expressed, respectively, by [62]

$$s_{t1} = A_{t1}\delta(x - X_{t1})\delta(y - Y_{t1}) \exp\left(j4\pi\frac{R_{t1}}{\lambda}\right) \qquad (3.96)$$

$$s_{t2} = A_{t2}\delta(x - X_{t2})\delta(y - Y_{t2}) \exp\left(j4\pi\frac{R_{t2}}{\lambda}\right) \qquad (3.97)$$

where A_{ti} is the combined gain with azimuth-compression gain and backscatter coefficient, (X_{ti}, Y_{ti}) is the target position in the imaged images, and R_{ti} is the instantaneous slant range modified by convolving it with the azimuth reference function. Note that in an actual system the moving point target cannot be well focused with the stationary-terrain matched filter. Similarly, the signals reflected from the stationary target and received by the two antennas are

$$s_{c1} = A_{c1}\delta(x - X_{c1})\delta(y - Y_{c1}) \exp\left(j4\pi\frac{R_{c1}}{\lambda}\right) \qquad (3.98)$$

$$s_{c2} = A_{c2}\delta(x - X_{c2})\delta(y - Y_{c2}) \exp\left(j4\pi\frac{R_{c2}}{\lambda}\right) \qquad (3.99)$$

Suppose the two point targets are overlapped with each other

$$X_{t1} = X_{t2} \qquad (3.100a)$$

$$Y_{t1} = Y_{t2} \qquad (3.100b)$$

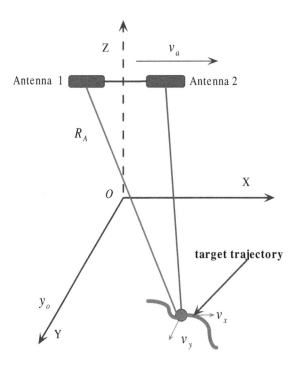

FIGURE 3.25: Geometry of an ATI SAR system.

Interferometric phase processing yields

$$
\begin{aligned}
s_{\text{ATI}} =& (s_{t1} + s_{c1}) \cdot (s_{t2} + s_{c2})^* \\
=& A_{t1} A_{t2} \exp\left[j4\pi\left(\frac{R_{t1} - R_{t2}}{\lambda}\right)\right] + A_{c1} A_{c2} \exp\left[j4\pi\left(\frac{R_{c1} - R_{c2}}{\lambda}\right)\right] \\
& + A_{t1} A_{c2} \exp\left[j4\pi\left(\frac{R_{t1} - R_{c2}}{\lambda}\right)\right] + A_{c1} A_{t2} \exp\left[j4\pi\left(\frac{R_{c1} - R_{t2}}{\lambda}\right)\right]
\end{aligned}
$$

$$(3.101)$$

The first term is the desired moving target's interferogram. The second term is the stationary target's interferogram. Its phase should be equal to zero because a stationary scene does not change with time, i.e., $R_{c1} = R_{c2}$. The remaining two terms are cross terms, which come from the clutter contamination at the SAR image formation stage. As the phase angle is 2π periodic, the two cross terms may have different phase values; hence, the effects of cross terms on the total along-track interferometric phase are not easily predictable.

As ATI SAR output is signal power, slowly moving targets will not be attenuated along with the stationary clutter when we utilize magnitude and phase information for target extraction. In the case of low signal-to-clutter ratio (SCR), ATI SAR will lose its ability to detect slowly moving targets and to correctly estimate their velocities because the system noise (additive thermal noise and multiplicative phase noise) scatters the stationary clutter signal around the real axis in the complex plane. If the clutter contribution is not negligible when compared to the signal power, the estimation of the target radial velocity from the contaminated interferometric phase may lead to erroneous results.

In poor SCR environment, this effect will become more serious for slowly moving targets and the targets will be indistinguishable from the clutter. Moreover, in this case the targets' impulse responses are not normal delta functions, particularly for the moving targets because they are poorly focused due to unmatched azimuth compression filter. This leads to a point target's response overlaps with several neighboring resolution cells. This also means a varying SCR across the target's response, which in turn affects its interferometric phase. Therefore, the ATI SAR is a clutter-limited GMTI detector, and it is necessary to apply some efficient clutter suppression or cancellation techniques.

In order to accurately estimate the target's true velocity, clutter contamination on the signal must be minimized. Precise knowledge of the interferogram's phase and amplitude statistics is very important for distinguishing the moving targets from the clutter. A straightforward approach to clutter cancelation is the DPCA technique, which synthesizes a static antenna system allowing cancelation of static returns on a pulse-to-pulse basis. [262]. When the DPCA condition is matched, the clutter cancellation can then be performed by subtracting the samples of the radar returns received by two-way phase centers in the same spatial position, which are temporally displaced. The

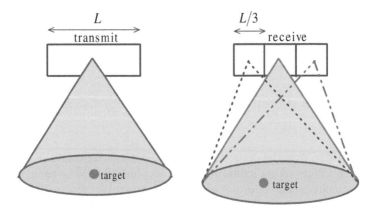

FIGURE 3.26: Three-antenna DPCA-based GMTI scheme.

radar returns corresponding to stationary objects like the clutter from natural scenes are cancelled, while the returns backscattered by moving targets have a different phase in the two acquisitions and remain uncancelled. Therefore, all static clutter scatterers are cancelled, leaving only moving targets and a much simplified target detection problem. However, in a two-antenna DPCA SAR there will be no additional degrees-of-freedom to estimate the moving targets' position information. Three-antenna SAR provides a potential solution to this problem.

3.6.2   Three-Antenna SAR

Figure 3.26 shows the scheme of a three-antenna DPCA-based moving target detection scheme. An antenna of length L_a is used as a single aperture in transmission and is split into three receiving subapertures in reception. In this case, the receive phase centers are displaced in the along-track direction by $L_a/3$. It is then effective to define the 'two-way' phase centers as the mid-point between the phase center of the whole transmitting antenna and the phase center of the receiving subapertures. It then goes as if the radar samples were collected by the transmitting and receiving antennas with phase centers colocated in the two-way phase centers. To reach this aim, it is assumed that the PRF is matched to the platform velocity and the distance between the three receiving subaperture phase centers. Under this condition, the two-way phase center of the trailing subaperture occupies the same location of the two-way phase center of the leading subaperture one PRI later.

We make an assumption of far-field, flat earth, free space, and single polarization for our model. Although three-antenna DPCA SAR can be realized with airborne and spaceborne platforms, we restrict ourselves to airborne

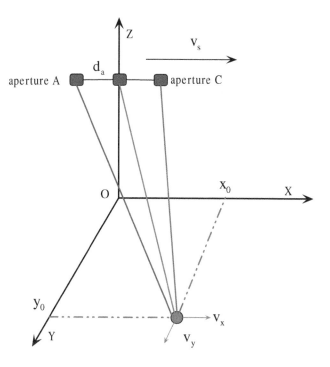

FIGURE 3.27: Geometry of a three-antenna DPCA-based GMTI system.

platform. Figure 3.27 shows the geometry of the three-antenna DPCA-based GMTI system.

The range history from the central aperture to a specific moving target located in (x_0, y_0) with velocity (v_x, v_y) can be represented by

$$
\begin{aligned}
R_c(\tau) &= \sqrt{(x_0 + (v_s - v_x)\tau)^2 + (y_0 + v_y\tau)^2 + h_s^2} \\
&\approx R_c + v_y\tau + \frac{(x_0 + (v_s - v_x)\tau)^2 + (v_y\tau)^2}{2R_c}
\end{aligned}
\tag{3.102}
$$

where τ is the azimuth time, and

$$
R_c = \sqrt{y_0^2 + h_s^2}
\tag{3.103}
$$

In an alike manner, the range histories of the left aperture and the right aperture can be represented, respectively, by

$$
R_l(\tau) \approx R_c + v_y\tau + \frac{(x_0 + d_a + (v_a - v_x)\tau)^2 + (v_y\tau)^2}{2R_c}
\tag{3.104}
$$

$$
R_r(\tau) \approx R_c + v_y\tau + \frac{(x_0 - d_a + (v_a - v_x)\tau)^2 + (v_y\tau)^2}{2R_c}
\tag{3.105}
$$

with $d_a = L_a/3$. Suppose the transmitted radar signal is

$$s(t) = \text{rect}\left[\frac{t}{T_p}\right] \exp\left[j2\pi\left(f_c t + \frac{1}{2}k_r t^2\right)\right] \tag{3.106}$$

After range compressing the three-antenna SAR data, we get

$$S_l(t,\tau) = T_p \exp\left[-j2\pi f_c \xi_l - j\pi k_r(t - \xi_l)^2\right] \cdot \frac{\sin\left(k_r\pi(t - \xi_l)T_p\right)}{k_r\pi(t - \xi_l)T_p} \tag{3.107a}$$

$$S_c(t,\tau) = T_p \exp\left[-j2\pi f_c \xi_c - j\pi k_r(t - \xi_c)^2\right] \cdot \frac{\sin\left(k_r\pi(t - \xi_c)T_p\right)}{k_r\pi(t - \xi_c)T_p} \tag{3.107b}$$

$$S_r(t,\tau) = T_p \exp\left[-j2\pi f_c \xi_r - j\pi k_r(t - \xi_r)^2\right] \cdot \frac{\sin\left(k_r\pi(t - \xi_r)T_p\right)}{k_r\pi(t - \xi_r)T_p} \tag{3.107c}$$

where

$$\xi_l = \frac{R_l(\tau) + R_c(\tau)}{c_0} \tag{3.108a}$$

$$\xi_c = \frac{2R_c(\tau)}{c_0} \tag{3.108b}$$

$$\xi_r = \frac{R_c(\tau) + R_r(\tau)}{c_0} \tag{3.108c}$$

To cancel the clutter, we perform

$$s_{cl}(t,\tau) = G_1 \cdot s_c(t,\tau) - s_l(t,\tau + t_d), \tag{3.109a}$$

$$s_{rc}(t,\tau) = s_r(t,\tau) - G_1 \cdot s_c(t,\tau + t_d) \tag{3.109b}$$

where $t_d = d_a/v_a$ is the relative azimuth delay between two apertures, and G_1 is used to compensate the corresponding phase shift

$$G_1 = \exp\left(-j\frac{2\pi d_a^2}{4R_c\lambda}\right) \tag{3.110}$$

Since

$$A_0 = T_p \frac{\sin(k_r\pi(t - \xi_l)T_p)}{k_r\pi(t - \xi_l)T_p} \approx T_p \frac{\sin(k_r\pi(t - \xi_c)T_p)}{k_r\pi(t - \xi_c)T_p} \approx T_p \frac{\sin(k_r\pi(t - \xi_r)T_p)}{k_r\pi(t - \xi_r)T_p} \tag{3.111}$$

Equation (3.109a) can be expanded into

$$S_{cl}(t,\tau) = A_0 \exp\left\{-j\frac{4\pi}{\lambda}(R_c + v_y\tau)\right\} \cdot \left[1 - \exp\left(-j\frac{2\pi}{\lambda}2v_y t_d\right)\right]$$

$$\times \exp\left\{-j\frac{4\pi}{\lambda}\left[\frac{x_0^2 + \frac{d_a^2}{4}[(v_a - v_x)\tau]^2 - 2x_0(v_a - v_x)\tau + (v_y\tau)^2}{2R_c}\right]\right\} \tag{3.112}$$

From Eq. (3.112) we can notice that, if $v_y = 0$, there is $|S_{cl}(t, \tau)| = 0$; hence, the clutter has been successfully cancelled by this method. The remaining problem is to detect the moving targets. It can be further derived that its Doppler frequency center and Doppler chirp rate are represented, respectively, by

$$f_{dc} = -\frac{2}{\lambda}\left[v_y - \frac{(v_a - v_x)x_0}{R_c}\right] \tag{3.113}$$

$$k_d = -\frac{2}{\lambda R_c}\left[(v_a - v_x)^2 + v_y^2\right] \tag{3.114}$$

Equation (3.112) can then be rewritten as

$$S_{cl}(t, \tau) = A_0' \exp\left[2j\pi\left(f_{dc}\tau + \frac{1}{2}k_d\tau^2\right) + \phi_1\right] \tag{3.115}$$

where

$$\phi_1 = -\frac{4\pi}{\lambda}\left(R_c + \frac{x_0^2 + \frac{d_a^2}{4}}{2R_c}\right) \tag{3.116a}$$

$$A_0' = A_0\left[1 - \exp\left(-j\frac{4\pi}{\lambda}v_y t_d\right)\right] \tag{3.116b}$$

Thus, once the Doppler parameters described in the Eq. (3.115) are estimated, the target velocity (v_x, v_y) can then be determined from the Eqs. (3.113) and (3.114).

To utilize the ATI technique, after compensating the phase terms caused by the relative aperture displace between $s_{cl}(t, \tau)$ and $s_{cr}(t, \tau)$ by multiplying

$$G_2 = \exp\left[j\frac{2\pi}{\lambda}\left(\frac{v_s\tau d_a}{R_c}\right)\right] \tag{3.117}$$

From Eq. (3.109b) we have

$$s_{rc}(t, \tau) = A_0' \exp\left[2j\pi\left(f_{dc}\tau + \frac{1}{2}k_d\tau^2\right) + \phi_2\right] \tag{3.118}$$

with

$$\phi_2 = -\frac{4\pi}{\lambda}\left(R_c + \frac{x_0^2 - x_0 d_a + \frac{d_a^2}{2}}{2R_c}\right) \tag{3.119}$$

Then $s_{cl}(t, \tau)$ and $s_{rc}(t, \tau)$ can be interferometric processed with each other by conjugate multiplication

$$s_{cl}(t, \tau)s_{rc}^*(t, \tau) = |A_0'|^2 \exp\left(-\frac{2\pi\left(x_0 d_a - \frac{d_a^2}{4}\right)}{\lambda R_c}\right) = |A_0'|^2 \exp\left(\Phi\right), \tag{3.120}$$

where Φ is the interferometric phase. The azimuth position of the moving target can be determined by

$$x_0 = \frac{\pi d_a^2 - 2R_c\lambda\Phi}{4\pi d_a} \tag{3.121}$$

As the velocity of the moving target is relative small, from Eq. (3. 102) R_c can be approximated by

$$R_c = \frac{c_0\hat{t}}{2} \tag{3.122}$$

where

$$\frac{\sin(k_r\pi(\hat{t} - \xi_l)T_p)}{k_r\pi(\hat{t} - \xi_l)T_p} = \max\left\{\frac{\sin(k_r\pi(t - \xi_l)T_p)}{k_r\pi(t - \xi_l)T_p}\right\} \tag{3.123}$$

As there is $-W_a/2 \le x_0 \le W_a/2, W_a \gg d_a$ with W_a being the synthetic aperture length in azimuth, the interferometric phase is limited by

$$|\Phi| \ll \pi\left|\frac{\pm W_a d_a - \frac{d_a^2}{2}}{R_c\lambda}\right| \approx |\pi\frac{W_a}{R_c\frac{\lambda}{d_a}}| \le \pi \tag{3.124}$$

The interferometric phase $\Phi \in [-\pi, \pi]$ is unambiguous; hence, the unambiguous x_0 can be obtained in this way. Once R_0 and x_0 are determined, the y_0 can then be derived from the Eq. (3.103) because the platform altitude h_a can be obtained from the inboard motion sensors.

To estimate the Doppler parameters, applying the simplified fractional Fourier transform (SFrFT) [409] to Eq. (3.115), we get

$$\chi_\alpha(\mu) = A_0'\exp\left(j\phi_1\right)\cdot$$
$$\int_{-T_p/2}^{T_p/2}\exp\left[j2\pi f_{dc}\tau + j\pi k_d\tau^2 + j\frac{\tau^2}{2}\cot(\alpha) - j\mu\tau\csc(\alpha) + j\phi_1\right]d\tau \tag{3.125}$$

It arrives its maximum at [406]

$$\{\hat{\alpha}_0, \hat{\mu}_0\} = \arg\max_{\alpha,\mu}|X_p(\mu)|^2 \tag{3.126}$$

$$\begin{cases} \hat{k}_d = -\cot(\hat{\alpha}_0), \\ \hat{\alpha}_0 = \frac{|X_{\hat{p}}(\hat{\mu}_0)|}{\Delta\tau}, \\ \hat{f}_{dc} = \hat{\mu}_0\csc(\hat{\alpha}_0) \end{cases} \tag{3.127}$$

This condition forms the basis for estimating the moving targets' parameters. In the SFrFT domain with a proper α, the spectra of any strong moving targets will concentrate to a narrow impulse, and that of the clutter will be spread. If we can construct a narrow band-stop filter in the SFrFT domain whose center frequency around at the center of the narrowband spectrum of a strong moving target, then the signal component of this moving target can be extracted from the initial signal. With this method the strong moving targets can be extracted iteratively, thereafter the weak moving targets may be detectable. This method can be regarded as an extension of the CLEAN algorithm [371] to the SFrFT.

In summary, after cancelling the stationary clutter, the identification of moving targets can be implemented with SFrFT in the following steps:

Step 1. Apply SFrFT to the SAR data in which the clutter has been cancelled with different α, and from the maximal peak get the numerical estimation of $(\hat{\mu}, \hat{\alpha})$.

Step 2. Apply $F_r^{\hat{\alpha}}$ to the same data, we have

$$X_{\hat{\alpha}}(\mu) = \chi_{\hat{\alpha}}(\mu) \tag{3.128}$$

Step 3. After identifying the first moving target, we then construct a narrow band-stop filter $M(\mu)$ to notch the narrow band-stop spectrum of this moving target

$$X'_{\hat{\alpha}}(\mu) = X_{\hat{\alpha}}(\mu)M(\mu) \tag{3.129}$$

Step 4. The filtered signals are then transformed back to time-domain by an inverse SFrFT.

Step 5. Repeat the operations from Step 1 to Step 4 until all the desired moving targets are identified.

Once the Doppler parameters of each target are obtained, substituting them into Eq. (3.113), we can get the v_x

$$v_y = \frac{(v_s - v_x)x_0}{R_c} - \frac{\lambda f_{dc}}{2} \approx \frac{v_s x_0}{R_c} - \frac{\lambda f_{dc}}{2} \tag{3.130}$$

In this step, since v_s, x_0, R_c, λ and f_{dc} are all known variables, the v_y can be determined successfully. In a like manner, substituting Eq. (3.130) into Eq. (3.114), we can get the v_x

$$v_x = v_a - \sqrt{-\frac{\lambda k_d R_c}{2} - v_y^2} \tag{3.131}$$

Note that, here $k_d \leq 0$ is assumed. Now, the parameters (v_x, v_y) and (x_0, y_0) are all determined successfully. Next, the moving targets can then be focused with a uniform image formation algorithm, such as range-Doppler (RD) and Chirp Scaling algorithms [449]. The corresponding processing steps are given in Figure 3.28.

3.6.3 Simulation Results

To evaluate the performance of the described processing algorithm, three-antenna DPCA stripmap SAR data from three point targets, one moving target and two stationary targets, are simulated using the parameters listed in Table 3.4. Figure 3.29 shows the processing results using the general range-Doppler imaging algorithm. It can be noticed that the imaged moving target

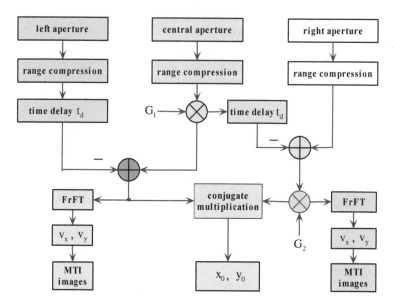

FIGURE 3.28: Illustration of the SFrFT and DPCA combined parameters estimation algorithm.

TABLE 3.4: Simulation parameters

Parameters	Values	Units
carrier frequency	1.25	GHz
pulse repeated frequency	360	Hz
flying altitude	7000	m
flying velocity	180	m/s
pulse duration	5	μs
range resolution	5	m
antenna length of each aperture	1	m
position of the target A	$(x = 50,\ y = 12000)$	m
position of the target B	$(x = 58,\ y = 12000)$	m
position of the target C	$(x = 50,\ y = 12250)$	m

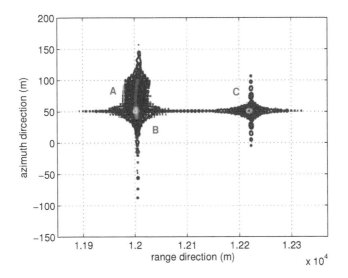

FIGURE 3.29: Processing results before clutter cancellation by DPCA operation.

B is overlapped with the stationary target A. The moving target cannot be identified from this figure, because we cannot discern which is the moving one and which is the stationary one. Figure 3.30 shows the processing results after clutter cancellation by the DPCA operation. The clutter and static returns have been cancelled successfully, leaving only moving targets and a much simplified target detection problem. However, the moving target is not focused due to the improper Doppler parameters used in the range-Doppler imaging algorithm. Moreover, the imaged target position is also drifted. To focus the moving target, accurate Doppler parameters are required.

Many algorithms considering linearly frequency modulated (LFM) signal detection, such as Wigner-Ville distribution (WVD) and Radon-Wigner transform, have been proposed. These algorithms are developed primarily for detecting single LFM signal in noise. However, they may generate cross terms, particularly in the presence of multiple moving targets. Since cross terms tend to oscillate more rapidly in the time-frequency plane than signal auto-components, two-dimensional smoothing suppresses cross-term artifacts at an expense of decreased localization. In contrast, since SFrFT is a linear transform, there will be no cross terms. We can easily locate the peak in the result of SFrFT shown in Figure 3.31. From the peak we get $(\mu = 0.2567, \alpha = 0.0092)$. Accordingly, the estimated Doppler parameters of the moving target are $(k_d = -17.34, f_{dc} = 4.45)$. Using these estimated parameters, Figure 3.32 gives the corresponding imaging results. It is shown that the moving targets can be successfully focused in this way.

In the previous discussions, the clutter cancellation is performed between

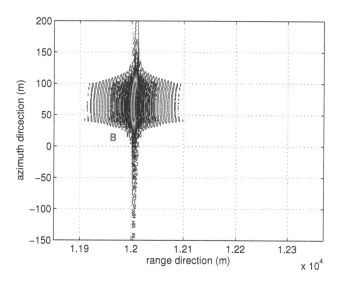

FIGURE 3.30: Processing results after clutter cancellation by DPCA operation.

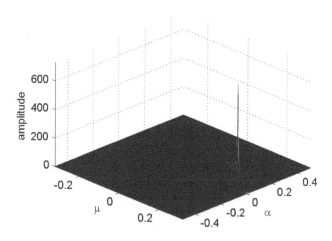

FIGURE 3.31: FrFT domain of the return of the single moving target.

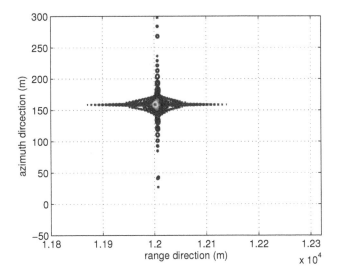

FIGURE 3.32: The focused image of the single moving target.

two DPCA antennas. Hence, the clutter cancellation performance depends mainly on the correlation characteristics of the signals from fore and aft antennas. But phase center offset and antenna deformation may cause decorrelation. It is thus necessary to analyze the decorrelation performance.

Suppose the clutter signals from the two DPCA antennas are represented by

$$s_1 = c_1 + n_1 \tag{3.132a}$$

$$s_2 = c_2 + n_2 \tag{3.132b}$$

where c_i $(i = 1, 2)$ denotes clutter signals from the i-th antenna, and n_i denotes additive noise in the the i-th antenna. The covariance matrix between the two DPCA antennas can then be represented by

$$
\begin{aligned}
\mathbf{R} =& E\left\{ \begin{bmatrix} s_1{}^* \cdot s_1 & s_1{}^* \cdot s_2 \\ s_2{}^* \cdot s_1 & s_2{}^* \cdot s_2 \end{bmatrix} \right\} \\
=& \begin{bmatrix} E\{c_1{}^* \cdot c_1 + n_1{}^* \cdot n_1\} & E\{c_1{}^* \cdot c_2\} \\ E\{c_2{}^* \cdot c_1\} & E\{c_2{}^* \cdot c_2 + n_2{}^* \cdot n_2\} \end{bmatrix}
\end{aligned}
\tag{3.133}
$$

The multiplicative random phase noise that decorrelates the antenna 2 from the antenna 1 can be modeled as

$$c_2 = c_1 \exp\left(j\varphi\right) \tag{3.134}$$

where c_1 is deterministic and φ is normally distributed. Suppose φ with a

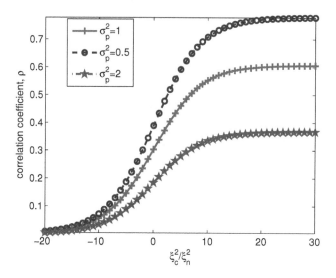

FIGURE 3.33: Example correlation coefficients as a function of clutter-to-noise ratio.

probability density function of $f(\varphi) = N(0, \sigma_\varphi^2)$, we can get

$$E\{c_1{}^* \cdot c_1\} = \xi_c^2, \qquad E\{n_1{}^* \cdot n_1\} = \xi_n^2$$
$$E\{c_1{}^* \cdot c_2\} = E\{c_2{}^* \cdot c_1\} = \xi_c^2 \exp\left(-\sigma_\varphi^2/2\right) \tag{3.135}$$

Then, Equation (3.133) can be further simplified to

$$\mathbf{R} = \begin{bmatrix} \xi_c^2 + \xi_n^2 & \xi_c^2 \exp\left(-\sigma_\varphi^2/2\right) \\ \xi_c^2 \exp\left(-\sigma_\varphi^2/2\right) & \xi_c^2 + \xi_n^2 \end{bmatrix} = \begin{bmatrix} \xi^2 & \xi^2 \cdot \rho \\ \xi^2 \cdot \rho & \xi^2 \end{bmatrix} \tag{3.136}$$

where ρ is the correlation coefficient, which can be determined by [63]

$$\rho = \frac{\xi_c^2/\xi_n^2}{1 + \xi_c^2/\xi_n^2} \exp\left(-\frac{\sigma_\varphi^2}{2}\right) \tag{3.137}$$

Figure 3.33 shows several example correlation coefficients. We can notice that the decorrelation characteristics depend on both additive noise and multiplicative phase noise, but the DPCA clutter cancellation performance depends only on multiplicative phase noise. Thus, DPCA SAR is a noise-limited detector. In contrast, ATI SAR is a clutter-limited detector. Here, the advantage of DPCA SAR and that of ATI SAR are combined; hence, the moving target detection performance can be improved.

3.7 Conclusion

In this chapter, we summarized SAR high-resolution and wide-swath imaging problems and discuss azimuth multi-antenna-based solutions. Two different approaches to implementing the DPCA, SPCM SAR and MPCM SAR, were introduced. The azimuth nonuniform sampling problem was discussed, followed by the azimuth signal reconstruction algorithm. Numerical simulation results were provided. Additionally, azimuth scanning multibeam SAR signal models and system performance were discussed. The use of azimuth multi-antenna SAR in GMTI was also discussed and a three-antenna-based ground moving targets imaging approach was presented.

4

Elevation-Plane Multi-Antenna SAR

In this chapter, we continue to discuss elevation-plane multi-antenna SARs for overcoming the fundamental tradeoff between swath width and range ambiguity. Various methods have been proposed to overcome this limitation. The depth-of-focus approach proposed in [44] can defocus the range ambiguities and thereby reduce their peak levels. However, the energy introduced by widening the elevation beam will result in degraded integrated sidelobe ratio (ISLR) performance. This chapter considers an alternative approach. The SAR operates with a pulse repetition frequency (PRF) appropriate to the desired resolution, but with a broader swath width than that implied by the minimum antenna area constraint. The SAR antenna on the receiver is an adaptive array, which forms multiple (N) separate receive patterns in the elevation direction, each with nulls in the directions of $N-1$ ambiguities, and a beam in the direction of the appropriate wanted echo.

This chapter is organized as follows. Section 4.1 makes a brief introduction of the null steering in range dimension. Section 4.2 introduces the elevation-plane multi-antenna SAR scheme at the receiver for wide-swath remote sensing, followed by several practical issues in Section 4.3. Section 4.4 presents the chirp scaling (CS)-based imaging algorithm for range multi-antenna SAR wide-swath remote sensing. Finally, system performance is analyzed in Section 4.5 and numerical simulation results are provided in Section 4.6.

4.1 Null Steering in the Elevation-Plane

The use of array nulls to achieve wide-swath SAR imaging has been proposed previously in [77, 78, 164]. In this technique, which we refer to as 'null steering', the PRF is increased for a higher azimuth resolution. The range ambiguities are controlled by steering the nulls of the receiving array at them. The receiving antenna is formed from an array of antennas, each has similar size as a conventional SAR antenna. Correct choice of element spacing in the receiving array ensures that when the nulls are correctly positioned on the ambiguities, the array main-lobe peak will coincide (approximately) with the imaging point. Under these conditions, the two-way voltage gain is never simultaneously high in the directions of the image point and its ambiguities,

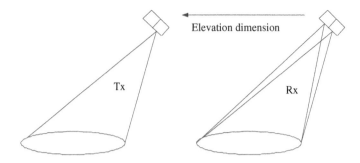

FIGURE 4.1: Two-element elevation-plane array allows for range ambiguity suppression.

even through the transmit footprint simultaneously includes the imaging point and its ambiguities.

The signal processing required for null steering is conceptually straightforward. Either range, azimuth or both range and azimuth [45] ambiguities can be suppressed using the null steering technique. In this chapter, we consider only the null steering in range dimension, which uses an elevation-plane array to steer nulls in the directions of the range ambiguities. Figure 4.1 illustrates this technique for a two-element elevation-plane array. It allows suppression of a single-range ambiguity and permits the imaging swath to be approximately doubled.

The swath illuminated by the transmitter is determined by

$$\frac{c_0}{2}\left(\frac{n}{\text{PRF}} + T_p + \Delta T\right) < R < \frac{c_0}{2}\left(\frac{n+2}{\text{PRF}} - T_p\right) \tag{4.1}$$

If the whole swath is divided into two subswaths, we then have

$$\frac{c_0}{2}\left(\frac{n}{\text{PRF}} + T_p + \Delta T\right) < R_1 < \frac{c_0}{2}\left(\frac{n+1}{\text{PRF}} - T_p\right) \tag{4.2a}$$

$$\frac{c_0}{2}\left(\frac{n+1}{\text{PRF}} + T_p + \Delta T\right) < R_2 < \frac{c_0}{2}\left(\frac{n+2}{\text{PRF}} - T_p\right) \tag{4.2b}$$

During each receiving interval a number of range profiles are recorded; one for each antenna in the receiving array. These range profiles are combined using the array weights that steer the array nulls to the ambiguities of the current imaging point, as shown in Figure 4.2. Next, conventional SAR processing can then be performed on the ambiguity-free range profiles.

When processing the null-steered SAR data, consideration must be given to the following two points. First, the element weights must be updated for each radar position. This is because the ambiguities of an imaging point remain at fixed positions on the terrain as the radar traverses the synthetic aperture. Second, the required element weights are different from each imaging point.

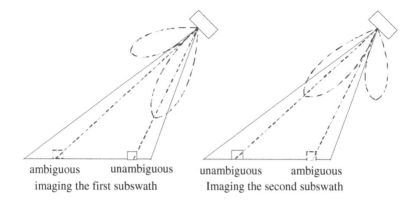

ambiguous   unambiguous   unambiguous   ambiguous

imaging the first subswath   Imaging the second subswath

FIGURE 4.2: Suppressing range ambiguity with the null steering technique.

This is because the positions of the ambiguities depend on the position of the imaging point. As SAR data are recorded digitally, the radar can be made to traverse the flight path (in a virtual sense) any number of times. This allows the nulls to be positioned differently for each imaging point.

The greatest impediment to accurate placement of the nulls is uncertainty in the platform altitude. Simulation results [43] have shown that the effects of roll estimate errors can be considered independently from those of yaw and pitch. (Roll is measured about the flight path; pitch is measured about the nominal to the plane containing the flight path and the nominal boresight.) Roll errors have an impact on the range ambiguity suppression, whereas yaw and pitch errors have an impact on the azimuth ambiguity suppression. The author concluded that a roll estimate accuracy of ±400 milli-degrees and a yaw estimate (combines both the pitch and yaw estimate errors) accuracy of ±10 milli-degrees are sufficient to obtain adequate ambiguity suppression. Uncertainty in the positions of the ambiguities must be included in these angles. Another disadvantage is that the resulting swathes are not continuous; blind ranges exist because the receiver must be switched off during transmission and reception.

Adaptive array techniques for steering range nulls have been investigated by Griffiths and Mancini [164]. They concluded that such techniques are feasible and that an angular error of approximately of 0.3° is achievable, which is adequate for the null steering technique.

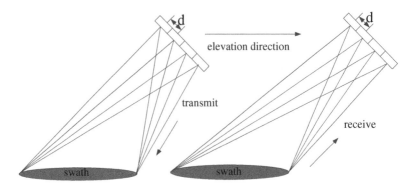

FIGURE 4.3: Illustration of the elevation-plane multi-antenna SAR.

4.2   Elevation-Plane Multi-Antenna SAR

Figure 4.3 shows the elevation-plane multi-antenna SAR. On transmit, activating all antenna elements generates a wide beam to illuminate the whole swath. During each receiving interval a number of range profiles are recorded by the multi-aperture receiver. That is to say, we divide the receiving antenna elements into multiple (M) groups in the elevation dimension.

In a single-aperture SAR, it is well known that the unambiguous slant range R is constrained by

$$\frac{c_0}{2}\left(\frac{i}{\text{PRF}} + T_p\right) < R_c < \frac{c_0}{2}\left(\frac{i+1}{\text{PRF}} - T_p\right) \tag{4.3}$$

If we divide the whole swath into M subswaths, the slant range constraint can be relaxed to

$$\frac{c_0}{2}\left(\frac{i+m-1}{\text{PRF}} + T_p\right) < R < \frac{c_0}{2}\left(\frac{i+m}{\text{PRF}} - T_p\right), 0 \le m \le M-1 \tag{4.4}$$

Considering the geometry shown in Figure 4.4, the looking-down angle from the first subaperture to the m-th subswath can be expressed as

$$\alpha_m = \alpha\left[\frac{c_0}{2}\left(t + \frac{m-1}{\text{PRF}}\right)\right], 0 \le m \le M-1 \tag{4.5}$$

with

$$\alpha(x) = \arccos\left[\frac{x^2 + h_s^2 + 2h_s R_e}{2x(h_s + R_e)}\right] - \beta \tag{4.6}$$

where x, R_e, t and β represent the target's azimuth position, Earth radius, sampling time and incidence angle, respectively.

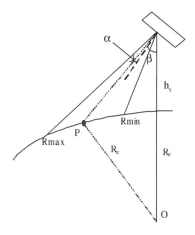

FIGURE 4.4: Geometry of the elevation-plane multiaperture SAR.

Suppose the transmitted signal is

$$s(t) = \text{rect}\left[\frac{t}{T_p}\right] \exp\left[j2\pi f_c t + j\pi k_r \left(t - \frac{T_p}{2}\right)^2\right] \tag{4.7}$$

where rect[·] is a window function, f_c is the carrier frequency, and k_r is the chirp rate. After matched filtering, the signals in the n-th ($n = 1, 2, \ldots, N$) receiving subaperture can be expressed as

$$
\begin{aligned}
y_n(t, \tau) = \sum_{m=0}^{M-1} & \int_{c_0 m/2\text{PRF}}^{c_0(m+1)/2\text{PRF}} \int_{v_s t - L_a/2}^{v_s t + L_a/2} \exp\left(-j\frac{4\pi R}{\lambda} - j\frac{m2\pi f_c}{\text{PRF}}\right) \\
& \times \exp\left[-j\frac{2\pi x^2}{\lambda(r + \frac{mc_0}{2\text{PRF}})}\right] \exp\left\{j2\pi(n-1)\frac{d_r}{\lambda}\sin\left[\alpha\left(R + \frac{mc_0}{2\text{PRF}}\right)\right]\right\} \\
& \times \text{sinc}\left\{B_r\left[\tau + \frac{m}{\text{PRF}} + \frac{nd_r \sin\left[\alpha(R + \frac{mc_0}{2\text{PRF}})\right] - 2r}{c_0} - \frac{x^2}{c_0\left(R + \frac{mc_0}{2\text{PRF}}\right)}\right]\right\} dxdr
\end{aligned}
\tag{4.8}
$$

where τ is the azimuth slow time, v_s is the platform flying speed, L_a is the synthetic aperture length, λ is the transmitted signal wavelength, d_r is the subaperture antenna length, and B_r is the transmitted signal bandwidth. Since

$$nd_r \sin\left[\alpha\left(R + \frac{mc_0}{2\text{PRF}}\right)\right] \ll 2R \tag{4.9}$$

Eq. (4.8) can be further simplified to

$$
y_n(t,\tau) \approx \sum_{m=0}^{M-1} \exp\left[j2\pi(n-1)\frac{d_r}{\lambda}\sin\left[\alpha\left(\tau + \frac{n+m}{\text{PRF}}\right)\frac{c_0}{2}\right]\right]
$$

$$
\times \int_{v_s t - L_a/2}^{v_s t + L_a/2} \exp\left(-j\frac{4\pi R_{m,n}(\tau,x)}{\lambda}\right)\exp\left(-j\frac{2\pi x^2}{R_{m,n}(\tau,x)\lambda}\right)dx
$$

(4.10)

where

$$
R_{m,n}(\tau,x) \approx \left(\tau + \frac{n+m}{\text{PRF}}\right)\frac{c_0}{2} - \frac{x^2}{\left(\tau + \frac{n+m}{\text{PRF}}\right)\frac{c_0}{2}}
$$

(4.11)

Eq. (4.10) can be rewritten as a matrix form

$$
\mathbf{Y}(t,\tau) = \mathbf{W}(t)\cdot\mathbf{S}(t,\tau)
$$

(4.12)

where $\mathbf{Y}(t,\tau) = [y_1(t,\tau), y_2(t,\tau), \ldots, y_N(t,\tau)]^T$, the sensing matrix $\mathbf{W}(t)$ is expressed as

$$
\mathbf{W}(t) = \begin{bmatrix} 1 & \cdots & 1 \\ \exp\left(j\frac{2\pi d_r \sin(\alpha_0)}{\lambda}\right) & \cdots & \exp\left(j\frac{2\pi d_r \sin(\alpha_{M-1})}{\lambda}\right) \\ \vdots & \ddots & \vdots \\ \exp\left(j\frac{2\pi(M-1)d_r \sin(\alpha_0)}{\lambda}\right) & \cdots & \exp\left(j\frac{2\pi(M-1)d_r \sin(\alpha_{M-1})}{\lambda}\right) \end{bmatrix}
$$

(4.13)

The signals $\mathbf{S}(\mathbf{t},\tau) = [s_1(t,\tau), \ldots, s_N(t,\tau)]^T$, reflected from the N subswaths, are expressed as

$$
s_m(t,\tau) = \int_{v_s t - L_a/2}^{v_s t + L_a/2} \exp\left(-j\frac{4\pi R_{m,n}(\tau,x)}{\lambda}\right)\exp\left(-j\frac{2\pi x^2}{R_{m,n}(\tau,x)\lambda}\right)dx \quad (4.14)
$$

It can be easily proved that the sensing matrix $\mathbf{W}(t)$ is a full rank matrix. The subswath returns $\mathbf{S}(t,\tau)$ can then be reconstructed from the radar returns by

$$
\mathbf{S}(t,\tau) = \mathbf{W}_0^{-1}(t)\cdot\mathbf{Y}(t,\tau)
$$

(4.15)

where $\mathbf{W}_0(t)$ is the reference matrix used to reconstruct the subswath returns, and $(\cdot)^{-1}$ denotes the inverse matrix operator. If $\mathbf{W}_0(t) = \mathbf{W}(t)$, the M subswath signals can be perfectly reconstructed from the aliasing returns for unambiguously wide-swath remote sensing.

4.3 Several Practical Issues

4.3.1 PRF Design

In an actual SAR system, to ensure a successful signal transmission, the minimum transmit time window T_{tw} should be

$$T_{\text{tw}} \geq T_p + \tau_{\text{rp}} \qquad (4.16)$$

where τ_{rp} is the receiver protecting window. Correspondingly, the minimum receive time window T_{rw} should be

$$T_{\text{rw}} = \frac{2(R_{\text{max}} - R_{\text{min}})}{c_0} + \tau_{\text{rp}} \qquad (4.17)$$

where R_{max} and R_{min} denote the farthest range and nearest slant range in the imaged swath, respectively. The allowable maximum PRF is

$$\text{PRF}_{\text{max}} = \frac{1}{T_{\text{tw}} + T_{\text{rw}}} \qquad (4.18)$$

On the other hand, to avoid possible azimuth ambiguities, the minimum PRF should be

$$\text{PRF}_{\text{min}} > \frac{v_s}{\rho_a} \qquad (4.19)$$

where ρ_a is the the azimuth resolution.

4.3.2 Ill-Condition of the Sensing Matrix

Ideal conditions are assumed in the previous discussions; however, there may be some undesired system errors in the actual SAR system. Suppose there are system errors $\Delta \mathbf{W}(t)$ in the sensing matrix $\mathbf{W}(t)$, we then have

$$\mathbf{S}(t, \tau) + \Delta \mathbf{S}(t, \tau) = [\mathbf{W}(t) + \Delta \mathbf{W}(t)]^{-1} \cdot \mathbf{Y}(t, \tau) \qquad (4.20)$$

where $\Delta \mathbf{S}(t, \tau)$ is the corresponding reconstruction signal errors caused by the system errors $\Delta \mathbf{W}(t)$. Since $\Delta \mathbf{W}(t) \ll \mathbf{W}(t)$, we have

$$\left\| \mathbf{W}^{-1}(t) \right\|_2 \left\| \Delta \mathbf{W}(t) \right\|_2 < 1 \qquad (4.21)$$

We can get

$$\begin{aligned}
\frac{\|\Delta \mathbf{S}(t, \tau)\|_2}{\|\mathbf{S}(t, \tau)\|_2} &\leq \frac{\|\mathbf{W}^{-1}(t)\|_2 \cdot \|\Delta \mathbf{W}(t)\|_2}{1 - \|\mathbf{W}^{-1}(t)\|_2 \cdot \|\Delta \mathbf{W}(t)\|_2} \\
&= \frac{\text{cond}[\mathbf{W}(t)] \cdot \frac{\|\Delta \mathbf{W}(t)\|_2}{\mathbf{W}(t)}}{1 - \text{cond}[\mathbf{W}(t)] \cdot \frac{\|\Delta \mathbf{W}(t)\|_2}{\mathbf{W}(t)}}
\end{aligned} \qquad (4.22)$$

The cond[·] denotes the condition number defined as

$$\text{cond}[\mathbf{W}(t)] = ||\mathbf{W}(t)||_2||\mathbf{W}^{-1}(t)||_2 = \sqrt{\frac{\lambda_{\max}}{\lambda_{\min}}} \qquad (4.23)$$

where $\lambda_{\max}$ and $\lambda_{\min}$ denote, respectively, the maximum and minimum eigenvalues of the matrix $\mathbf{W}(t)\mathbf{W}^H(t)$ with $(\cdot)^H$ being the conjugate transpose. If a matrix has a small condition number, it is a well-conditioned matrix; otherwise, it is an ill-conditioned matrix. It is noticed that the condition number of the sensing matrix will influence subsequent subswath signal processing performance. Therefore, a small condition number is desired for the sensing matrix $\mathbf{W}(t)$.

As a simple example, we consider a SAR with two receiving antennas. In this case, the sensing matrix is

$$\mathbf{W}(t) = \begin{bmatrix} 1 & 1 \\ \exp\left(j2\pi\frac{d_r}{\lambda}\sin\alpha_0\right) & \exp\left(j2\pi\frac{d_r}{\lambda}\sin\alpha_1\right) \end{bmatrix} \qquad (4.24)$$

We have

$$\mathbf{W}(t)\mathbf{W}^H(t) = \begin{bmatrix} 2, & \exp\left(-j2\pi\frac{d_r}{\lambda}\sin\alpha_0\right) + \exp\left(-j2\pi\frac{d_r}{\lambda}\sin\alpha_1\right) \\ \exp\left(j2\pi\frac{d_r}{\lambda}\sin\alpha_0\right) + \exp\left(j2\pi\frac{d_r}{\lambda}\sin\alpha_1\right), & 2 \end{bmatrix} \qquad (4.25)$$

The condition number of the sensing matrix $\mathbf{W}(t)$ is derived as

$$\text{cond}[\mathbf{W}(t)] = \sqrt{\frac{2 + \sqrt{2}\cdot\sqrt{1 + \cos\left(\frac{2\pi d_r(\sin\alpha_0 - \sin\alpha_1)}{\lambda}\right)}}{2 - \sqrt{2}\cdot\sqrt{1 + \cos\left(\frac{2\pi d_r(\sin\alpha_0 - \sin\alpha_1)}{\lambda}\right)}}} \qquad (4.26)$$

The antenna length d_r can be approximated as

$$d_r = \frac{0.886\lambda}{2\cdot\sin\left[\alpha(R_{\max}) - \alpha(R_{\min})\right]} \qquad (4.27)$$

If we want the condition number to be $\text{cond}[\mathbf{W}(t)] \leq 3$, the SAR geometry should be satisfactory with

$$\cos\left(\frac{0.886\pi\left[\sin\alpha(R_{\max} - \frac{c_0}{2\text{PRF}}) - \sin\alpha(R_{\max})\right]}{2\cdot\sin\left[\alpha(R_{\max}) - \alpha(R_{\min})\right]}\right) \leq 0.28 \qquad (4.28)$$

If three and more receiving subswathes are employed in the SAR system, the corresponding condition number can be analyzed by numerical computation.

4.3.3   Interferences of Nadir Echoes

To implement efficient wide-swath imaging, the echoes reflected from the nadir points must be avoided. Otherwise, the nadir echoes will bring a catastrophic

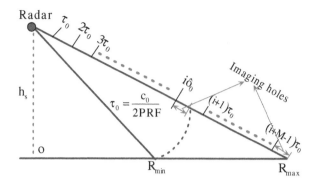

FIGURE 4.5: Illustration of imaging holes in the farthest and nearest swath.

interference on the desired echoes due to their specular reflection. To avoid this problem, the platform altitude h_s and operating PRF should be satisfied with the relation given in Eq. (4.1). Since it is a very rigorous constraint, we introduce a novel method to suppress the interferences of nadir echoes.

Consider the geometry of interaction of the radar pulse shown in Figure 4.5, there will be imaging hole in the farthest and nearest swath, hence we have

$$\begin{cases} \frac{c_0}{2}(\frac{i+m-1}{\text{PRF}} + T_1) < R_m < \frac{c_0}{2}(\frac{i+m}{\text{PRF}} - T_p), & m = 1 \\ \frac{c_0}{2}(\frac{i+m-1}{\text{PRF}} + \Delta T) < R_m < \frac{c_0}{2}(\frac{i+m}{\text{PRF}}), & 2 \le m \le M-1 \\ \frac{c_0}{2}(\frac{i+m-1}{\text{PRF}} + \Delta T) < R_m < \frac{c_0}{2}(\frac{i+m}{\text{PRF}} + T_2), & m = M \end{cases} \quad (4.29)$$

where

$$T_1 = \frac{2R_{\min}}{c_0} - \frac{i}{\text{PRF}}, \quad (4.30a)$$

$$T_2 = \frac{2R_{\max}}{c_0} - \frac{i'}{\text{PRF}} \quad (4.30b)$$

with

$$T_1 \ge T_p + \Delta T \quad (4.31a)$$

$$T_2 \ge T_p \quad (4.31b)$$

$$T_1 + T_2 < \frac{1}{\text{PRF}} \quad (4.31c)$$

Note that i' is also an integer, but it is not equal to i.

From Eq. (4.29) we know that

1) when $T_p + \Delta T < t < T_1$ with t being the fast time, the first subswath has no returns.

2) when $T_1 < t < \frac{1}{\text{PRF}} - T_2$, all the subswaths have returns.

3) when $\frac{1}{\text{PRF}} - T_2 < t < \frac{1}{\text{PRF}} - T_p$, the last subswath has no returns.

This information provides a potential solution to suppress or avoid nadir echoes. If we made the nadir echoes arrive in the receiver at $T_p + \Delta T < t < T_1$ and lie in the first subswath, or arrive in the receiver at $\frac{1}{\text{PRF}} - T_2 < t < \frac{1}{\text{PRF}} - T_p$ and lie in the last subswath, the nadir echoes can then be suppressed with digital beamforming on receiver [221]. To made the nadir echoes lie in the first subswath, the radar platform altitude should be

$$h_{s1} = \frac{c_0}{2}\left(\frac{i}{\text{PRF}} + t_d\right), \quad t_d < T_1 \tag{4.32}$$

In this case, the nadir echoes can be suppressed by changing the sensing matrix expressed in Eq. (4.13) into Eq. (4.33)

$$\begin{bmatrix} 1 & 1 & \cdots & 1 \\ \exp\left(j\frac{2\pi d_r \sin(\alpha_{d1})}{\lambda}\right) & \exp\left(j\frac{2\pi d_r \sin(\alpha_2)}{\lambda}\right) & \cdots & \exp\left(j\frac{2\pi d_r \sin(\alpha_M)}{\lambda}\right) \\ \vdots & \vdots & \ddots & \vdots \\ \exp\left(j\frac{2\pi(M-1)d_r \sin(\alpha_{d1})}{\lambda}\right) & \exp\left(j\frac{2\pi(M-1)d_r \sin(\alpha_2)}{\lambda}\right) & \cdots & \exp\left(j\frac{2\pi(M-1)d_r \sin(\alpha}{\lambda}\right) \end{bmatrix} \tag{4.33}$$

with

$$\alpha_{d1} = \alpha[h_{s1})] \tag{4.34}$$

Similarly, if we made the nadir echoes lie in the last subswath, the platform altitude should be

$$h_{s2} = \frac{c_0}{2}\left(\frac{i'}{\text{PRF}} + t'_d\right), \quad T_2 < t'_d < \frac{1}{\text{PRF}} \tag{4.35}$$

The nadir echoes can also be suppressed by changing the sensing matrix expressed in Eq. (4.13) into Eq. (4.36) with $\alpha_{d2} = \alpha[h_{s2})]$. Certainly the platform altitudes expressed in Eqs. (4.32) and (4.35) have a wider range than that expressed in Eq. (4.1). More importantly, the nadir echoes can be avoided or suppressed in this way.

$$\begin{bmatrix} 1 & 1 & \cdots & 1 \\ \exp\left(j\frac{2\pi d_r \sin(\alpha_1)}{\lambda}\right) & \exp\left(j\frac{2\pi d_r \sin(\alpha_2)}{\lambda}\right) & \cdots & \exp\left(j\frac{2\pi d_r \sin(\alpha_M)}{\lambda}\right) \\ \vdots & \vdots & \ddots & \vdots \\ \exp\left(j\frac{2\pi(M-1)d_r \sin(\alpha_1)}{\lambda}\right) & \exp\left(j\frac{2\pi(M-1)d_r \sin(\alpha_2)}{\lambda}\right) & \cdots & \exp\left(j\frac{2\pi(M-1)d_r \sin(\alpha}{\lambda}\right) \end{bmatrix} \tag{4.36}$$

To analyze the performance of this nadir echoes suppression approach, first the T_1 and T_2 defined in Eq. (4.31) must be determined. Once we made the nadir echoes lie in the first or the last subswath, the nadir echoes can then be suppressed with digital beamforming on receiver. Suppose there is one point target at a slant range of 68.5 km, Figure 4.6 gives the comparative pulse compression results before and after suppressing the nadir echoes. Due to range ambiguities, there is a ambiguous target located at 67.5 km in Figure 4.6(a). After applying the nadir echoes suppression method, the ambiguous target is significantly suppressed, as shown in Figure 4.6(b). Therefore, the nadir echoes can be suppressed in this way.

4.3.4 Blind Range Problem

Blind ranges occur when the return arrives at the receiver while the radar is transmitting a new pulse and receiver is blanked. The blind ranges are given by

$$r_n \approx n \frac{c_0}{2 \cdot \text{PRF}} \tag{4.37}$$

where r_n is the blind range for a given value of n, n is a positive integer that indicates which of the blind ranges is being determined [151]. The blind ranges are independent of the radar carrier frequency.

The blind range problem can be resolved by variable pulse repetition intervals (PRIs), as shown in Figure 4.7. For simplicity, the PRIs are designed to be an arithmetic progression

$$T_{k+1} - T_k = T_0, 1 \leq k \leq K - 1, T_0 \gg 2T_p \tag{4.38}$$

where T_k is the k-th PRI, and T_0 is a given constant satisfying $T_0 \gg 2T_p$. For the first pulse, its n-th blind range is determined by

$$c_0 \left(\sum_{i=1}^{n} T_i - T_p \right) < r_n < c_0 \left(\sum_{i=1}^{n} T_i + T_p \right), 1 \leq n \leq K - 1 \tag{4.39}$$

Similarly, for the other subsequent pulses the n-th blind range can be derived as

$$\begin{cases} c_0 \left(\sum_{i=k}^{n+k-1} T_i - T_p \right) < r_n < c_0 \left(\sum_{i=k}^{n+k-1} T_i + T_p \right), \\ \qquad 1 \leq n \leq K - k + 1 \\ c_0 \left(\sum_{i=k}^{K} T_i + \sum_{i=1}^{k-1} T_i - T_p \right) < r_n < c_0 \left(\sum_{i=k}^{n+k-1} T_i + \sum_{i=1}^{k-1} T_i + T_p \right), \\ \qquad 1 \leq n \leq K - k + 1 \end{cases} \tag{4.40}$$

It can be concluded from Eqs. (4.39) and (4.40) that, there will be no overlapped blind ranges. That is to say, the blind-range problem has been resolved in this way.

However, the variable PRIs may results in nonuniform azimuth (Doppler) sampling signal. After matched filtering we can get the following discrete azimuth signal [428]

$$x[l] = s_r \left[\left\lfloor \frac{l}{K} \right\rfloor \sum_{k=1}^{K} T_k + \sum_{k=1}^{l - \lfloor \frac{l}{K} \rfloor K} T_k \right] \tag{4.41}$$

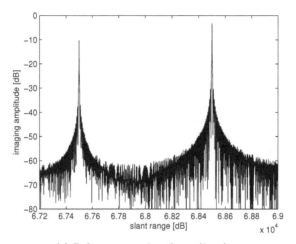

(a) Before suppressing the nadir echoes.

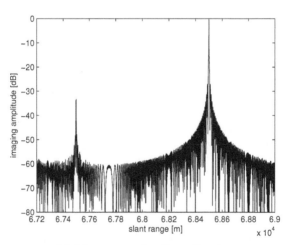

(b) After suppressing the nadir echoes.

FIGURE 4.6: Comparative pulse compression results before and after suppressing the nadir echoes.

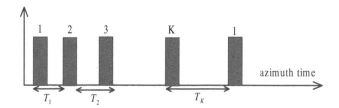

FIGURE 4.7: Illustration of the blind range suppression via variable PRIs.

where $s_r[t]$ is the continuous azimuth signal that has been processed by matched filters, and $\lfloor \cdot \rfloor$ is the minimum integer. Although the $x[l]$ is a nonuniform signal, we can divide $x[l]$ into K groups

$$x[l] = [x_1[K \cdot l_s], x_2[K \cdot l_s + 1], \ldots, x_K[K \cdot l_s + K - 1]] \tag{4.42}$$

where $0 \leq l_s \leq L_s = L/K$ with L being the length of the sampled signal sequence. The signals in each group are uniform and can be processed by the general Fourier transform algorithm

$$S(\omega) = \text{FFT}\{x_1[l_s]\} + \sum_{k=2}^{K-1} \left\{ e^{(-jl_s\omega \sum_{i=1}^{k-1} T_i)} \cdot \text{FFT}[x_k[l_s]] \right\} \tag{4.43}$$

Next, general algorithms can be employed for subsequent signal processing.

4.4   Multi-Antenna Chirp Scaling Imaging Algorithm

Like general SAR imaging processor, the task of the elevation-plane multi-antenna SAR imaging processor is to implement a filter, matched in both range and azimuth dimensions, to focus the raw data. In the following, we develop a CS-based image formation algorithm.

The i-th subaperture received signals that can be represented by

$$s_i(t, \tau) = \sum_{m=0}^{M-1} \sigma_m \exp\left[-j\pi k_r \left(\tau + \frac{n+m}{\text{PRF}} - \frac{2R_m(\tau)}{c_0}\right)^2\right]$$
$$\times \psi_{m,i}\left(R - \frac{\lambda}{2}f_d\tau\right) \exp\left(-j\frac{4\pi}{\lambda}R_m(\tau)\right) \tag{4.44}$$

where

$$\psi_{m,i}\left(R-\frac{\lambda}{2}f_d\tau\right) = \exp\left[j2\pi i\frac{d_r}{\lambda}\sin(\alpha_m)\right] \tag{4.45a}$$

$$f_d = \frac{2v_r}{\lambda} \tag{4.45b}$$

$$k_{a,m} = -\frac{2v_s^2}{\lambda\left(R+\frac{mc_0}{2\text{PRF}}\right)} \tag{4.45c}$$

with v_r being the relative radial velocity. The subsequent processing steps can then follow the general CS algorithm detailed in Chapter 2.

Step 1: Azimuth FFT According to the principle of stationary phase, Eq. (4.44) can be represented in range-Doppler domain by

$$
\begin{aligned}
S1_i(t, f_a) = \sum_{m=0}^{M-1} & \sigma_m \exp\left[-j\pi K_s\left(f_a, R+\frac{mc_0}{2\text{PRF}}\right)\right. \\
& \times \left.\left(t+\frac{m}{\text{PRF}} - \frac{2R_{s,m}\left(f_a, R+\frac{mc_0}{2\text{PRF}}\right)}{c_0}\right)^2\right] \\
& \times \psi_{m,i}\left[R-\frac{\lambda}{2}f_d\left(\frac{f_a-f_d}{k_{a,m}}\right)\right] \\
& \times \exp\left[-j\frac{4\pi f_0}{c_0}\left(R+\frac{\lambda(f_a-f_d)^2}{4k_{a,m}}\right)\right]
\end{aligned}
\tag{4.46}
$$

The equivalent chirp rate K_s, the range curvature factor C_s and the range mitigation $R_{s,m}$ can be expressed, respectively, by

$$K_s\left(f_a, R+\frac{mc_0}{2\text{PRF}}\right) = \frac{k_r}{1-k_r\frac{f_a^2}{k_{a,m}f_0^2}} \tag{4.47}$$

$$C_s(f_a) = \frac{\lambda(f_d^2-f_a^2)}{4k_{a,m}\left(R+\frac{mc_0}{2\text{PRF}}\right)} = \frac{\lambda^2}{8v_s^2}(f_a^2-f_d^2) \tag{4.48}$$

$$R_{s,m}\left(f_a, R+\frac{mc_0}{2\text{PRF}}\right) = \left(R+\frac{mc_0}{2\text{PRF}}\right)[1+C_s(f_a)] \tag{4.49}$$

Step 2: Chirp Scaling Phase Processing The range-Doppler domain signal $S1_i(t, f_a)$ is multiplied by the following chirp scaling factor

$$H_{m,1}(t, f_a) = \exp\left\{-j\pi K_s\left(f_a, R_{\text{ref1}}\right)C_s(f_a)\left[t+\frac{n}{\text{PRF}}-\frac{2}{c_0}R_{s,0}\left(f_a, R_{\text{ref2}}\right)\right]^2\right\} \tag{4.50}$$

whose phase term is chosen so that the range mitigation of each and every scatterer in each subswath is equalized to that of the reference range. Note that the R_{ref1} and R_{ref2} are the slant ranges to the whole swath center and the subswath center, respectively. Multiplying $S1_i(t, f_a)$ by $H_{m,1}(t, f_a)$ yields

$$
\begin{aligned}
S2_i(t, f_a) = \sum_{m=0}^{M-1} \psi_{m,i} &\left[R - \frac{\lambda}{2} f_d \left(\frac{f_a - f_d}{k_{a,m}} \right) \right] \exp\left(-j\Theta_m\right) \\
&\times \exp\left[-j\frac{4\pi}{\lambda} \left(R + \frac{mc_0}{2\text{PRF}} + \frac{\lambda(f_a - f_d)^2}{4k_{a,m}} \right) \right] \\
&\times \exp\left\{ -j\pi K_s\left(f_a, R_{\text{ref1}}\right) \left[1 + C_s(f_a)\Gamma^2(t, f_a) \right] \right\}
\end{aligned}
\tag{4.51}
$$

where

$$
\Theta_m = \pi K_s(f_a, R_{\text{ref1}})C_s(f_a)\left[1 + C_s(f_a)\right] \frac{4}{c_0^2} \left[R + \frac{c_0}{2} \frac{mC_s(f_a)}{\text{PRF}\left(1 + C_s(f_a)\right)} - R_{\text{ref2}} \right]^2
$$

$$
\tag{4.52a}
$$

$$
\Gamma(t, f_a) = \left[t - \frac{2}{c_0} \left(R + \frac{mc_0}{2\text{PRF}} \frac{C_s(f_a)}{1 + C_s(f_a)} + C_s(f_a)R_{\text{ref2}} \right) \right]^2
\tag{4.52b}
$$

This multiplication results in that the range mitigation terms in each subswath follow the same reference curvature trajectory

$$
\Delta R_m = \frac{mc_0}{2\text{PRF}\left(1 + C_s(f_a)\right)} + C_s(f_a)R_{\text{ref2}}, \quad 0 \le m \le M - 1
\tag{4.53}
$$

Step 3: Range FFT After applying the range FFT, we get

$$
\begin{aligned}
S3_i(f_r, f_a) = \sum_{m=0}^{M-1} \psi_{m,i} &\left[R - \frac{\lambda}{2} f_d \left(\frac{f_a - f_d}{k_{a,m}} \right) \right] \exp\left[-j\frac{4\pi}{c_0}\left(R + \Delta R_m\right)f_r \right] \\
&\times \exp\left\{ j\pi \frac{f_r^2}{K_s(f_a, R_{\text{ref1}})\left[1 + C_s(f_a)\right]} \right\} \Psi_{m,i}(R)
\end{aligned}
\tag{4.54}
$$

where

$$
\Psi_{m,i}(R) = \exp\left\{ -j\frac{4\pi}{\lambda}\left[R + \frac{mc_0}{2\text{PRF}} + \frac{\lambda(f_a - f_d)^2}{4k_{a,m}} \right] \right\} \exp\left(-j\Theta_m\right)
\tag{4.55}
$$

Step 4: Range Compression Range compressing $S3_i(f_r, f_a)$ with the range reference function

$$
H_2(f_r, f_a) = \exp\left\{ j\pi \frac{f_r^2}{K_s(f_a, R_{\text{ref1}})\left[1 + C_s(f_a)\right]} \right\}
\tag{4.56}
$$

yields

$$S4_i(f_r, f_a) = \sum_{m=0}^{M-1} \psi_{m,i} \left[R - \frac{\lambda}{2} f_d \left(\frac{f_a - f_d}{k_{a,m}} \right) \right]$$
$$\times \exp\left[-j\frac{4\pi}{c_0} (R + \Delta R_m) f_r \right] \Psi_{m,i}(R) \tag{4.57}$$

Step 5: Range Mitigation Correction Since each subswath has a different range curvature, the range mitigation correction should be applied separately for each subswath. The range mitigation correction function for the p-th subswath is

$$H_{3,p}(f_r, f_a) = \exp\left\{ j\frac{4\pi f_r}{c_0} \left[\frac{mc_0}{2\text{PRF}\left[1 + C_s(f_a)\right]} + C_s(f_a)R_{\text{ref2}} \right] \right\} \tag{4.58}$$

We then have

$$S5_{i,p}(f_r, f_a) = S4_i(f_r, f_a) \cdot H_{3,p}(f_r, f_a) \tag{4.59}$$

Step 6: Range IFFT The range IFFT collapses to the focused range envelope at the correct range position, leaving only azimuth phase terms. The output signal in range-Doppler domain is

$$S6_{i,p}(t, f_a) = \sum_{m=0}^{M-1} \psi_{m,i} \left[R - \frac{\lambda}{2} f_d \left(\frac{f_a - f_d}{k_{a,m}} \right) \right] \Psi_{m,i}(R)$$
$$\times \text{sinc}\left[\frac{t - \frac{2}{c_0}\left(R + \frac{(p-m)c_0 C_s(f_a)}{2\text{PRF}(1+C_s(f_a))} \right) + \frac{nc_0}{2\text{PRF}}}{B_r} \right] \tag{4.60}$$

Step 7: Subswath Data Fusion Using the sensing matrix

$$\mathbf{W}(t, f_a)_{m,i} = \exp\left\{ \frac{j2\pi(i-1)d_r}{\lambda} \sin\alpha \left[\frac{c_0}{2}\left(t + \frac{m}{\text{PRF}} \right) \right. \right.$$
$$\left. \left. -\frac{\lambda}{2} f_d \left(\frac{f_a - f_d}{k_{a,m}} \right) - \frac{(i-m)C_s(f_a)}{\text{PRF}(1+C_s(f_a))} \right] \right\} \tag{4.61}$$

we can reconstruct the subswath data as

$$S7_p(t, f_a) = \mathbf{W}_{p,:}^{-1}\mathbf{S}_M = \text{sinc}\left[\frac{1}{B_r}\left(t - \frac{2R}{c_0} \right) \right] \Psi_{m,i}(R) \tag{4.62}$$

where $\mathbf{S}_M = [S6_{i,1}, S6_{i,2}, \ldots, S6_{i,M}]^T$.

Step 8: Azimuth Filtering and Phase Residual Compensation Compensating the phase residual $\Psi_{m,i}(R)$ yields

$$S8_p(t, f_a) = \text{sinc}\left[\frac{1}{B_r}\left(t - \frac{2R}{c_0} \right) \right] \tag{4.63}$$

Step 9: Azimuth IFFT

The subswath image is obtained by an azimuth IFFT

$$S9_p(t, \tau) = \text{sinc}\left[\frac{1}{B_r}\left(t - \frac{2R}{c_0}\right)\right]\text{sinc}\left(\frac{\tau}{\rho_a}\right) \tag{4.64}$$

where ρ_a is the azimuth resolution.

Finally, combing the subswath images yields the whole swath image

$$S10(t, \tau) = \sum_{m=0}^{M-1}\text{sinc}\left[\frac{1}{B_r}\left(t - \frac{m+n}{\text{PRF}} - \frac{2R}{c_0}\right)\right]\text{sinc}\left(\frac{\tau}{\rho_a}\right) \tag{4.65}$$

The logical flow diagram of the image formation algorithm is shown in Figure 4.8.

4.5 System Performance Analysis

4.5.1 RASR Analysis

It is well known that the single-antenna SAR RASR is defined as [76]

$$\text{RASR}_{\text{sa}} = \max_i \left[\frac{\displaystyle\sum_{k=-n_h}^{n_h} \frac{\sigma_{i,k}G_{i,k}^2}{R_{i,k}^3 \sin(\theta_{i,k})}}{\frac{\sigma_{i,0}G_{i,0}^2}{R_{i,0}^3 \sin(\theta_{i,0})}}\right], \tag{4.66}$$

$$k \neq 0, i = 1, 2, \ldots, n, k = \pm 1, \pm 2, \ldots, \pm n_h$$

where $\sigma_{i,k}$ ($\sigma_{i,0}$), $G_{i,k}$ ($G_{i,0}$), $R_{i,k}$ ($R_{i,0}$) and $\theta_{i,k}$ ($\theta_{i,0}$) denote respectively the radar cross section (RCS) coefficient, antenna gain, slant range and incidence angle for the i-th ambiguous range cell. The k, which is the pulse index ($k = 0$ for the desired pulse), is positive for the preceding pulse and negative for the succeeding ones. Note that $k = n_h$ is is the number of pulses to the horizon.

Different from the single-antenna SAR, for the elevation-plane multi-antenna SAR the range ambiguous signals will arrive from the ranges of

$$R_{m,i,k} = R_{m,i} + \frac{c_0}{2}\frac{k}{\text{PRF}}, m = 1, 2, \ldots, M, k = \pm 1, \pm 2, \ldots, \pm n_h \tag{4.67}$$

where $R_{m,k}$ is the slant range for the m-th subaperture. The corresponding RASR for the m-th subswath can be expressed as

$$\text{RASR}_{\text{ma},m} = \max_i \left[\frac{\displaystyle\sum_{k=-n_h}^{n_h} \frac{\sigma_{m,i,k}G_{m,i,k}^2}{R_{m,i,k}^3 \sin(\theta_{m,i,k})}}{\frac{\sigma_{m,i,0}G_{m,i,0}^2}{R_{m,i,0}^3 \sin(\theta_{m,i,0})}}\right], k \neq 0 \tag{4.68}$$

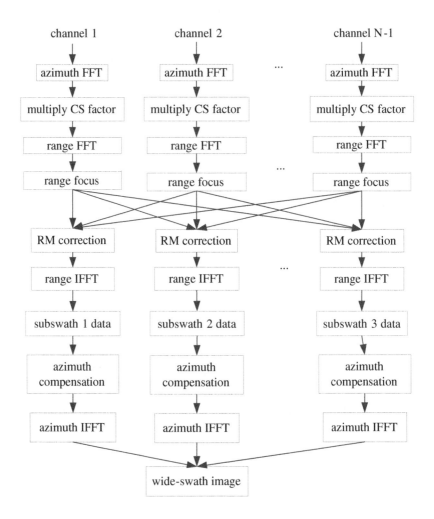

FIGURE 4.8: The logical flow diagram of the image formation algorithm.

The variables are defined in a similar manner as the Eq. (4.66). Considering all the subswaths, the final RASR of the elevation-plane multi-antenna SAR can be evaluated by

$$
\text{RASR}_{\text{ma}} = \max_i \left[\frac{\sum_{m=1}^{M} \sum_{k=-n_h}^{n_h} \frac{\sigma_{\text{m,i,k}} G_{\text{m,i,k}}^2}{R_{\text{m,i,k}}^3 \sin(\theta_{\text{m,i,k}})}}{\sum_{m=1}^{M} \frac{\sigma_{\text{m,i,0}} G_{\text{m,i,0}}^2}{R_{\text{m,i,0}}^3 \sin(\theta_{\text{m,i,0}})}} \right], k \neq 0 \qquad (4.69)
$$

Comparing Eqs. (4.66) and (4.69), we can notice that the elevation-plane multi-antenna SAR has different RASR from conventional single-antenna SAR. Their differences will be further investigated by simulation results in next section.

4.5.2 SNR Analysis

The target reflected per-pulse power available at the n-th receiving subaperture is determined by

$$
P_r = \frac{P_t \cdot G_t}{4\pi R_0^2(\theta_m)} \cdot \frac{\sigma_s}{4\pi R_0^2(\theta_m)} \cdot \frac{\lambda^2 G_r}{4\pi} \qquad (4.70)
$$

where P_t is the transmitting peak power, G_t and G_r are the transmitting and receiving antenna gain respectively, θ_m is the incidence angle for the m-th subswath, σ_s is the target RCS coefficient, and $R_0(\theta_m)$ is the corresponding slant range. As the total data samples are processed coherently to produce a single imaging resolution cell, the receiver thermal noise samples can be taken as independent from pulse to pulse. After coherent range and azimuth compression, the final image SNR is represented by

$$
\text{SNR}_{\text{image}} = \frac{M P_t G_t(\theta_m) G_r(\theta_m) \lambda^3 c_0 T_p \sigma_s \cdot \text{PRF}}{256\pi^3 v_s R_0^3(\theta_m) \sin(\theta_m) K_0 T_{\text{sys}} B_n L_s} \qquad (4.71)
$$

where M is the gain offered by jointly processing the total M subswath data, K_0 ($K_0 = 1.38 \times 10^{-23}$) is the Boltzmann constant, T_{sys} is the system noise temperature, B_n is the noise bandwidth and L_s is the loss factor. Ideally, the SAR with an elevation-plane multi-antenna receiver as compared to conventional single-antenna SAR can obtain an additional SNR gain with a factor of M due to the coherently subswath data processing.

4.6 Numerical Simulation Results

Suppose the following system parameters: carrier frequency is $f_c = 1.2$ GHz, PRF $= 3500$ Hz, $M = 2$, and subswath width is $\Delta W = 50$ km, Figure

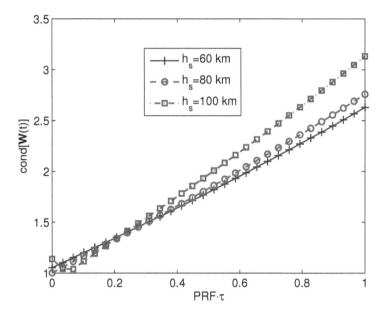

FIGURE 4.9: The sensing matrix condition number for different platform altitude.

4.9 shows the condition number of the sensing matrix for different platform altitudes. Note that the same subswath width is assumed in the simulations. It can be noticed that the condition number for the same swath width increases with the increase of the platform altitude. Figure 4.10 shows the condition number of the sensing matrix for different number of subswaths at an altitude of 60 km. Note that an equal subswath width of 50 km is assumed in the simulations. Also notice that the condition number increases with the increase of the subswath number.

To evaluate the quantitative performance, we design a conceptual elevation-plane multi-antenna SAR system which operates at a carrier frequency of 1.2 GHz. Table 4.1 gives the system parameters. Note that, to obtain a wide swath, a relatively large incidence angle is employed in the system. The minimum and maximum incidence angles can be derived as 41° and 70°, respectively. The swath width is 100 km which is divided into two subswaths, each with a width of $W_{\text{sub}} = 50$ km. According to the elevation-plane multi-antenna SAR scheme described previously, the system operating PRF is designed to be 3500 Hz. We consider a comparative single-antenna SAR with an operating PRF of 1750 Hz which is an half of the operating PRF for the elevation-plane multi-antenna approach. Figure 4.11 gives the comparative RASR performance as a function of the slant range. It can be noticed that, as compared to the conventional single-antenna approach, the SAR with

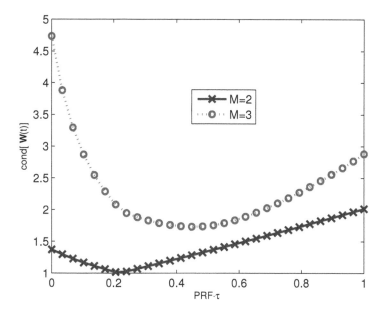

FIGURE 4.10: The sensing matrix condition number for different number of subswaths.

an elevation-plane multi-antenna receiver can operate at a higher (ideally, M times) PRF, however, without degrading the RASR performance. This means that a wider Doppler bandwidth and a higher azimuth resolution are allowed, which is particularly valuable for high-resolution wide-swath SAR imaging.

Numerical simulations are also performed to evaluate the chirp-scaling-based image formation algorithm. Using the simulation parameters listed in Table 4.2, Figure 4.12 shows the comparative imaging results. It can be noticed that the two targets are not successfully focused by the conventional Chirp-Scaling algorithm due to range ambiguity. In contrast, they are well focused by dividing the raw data to two subswaths and processing them with the multi-antenna chirp-scaling algorithm one by one. The final image is obtained by synthesizing the subswath images.

4.7 Conclusion

This chapter discussed several elevation-plane multi-antenna SARs for high-resolution wide-swath remote sensing. The main problem in high-resolution wide-swath SAR imaging is to suppress azimuth and/or range ambiguities.

TABLE 4.1: System parameters using for simulating the sensing matrix.

Parameters	Values	Units
carrier frequency f_c	1.2	GHz
platform velocity v_s	500	m/s
platform altitude h_s	60	km
transmit peak power P_t	2000	W
transmit pulse duration T_p	8	μs
transmit signal bandwidth B_r	100	MHz
transmit/receive antenna gain G_t/G_r	20	dB
minimum slant range $R_{\min}$	80	km
maximum slant range $R_{\max}$	180	km
Earth radius R_e	6370	km
target RCS coefficient σ_s	-15	dB
number of subswath M	2	$--$
system operating PRF	3500	Hz

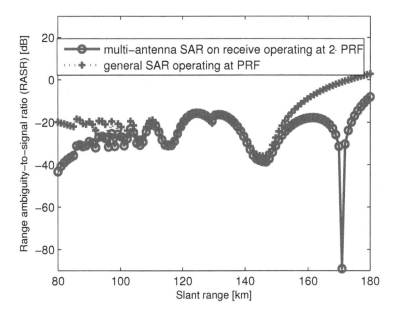

FIGURE 4.11: Comparative RASR results between the elevation-plane multi-antenna SAR and a general single-antenna SAR.

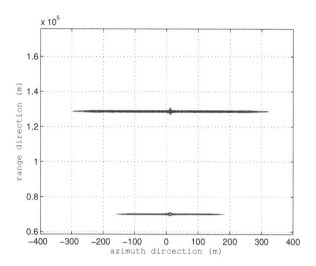

(a) Conventional CS algorithm.

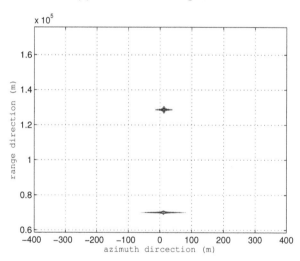

(b) Multi-antenna CS algorithm.

FIGURE 4.12: The comparative imaging results.

TABLE 4.2: System parameters using for simulating the imaging algorithm.

Parameters	Values	Units
carrier frequency f_c	5.3	GHz
platform velocity v_s	500	m/s
platform altitude h_s	20	km
operating PRF f_p	2560	Hz
total swath width	117.2	km
minimum slant range R_{min}	58.6	km
maximum slant range R_{max}	175.8	km
incidence angle θ	76	°
Earth radius R_e	6370	km
target 1 position	(70000, 10)	m
target 2 position	(128694, 100)	m
number of subswath M	2	--

The null steering in elevation-plane technique can suppress range ambiguities, but it has an impediment for unstable motion platform. Elevation-plane multi-antenna SAR is a representative high-resolution wide-swath imaging technique, but it has several practical issues such as ill-condition of the sensing matrix, interferences of the nadir echoes, and blind range problems. We present two methods to resolve the interferences of the nadir echoes and blind range problem, respectively. The ill-condition of the sensing matrix can be overcome by using multiple-input multiple-output (MIMO) SAR. This topic is discussed in Chapter 6. If the sensing matrix is well-conditioned, the elevation-plane multi-antenna SAR data can be efficiently processed by the CS-based imaging algorithm.

5

MIMO SAR Waveform Diversity and Design

All the multi-antenna SARs discussed in previous chapters are single-input multiple-output (SIMO) operation mode. It is necessary to extend them to multiple-input multiple-output (MIMO) operation mode. In this chapter, we first discuss waveform diversity and design for MIMO SAR imaging. MIMO is a technique used in wireless communications to increase data throughout and link range without additional bandwidth or transmitting power. MIMO radar has received vibrant attention in recent years; however, the MIMO SAR discussed in this book is different from general MIMO radars because aperture synthesis with moving platform is usually employed in MIMO SAR but stationary platform without aperture synthesis is employed in MIMO radar. Given that MIMO SAR is in its infancy, it is generally assumed that independent signals are transmitted through multiple antennas placed in moving platforms, and that these signals, after propagating the environment, are received by multiple antennas. Generally speaking, when compared with traditional SARs, MIMO SAR has two advantages, increased degrees-of-freedom and spatial diversity gain.

In MIMO SAR systems, each antenna should transmit a unique waveform, orthogonal to the waveforms transmitted by other antennas. This is similar to the waveform diversity discussed in MIMO radars. However, the waveforms applied in normal MIMO radar for target detection may be not suitable for MIMO SAR imaging. The waveforms used in MIMO SAR systems should have a large bandwidth, so that a high range resolution can be obtained. Another property is that SAR is usually placed inside airplanes or satellites; a high transmit power is thus required for the transmitted waveforms. For these two reasons, the waveforms used in MIMO SARs should have a large time-bandwidth product and a constant modulus, so as to achieve a high range resolution and reduce the required peak transmitting power. Certainly the waveforms should also have good ambiguity characteristics such as range resolution, Doppler tolerance, adjacent-band interferences caused by the waveform itself, and matched filtering sidelobe performance.

This chapter is organized as follows. Section 5.1 makes a critical overview of waveform diversity and design. It is not intended to cite all work performed in the related areas, but the important accomplishments relevant to the development of MIMO SAR waveform diversity are summarized. Section 5.2 introduces the orthogonal polyphase-coded waveform, which has some advantages over binary-coded waveforms. Section 5.3 introduces the orthogonal discrete frequency-coding waveform (DFCW). Its correlation and Doppler performance are analyzed by the ambiguity function. Section 5.4 discusses the random stepped frequency (RSF) waveform. Next, orthogonal frequency diversion multiplexing (OFMD) phase modulation waveform is introduced in Section 5.5, followed by the OFDM chirp modulation waveform in Section 5.6. Finally, constant-envelope OFDM waveform is discussed in Section 5.7.

5.1 Introduction

For many years, conventional radars transmitted and received the same waveform, and processed that waveform identically on very pulse or burst within a coherent processing interval. In fact, prior to the 1990s, waveform diversity did not exist as a separate research area, but instead activities occurred as a part of a variety of other efforts. Interest in high-power microwave effects accelerated interest in waveform manipulation during the 1950s. In 1953, Dicke [94] became the first author to suggest the use of linear frequency modulation (LFM) to achieve pulse compression. The 1960s then saw an explosion of interest in waveform design for clutter rejection, electromagnetic compatibility, and spread spectrum techniques for communication and radar. A number of people began to investigate optimum transmit waveforms in the presence of clutter [12, 88, 331, 364, 379, 359].

In 2000, the U.S. Air Force, Army, and Navy organized as tri-service waveform diversity working group to address single-aperture sensors performing other functions, including navigation and communications. This working group led to the establishment of an annual international conference on waveform diversity and design. A variety of other efforts were also launched, including a U.S. Air Force Office of Scientific Research Multidisciplinary University Research Initiative, and demonstrations at the U.S. Air Force Research Laboratory Rome Research Site and the Naval Research Laboratory. The IEEE has given a standard definition of the term *waveform diversity*: "Adaptivity of the radar waveform to dynamically optimize the radar performance for the particular scenario and task. May also exploit adaptivity in other domains, including the antenna radiation pattern (both on transmit and receive), time domain, frequency domain, coding domain and polarization domain".

Many of today's waveform and spatially diverse capabilities are made possible due to the advent of lightweight digital electronics. Since Moore [287] published his famous monograph in 1965, the number of components per device, and hence processing power, has doubled roughly every two years. This review revealed a number of important contributions to waveform diversity, but the early pioneers were limited by vacuum tube semiconductors, mechanical switches, hardware signal processors, and slide rule computers. The state of technology limited engineers from realizing the kernels of waveform diversity imaged at that time. But, today's technology offers programmable waveform generators. Moreover, the individual oscillators can be tied together to achieve coherency. Today it is possible to generate more complex waveforms than that can currently be analyzed. This fact allows radar system designers to explore the notion of waveform diversity, where different radar platforms transmit different waveforms. Modern radars are thus increasingly being equipped with arbitrary waveform generators that enable simultaneous transmission of different waveforms from different polarimetric antennas, even on a pulse-to-pulse basis. The available design space encompasses spatial location, polarization, time, and frequency. Thus, although we must respect time and bandwidth constraints, the number of possibilities is vast [42].

Recent works in radar adaptive processing have suggested that using distributed apertures in conjunction with frequency or waveform diversity leads to significantly narrower main beams with attendant gains in performance [3]. However, because of the large distance between receiving elements, usually in the orders of thousands of wavelengths apart, grating lobes have became a major issue in the application. The authors proposed frequency diversity as an effective scheme to reduce the grating lobes. All elements in the array are both transmitters and receivers. In such a system, each element transmits at a different center frequency and each transmission is isolated by band-pass filters at each element. An array of N elements, therefore, would result in N^2 spatial degrees-of-freedom. One disadvantage of such a method is that the frequency range of the carrier frequencies used were very large and this results in a very large bandwidth being occupied. There exists quite a large spacing between carrier frequencies while the signal does not occupy the whole bandwidth over adjacent carrier frequencies. This is analogous to the frequency division multiplexing case used in wireless communications where a guard frequency band is given to each signal at each carrier frequency to prevent spectrum overlap. Possible solutions to these problem faced in radar systems can be sought through similar solutions used in wireless communications.

In [347], waveform diversity was studied for the application of target tracking in the presence of clutter. A general frequency modulation (FM) structure was selected and the waveform parameters, such as the FM waveform type, the FM rate and the waveform duration, are selected to minimize a cost function involving the actual target position and the estimated target position. The simulation results showed clear benefits in adapting the waveform to the scenario because the mean squared error (MSE) of the target tracking was significantly reduced. This work set the stage for and motivate further research in other means of extending waveform diversity to the important problem of weak target detection in interference. Crucially, the authors ignore the interaction between the transmitted FM waveforms. In [227], the authors proposed using varying FM rates for distributed apertures in a time domain fashion. This, however, produces significant delay in detection, an issue of importance in higher-speed target detection.

Waveform diversity can also be accomplished via the modulation of certain types of codes with the transmission signal or the selection of a particular waveform on transmission. Fitzgerald [124] demonstrated the inappropriateness of selection of waveform based on measurement quality alone: the interaction between the measurement and the track can be indirect, but must be accounted for. Sowelam and Tewfik [354] developed a signal selection strategy for radar target classification, and a sequential classification procedure was proposed to minimize the average number of necessary signal transmissions. In [340], the use of polyphase codes, in particular the Golay complementary codes, to be modulated with the transmitted signal pulse have provided excellent sidelobe suppression. It is shown in [48, 137] that through the selection of transmission waveforms, waveform diversity can be achieved. The work in [137] developed algorithms to match transmission waveforms to target and clutter statistics provided that they are known in advance, where an optimized waveform vector is used to maximize the corresponding SNR. A different approach to the waveform selection problem was taken in [48], where the selection procedure is based on the formulation of the multi-static ambiguity function. The results show that certain waveforms will provide better performance under certain scenarios, which suggested that the advantage of a system allowing for waveform diversity can greatly improve the resolution.

In recent years, waveform diversity for MIMO radar has received vibrant attention [42, 136]. Stoica et al. [356] performed the design of the covariance matrix to control the spatial power. Yang et al. [459, 460, 462] considered waveform diversity design mainly for the estimation of extended tatgets. Leshem et al. [234] performed quantitative analysis demonstrating the relationship between the information theoretic and estimation criteria. These methods often assume some prior knowledge of the impulse response of the target and use this knowledge to choose the waveforms, which optimize the mutual information between the received signals and the impulse response of the target. Additionally, Li et al. [248] proposed a method of MIMO radar range compression, waveform optimization and waveform synthesis. In [247], a cyclic transmit signal with quadratically phase shift keyed (QPSK) Hadamard codes scrambled by a pseudo-noise sequence was proposed for MIMO SAR imaging, but high-resolution SAR imagery cannot be obtained by this method because of its limited frequency bandwidth. In [68, 203], one digital beam forming SAR system is extended into the MIMO SAR concept with waveform diversity, but only the simple up- and down-chirp signals are used. Consequently only a limited number of orthogonal waveforms can be employed in the system.

Another waveform diversity technique is the OFDM which is a popular choice for common radar and communication signal because OFDM offers advantages such as robustness against multipath fading and relatively simple synchronization. OFDM-like signals has been shown to be suitable for radar applications [235] and the feasibility of integrating communication functions in radar networks [142, 231] have also been explored. For the case of the radar function, it has also been pointed out in another study [134] that OFDM-coded radar signals are comparable with LFM signals and, furthermore, experiences no range-Doppler coupling. Wang [414, 418] extended conventional OFDM signal to OFDM chirp waveform. The designed OFDM chirp waveform has a good peak-average performance, a large time-bandwidth product, and no visible gaps in the covered bandwidth. However, the spectra is not uniform across the bandwidth like the conventional chirp waveforms.

5.2 Polyphase-Coded Waveform

Pulse compression techniques have been widely used in many modern radar systems. One of the early methods for pulse compression is by phase coding. We start from a pulse of duration T_p. The pulse is divided into M bits of identical duration $t_b = T_p/M$. The complex envelope of the phase-coded pulse is given by [237]

$$u(t) = \frac{1}{\sqrt{T_p}} \sum_{m=1}^{M} u_m \text{rect} \left[\frac{t - (m-1)t_b}{t_b} \right] \tag{5.1}$$

where $u_m = \exp(j\phi_m)$ and the set of M phases $\{\phi_1, \phi_2, \ldots, \phi_M\}$ is the phase code associated with $u(t)$. The phase codes are chosen so that the autocorrelation function of the waveform has the largest peak signal to sidelobe ratio for a certain code length. Binary phase codes were originally developed in which the phase elements ϕ_i are restricted to 0 or π. The main drawback of binary codes such as Barker code and m-sequences is their sensitivity to Doppler shift.

Polyphase codes are normally derived from the phase history of frequency-modulated pulse. The Frank code and P1 and P2 codes, the modified version of Frank code, are derived from the frequency stepped pulses. These three codes are only applicable for perfect square length and can be expressed as [239]

$$\text{Frank}: \phi_{i,j} = (2\pi/L)\,(i-1)(j-1) \tag{5.2}$$

$$\text{P1}: \phi_{i,j} = (\pi/L)\,[L - (2j-1)]\,[(j-1)L + (i-1)] \tag{5.3}$$

$$\text{P2}: \phi_{i,j} = (\pi/L)\,[(L+1)/2 - j]\,[(L+1)/2 - i] \tag{5.4}$$

Two other well-known polyphase codes are P3 and P4 codes derived from the linear frequency-modulated pulse. Unlike Frank, P1 and P2 codes, the length of P3 and P4 codes can be arbitrary. P3 and P4 codes can be expressed as [240]

$$\text{P3}: \phi_i = \pi(i-1)^2/M \tag{5.5}$$

$$\text{P4}: \phi_i = \pi(i-1)(i-1-M)/M \tag{5.6}$$

It is known that Frank, P1 and P2 codes are more Doppler-tolerant than binary phase codes and P3 and P4 codes are even better [456].

Assuming that an orthogonal polyphase code set consists of L signals with each signal containing N subpulses represented by a complex number of sequence, the signal can be expressed as

$$s_l[n] = e^{j\phi_l[n]}, n = 1, 2, \ldots, N; l = 1, 2, \ldots, L \qquad (5.7)$$

where $\phi_l[n]$ $(0 \leq \phi_l[n] < 2\pi)$ is the phase of subpulse n of signal l in the signal set. If the number of the distinct phases available to be chosen for each subpulse in a code sequence is M, the phase for a subpulse can only be selected from

$$\phi_l[n] \in \left\{ 0, \frac{2\pi}{M}, 2 \cdot \frac{2\pi}{M}, \ldots, (M-1) \cdot \frac{2\pi}{M} \right\} \qquad (5.8)$$

Using this equation for a value of 4, the choice of phase values would be 0, $\frac{\pi}{2}$, π and $\frac{3\pi}{2}$ and as the value of admissible phase values M increases so does the complexity of the receiver.

Considering a polyphase code set $\mathbf{S}$ with code length of N, set size of L, and distinct phase number M, the phase values of $\mathbf{S}$ can be represented by the following $L \times N$ phase matrix [91]

$$\mathbf{S}(L, N, M) = \begin{bmatrix} \phi_1[1] & \phi_1[2] & \cdots & \phi_1[N] \\ \phi_2[1] & \phi_2[2] & \cdots & \phi_2[N] \\ \vdots & \vdots & \ddots & \vdots \\ \phi_L[1] & \phi_L[2] & \cdots & \phi_L[N] \end{bmatrix} \qquad (5.9)$$

where the phase sequence in row l $(1 \leq l \leq L)$ is the polyphase sequence of signal L, and all the elements in the matrix can only be chosen from the phase set. The aperiodic auto-correlation and cross-correlation properties should satisfy the following relations

$$A(s_l, k) = \begin{cases} \frac{1}{N} \sum_{n=1}^{N-k} s_l[n] s_l^*[n+k] = 0, 0 < k < N \\ \frac{1}{N} \sum_{n=-k+1}^{N} s_l[n] s_l^*[n+k] = 0, -N < k < 0 \end{cases} \qquad (5.10)$$

$$C(s_p, s_q, k) = \begin{cases} \frac{1}{N} \sum_{n=1}^{N-k} s_p[n] s_q^*[n+k] = 0, 0 \leq k < N \\ \frac{1}{N} \sum_{n=-k+1}^{N} s_p[n] s_q^*[n+k] = 0, -N < k < 0 \end{cases} \qquad (5.11)$$

where $A(s_l, k)$ and $C(s_p, s_q, k)$ are the aperiodic autocorrelation function of polyphase sequence s_l and the cross-correlation function of sequences s_p and s_q, respectively. The design of an orthogonal polyphase code set is equivalent to the construction of a polyphase matrix with the above auto-correlation and cross-correlation constraints.

Algebraic methods have been used to design binary and polyphase sequences with low aperiodic autocorrelation sidelobes [69, 133, 160, 251, 383]. Even though there are a few reports in the literature about algebraic construction of two sequences with low cross-correlation [99, 291, 351], it seems to be very difficult to algebraically design a set of three or more sequences with low cross-correlation between any two sequences in the set. Alternately, a more practical approach to designing a polyphase code set is to numerically search the best polyphase sequence by minimizing a cost function that measures the degree to which a specific result meets the design requirements. Given the predetermined values of L, M, and N, a group of polyphase sequences that are automatically constrained by the two equations can be generated by minimizing the following equation

$$E = \sum_{l=1}^{L} \sum_{k=1}^{N-1} |A(s_l, k)|^2 + \lambda \sum_{p=1}^{L-1} \sum_{q=p+1}^{L} \sum_{k=-(N-1)}^{N-1} |C(s_p, s_q, k)|^2 \qquad (5.12)$$

The author of [91] described some numerically optimized sequences based on aperiodic auto-correlation peak-to-sidelobe ratio and cross-correlation energies, but didn't consider the Doppler problem. This method was further improved in [197]. Figure 5.1 compares the correlation performances of these two kinds of polyphase sequence with the same length of 40.

A simulated annealing algorithm-based method to determine the each phase value in the phase sequence is proposed in [204]. This method is processed in the following four basic steps:

- Choose the initial values of L, M, and N.

- Choose the annealing temperature. The initial starting temperature is based on the standard deviation of the initial cost distribution.

- Start the annealing process by "perturbing" the phase coding by randomly choosing a coding sequence from the waveform set. If the cost is reduced by the perturbation, it will be accepted based on the cost change and the current annealing temperature. This process is repeated until the standard deviation of the cost values becomes stable at the annealing temperature. Then the annealing temperature is changed until the standard deviation of the cost values becomes stable again.

- Stop the process when the cost is not reduced over the period of three consecutive temperature reductions.

5.3 Discrete Frequency-Coding Waveform

DFCW has a relative large compression ratio when compared with polyphase-coded waveform. DFCW can be expressed as [69, 256]

$$s_p(t) = \sum_{n=0}^{N-1} a(t) \exp\left(j 2\pi f_n^p t\right) \qquad (5.13)$$

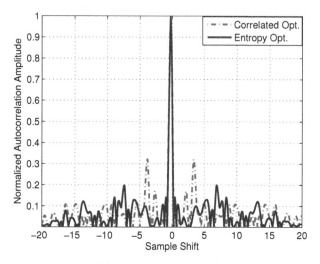

(a) Auto-correlation results.

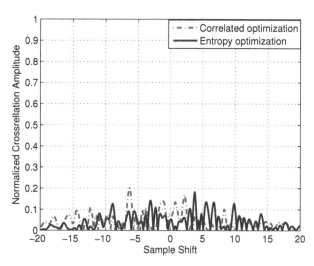

(b) Cross-correlation results.

FIGURE 5.1: Comparison of correlation performance between the correlated optimization and entropy optimization.

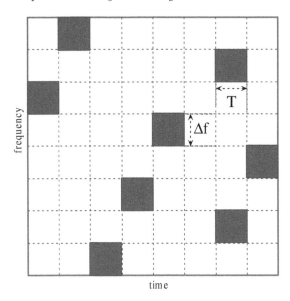

FIGURE 5.2: An example DFCM waveform set.

where

$$a(t) = \begin{cases} \frac{1}{T}, & (n-1)T \le t \le nT \\ 0, & \text{otherwise} \end{cases} \tag{5.14}$$

T is the subpulse duration, $p = 1, 2, \ldots, L$ with L being the number of frequency-hopping steps, N is the number of subpulse, $f_n^p = f_0^p + (n-1) \cdot \Delta f$ is the coding frequency of the n-th subpulse of the p-th waveform in the DFCWs, and $\Delta f = 1/T$. The coding frequency sequence $\{n_1 \Delta f, n_2 \Delta f, n_3 \Delta f, \ldots, n_N \Delta f\}$ can be simply represented by the firing sequence $\{n_1, n_2, n_3, \ldots, n_N\}$, which is a unique permutation of canonic sequence $\{0, 1, 2, \ldots, N-1\}$.

The delay-Doppler ambiguity function for the p-th waveform in the DFCW waveform set can be defined as

$$\chi_{\text{dfcw}}(\tau, f) = \frac{1}{NT} \int_{-\infty}^{\infty} s_p^*(t) s_p(t - \tau) e^{j2\pi ft} dt \tag{5.15}$$

As the cross-ambiguity function between $\exp(j2\pi f_n^p t)$ and $\exp(j2\pi f_n^q t)$ can be easily derived as

$$
\begin{aligned}
\psi_{\text{pq}}(\tau, f) &= \frac{1}{T} \int_{-\infty}^{\infty} e^{-j2\pi f_n^p t} e^{j2\pi f_n^q t} dt \\
&= \begin{cases} \frac{T-|\tau|}{\pi\alpha(T-|\tau|)} \sin[\pi\alpha(T-|\tau|)] \exp\left[-j\pi\alpha(T+\tau) - j2\pi f_q \tau\right], \\ 0, \qquad\qquad\qquad\qquad\qquad\qquad\qquad\qquad \text{otherwise} \end{cases}
\end{aligned} \tag{5.16}
$$

where $\alpha = f_n^p - f_n^q - f$, substituting the above equation to Eq. (5.15) yields

$$\chi_{\text{dfcw}}(\tau, f) = \frac{1}{N} \sum_{n=0}^{N-1} e^{j2\pi nft} \left[\psi_{\text{pp}}(\tau, f) + \sum_{m=0,m\neq n}^{N-1} \psi_{\text{pq}}(\tau - (p-q)T, f) \right]$$

(5.17)

Note that, if $|\tau| \geq T$, we then have

$$\psi_{\text{pq}}(\tau, f) = \psi_{\text{pp}}(\tau, f) = 0 \tag{5.18}$$

It can be shown that the $\sum_{m=0,m\neq n}^{N-1} \psi_{\text{pq}}(\tau - (p-q)T, f)$ has little effect on the desired $\psi_{\text{pp}}(\tau, f)$ [254]. We can consider only the terms containing $\psi_{\text{pp}}(\tau, f)$. In this case, Eq. (5.17) can be simplified to

$$\chi_{\text{dfcw}}(\tau, f) = \begin{cases} \frac{T-|\tau|}{T} e^{j\pi(N-1)(\tau\Delta f)} \frac{\sin(\pi N\tau\Delta f)}{N\sin(\pi\tau\Delta f)}, & |\tau| \leq T, \\ 0, & \text{otherwise} \end{cases} \tag{5.19}$$

We then have

$$|\chi_{\text{dfcw}}(\tau, f)| = \left| \frac{T - |\tau|}{T} \frac{\sin(\pi N\tau\Delta f)}{N\sin(\pi\tau\Delta f)} \right| \tag{5.20}$$

To avoid possible grating lobes, it should be satisfactory with $T\Delta f = 1$.

Notice that the auto-correlation sidelobe peaks are almost independent of the firing frequency order, at around -13.2 dB [348]. Such a large sidelobe peak is a disadvantage for detecting weak targets. Replacing the fixed frequency in each subpulse with linearly modulated frequency can lower the auto-correlation sidelobe peaks, but this will result in grating lobes if $T\Delta f > 1$ [255]. A simulated annealing algorithm was proposed in [90] to optimally design DFCW sequences. Genetic algorithms [255] and modified genetic algorithms [256] are also used to design orthogonal DFCW sequences with good aperiodic correlation performance. Suppose $N = 32$ and $L = 3$, Table 5.1 gives an example DFCW set optimized by the genetic algorithms. Furthermore, generation of DFCW using accelerated particle swarm optimization algorithms were investigated in [323].

5.4   Random Stepped-Frequency Waveform

A basic stepped-frequency waveform is a signal that breaks up the pulse width in time with a different frequency across each subpulse. It means that the subpulse's frequency is uniform or fixed across the subpulse. Stepped-frequency waveform can imitate LFM signal by linearly ordering the subpulse frequencies. This waveform is a discrete approximation to LFM, but would result in an ambiguity region similar to the continuous LFM ambiguity function

TABLE 5.1: An example DFCW set optimized by the genetic algorithm.

No.	Code 1	Code 2	Code 3	No.	Code 4	Code 5	Code 6
1	11	4	2	17	15	23	16
2	9	27	20	18	12	8	17
3	28	15	12	19	20	19	24
4	22	20	5	20	7	0	19
5	5	16	6	21	18	2	9
6	16	3	22	22	4	25	3
7	1	21	26	23	29	31	14
8	19	30	11	24	27	11	18
9	2	9	23	25	10	22	29
10	30	29	4	26	25	13	0
11	31	1	8	27	23	10	27
12	8	26	31	28	24	7	7
13	21	12	10	29	14	28	21
14	17	6	15	30	6	5	1
15	0	17	28	31	26	14	25
16	13	18	13	32	3	24	30

diagram. Although stepped-frequency waveform does not require specific subpulse frequency ordering, there are advantages to specifying a subpulse frequency order. Specifically, Costas codes apply a predetermined ordering to the subpulse frequencies to achieve reduction in sidelobe levels [349]. The predetermined order in Costas codes reduces the height of range and Doppler sidelobe levels in the ambiguity function because the subpulse frequency order of Costas codes stay at the same on a pulse-by-pulse basis. If the subpulse frequency order changes randomly pulse-by-pulse, then the $B_r \times T_p$ region will be completely filled as the SAR platform flies along the synthetic aperture. Pulse-by-pulse means each synthetic aperture array location. For each array location, the subpulse order is a random selection of the N^2 available squares in the $B_r \times T_p$ region. According to [349], random frequency selection in a stepped-frequency waveform represents a noise-like waveform which has an ambiguity function that represents a thumbtack.

5.4.1   Basic RSF Waveforms

A stepped-frequency waveform of duration T_p is formed by combining N subpulses of duration $T = T_p N$, each with a different frequency f_n. The frequencies are uniformly spaced within a bandwidth B_r and can be applied to the subpulse in any order. The entire transmitting waveform can be written as [264]

$$s(t) = \frac{1}{T_p} \sum_{n=0}^{N-1} \Pi \left(\frac{t - nT}{T} \right) \text{Re} \left\{ e^{j2\pi f_n t} \right\}, \qquad -\frac{T_p}{2} \leq t \leq \frac{T_p}{2} \qquad (5.21)$$

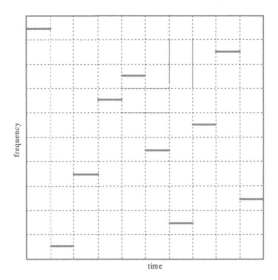

FIGURE 5.3: An example RSF time-frequency grid.

where $\Pi(x)$ is the rectangle function defined as

$$\Pi(x) = \begin{cases} 0, & |x| > \frac{1}{2} \\ 1, & |x| < \frac{1}{2} \end{cases} \tag{5.22}$$

RSF waveforms are generated by using a random frequency on each subpulse. It is common to represent the frequency ordering using a time-frequency grid. The grid is divided into N^2 regions of size $B_r/N \times T_p/N$. Figure 5.3 shows a single random frequency ordering, where the actual frequency values which are transmitted during a particular subpulse are shown. Only a single frequency is transmitted during each subpulse.

The transmission of N distinct tones by itself would not normally be enough to adequately cover the entire bandwidth. Given N subpulses a single RSF waveform covers only N of the N^2 possible sections of the time-frequency grid. The ambiguity function for a single RSF waveform is a thumbtack with noise-like sidelobes. At each synthetic aperture location the frequency ordering is randomly chosen, eventually resulting in the complete filling of the time-frequency region when the pulses are coherently combined. It is derived in [264] that coherent combination of many RSF waveforms results in an ambiguity function

$$|\chi(\tau, f_d)|^2 = |\text{sinc}(B_r\tau)\text{sinc}(\tau f_d)|^2 \tag{5.23}$$

Different from LFM waveforms, RSF waveforms exhibit a strong response only when the received signal is closely matched by the filter. When a shift in frequency exists, the amplitude of the response tapers off as $\text{sinc}(\tau f_d)$ which has nulls at $1/T_p$.

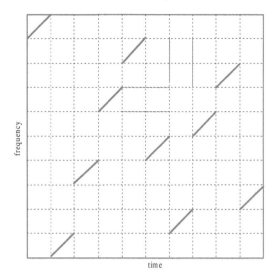

FIGURE 5.4: An example RSF-LFM time-frequency grid.

5.4.2 RSF-LFM Waveforms

Additional improvement to the basic RSF waveform can be obtained with more uniform frequency coverage across the designated bandwidth. If the time-bandwidth region is uniformly filled, the approximation to Eq. (5.23) and the nulling of aliased scatterers will be improved. One approach which has been taken involves linearly modulating the RSF subpulses (called RSF-LFM), as shown in Figure 5.4. It is evident that the frequency across each subpulse is no longer constant, but linearly sweeps across the subpulse bandwidth.

The transmitted RSF-LFM signal can be expressed as

$$s(t) = \frac{1}{T_p} \sum_{n=0}^{N-1} \Pi\left(\frac{t-nT}{T}\right) \operatorname{Re}\left\{e^{j2\pi f_n t} e^{j\pi \frac{B_r t^2}{NT}}\right\}, \qquad -\frac{T_p}{2} \le t \le \frac{T_p}{2} \quad (5.24)$$

The LFM coding on the RSF subpulses theoretically allows the time-bandwidth region to be filled more completely. Better frequency coverage can be obtained from RSF-LFM by changing the underlying stepped-frequency waveform by reducing the number of subpulses and changing the way the LFM is applied to the subpulses. This method changes the $B_r T_p = N^2$ relationship of the time-frequency region to $B_r T_p > N^2$. Results for this type of waveform modification do show some improvement in aliased energy reduction.

5.4.3   Phase-Modulated RSF Waveforms

One way to accomplish more uniform coverage is phase modulation of the individual RSF subpulses. Phase modulations may be classified as either binary phase or polyphase.

The most representative binary phase modulation is the Barker codes. Barker codes exist for various lengths of up to a maximum of 13. The length 13 Barker code is

$$[+++++--++-+-+] \tag{5.25}$$

where the $+$ and $-$ indicate the zero phase shift and 180°, respectively. The length 13 Barker code's autocorrelation function has a peak-to-sidelobe ratio of -22.3 dB [349]. In order to apply the length 13 Barker code to a RSF waveform, each of the N subpulses of Eq. (5.21) is divided further into 13 sub-subpulses. The phase of each sub-subpulse is coded according to Eq. (5.25). The effect of subdividing the subpulse into sub-subpulses results in frequency spreading related to the time duration of the sub-subpulses. It is shown in [182] that the mainlobe frequency response of the RSF-Barker coded subpulse is approximately 13 times wider than the RSF subpulse. While the frequency coverage of the subpulse is increased, the sidelobe levels have increased substantially and do not taper off nearly as much as the RSF subpulse sidelobe levels.

The Frank codes can also be applied to each of the N RSF subpulses of Eq. (5.21) in a similar manner as they are applied for the RSF-Barker waveform. Each subpulse of the RSF waveform is divided further into N^2 sub-subpulses, where N^2 represents the length of the code. For a RSF subpulse comprised of 200 time samples, a length 25 Frank code will have sub-subpulses that are 8 samples long. The phase of each sub-subpulse is coded according to the following matrix. The phases of the length N^2 code are taken starting at the top row, left to right, with each successive row following in like manner. It is shown in [182] that the mainlobe frequency response of the RSF-Frank coded subpulse is approximately 25 times wider than that of the RSF subpulse. The frequency coverage of each subpulse is increased, and sidelobe levels have increased similarly to the RSF-Barker coded subpulses.

$$\frac{2\pi}{M} \begin{bmatrix} 0 & 0 & 0 & \cdots & 0 \\ 0 & 1 & 2 & \cdots & (N-1) \\ 0 & 2 & 4 & \cdots & 2(N-1) \\ 0 & 3 & 6 & \cdots & 3(N-1) \\ \vdots & \vdots & \vdots & \ddots & \vdots \\ 0 & (N-1) & 2(N-1) & \cdots & (N-1)^2 \end{bmatrix} \tag{5.26}$$

When the P4 codes are applied, each subpulse of the RSF waveform is divided further into N sub-subpulse. For a RSF subpulse comprised of 200 time samples, a length 25 P4 code will have sub-subpulses that are 8 samples long. The phase of each sub-subpulse is coded according to

$$\phi_i = \left[\frac{\pi(i-1)^2}{N} \right] - \pi(i-1), \quad 1 \leq i \leq N \tag{5.27}$$

It is shown in [182] that the mainlobe frequency response of the RSF-P4 coded subpulse is approximately 25 times wider than that of the RSF subpulse. The frequency coverage of each subpulse is increased, and sidelobe levels have increased similarly to the RSF-Barker and RSF-Frank coded subpulses. The frequency coverage of each subpulse is increased, and it is obvious that sidelobe levels have increased substantially and do not taper off nearly as much as the RSF subpulse sidelobe levels.

Note that phase modulated waveforms no longer have nicely defined waveform bandwidths but significantly frequency content extends beyond the bandwidth of the original RSF waveform. Hardware limitations such as limited amplifier bandwidths will cut off the frequency spreading. Such bandwidth limitations will result in higher sidelobe levels and reduced gain from the matched filter. Another note is that, many other phase coded waveforms exist. Pseudonoise sequences, complementary codes (such as Golay or Welti codes) and Huffman codes are some examples.

5.5 OFDM Waveform

OFDM was originally developed as a multicarrier, wideband digital communications modulation technique in which high-rate data is transmitted in parallel at a slower rate over a frequency-selective channel via multiple closely-spaced narrowband orthogonal subcarriers [64], as shown in Figure 5.5. With a large number of OFDM subcarriers, each frequency segment, or subband, in the signal spectrum spanned by a single modulated subcarrier can be viewed as being a frequency-flat, as opposed to a frequency-selective, channel. This allows channel equalization in the frequency domain on a subband-by-subband basis, though simple amplitude and phase compensation [258]. Time and frequency synchronization is crucial in OFDM communication systems to preserve subcarrier orthogonality. For radar, however, sensitivity to synchronization is beneficial since the radar receiver uses a stored version of the transmitted signal and measures the time-delay and frequency offsets between the transmitted signal and the received echo to determine a target's range and closing velocity, or range gate.

Although OFDM has been elaborately studied and commercialized in digital communication field, it has not so widely been studied by the radar community apart from a few recent efforts [26, 134, 141, 342, 357, 376, 430]. OFDM radar has the ability to tailor the transmitted pulse spectrum that is not possible for conventional radar pulse compression modulations. An OFDM radar

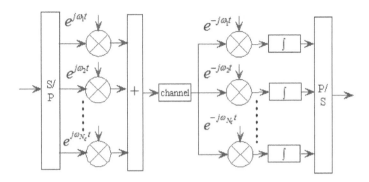

FIGURE 5.5: Basic diagram of OFDM modulation and demodulation system, where ω_i is the subcarrier frequency.

could be designed to have subcarriers that span its entire operating band. By turning off particular subcarriers by simply weighting them with zero in the transmitter, the transmitted signal can be spectrally tailored to have notches corresponding to other in-band communications or radar systems. This allows an OFDM-based radar to coexist with friendly narrowband radars and communication links that are in-band, or to avoid narrowband interference and jamming [27].

The advantages of using OFDM in radar can be summarized as:

- Waveforms are generated digitally with possibility of pulse-to-pulse shape variation.
- It offers high resolution and good multi-path potential.
- Noise-like waveforms offer increased low probability of intercept/detection.
- Ease of narrowband jamming/interference mitigation by simply turning off certain subbands.
- Multiband approach affords benefits of frequency hopping.
- Same architecture can be used to transmit large amounts of data in real time.
- Flexible usage of subbands such as Doppler, location and communication.
- Current technology allows for relatively inexpensive implementation.

5.5.1   OFDM Single-Pulse Waveform

An OFDM pulse of duration T_p consists of N_s code sequences, each with N_c symbols or chips, transmitted simultaneously on N_s complex sinusoidal

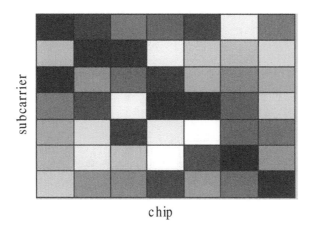

FIGURE 5.6: Structure of an OFDM code matrix.

subcarriers. Each OFDM symbol has duration T_c. The overall pulse compression code can be represented as an $N_s \times N_c$ matrix. The rows represent the subcarriers and the columns represent the chips. The matrix elements can be complex-valued. This matrix can be also viewed as a N_b-element code sequence, where each element has a duration T_b, demultiplexed onto the N_s subcarriers such that each subcarrier has a code of N_c elements, where $N_b = N_s N_c$ and $T_c = N_s T_b$.

The baseband OFDM signal can be expressed as [341]

$$s_{\text{OFDM}}(t) = \frac{1}{\sqrt{\mathbf{w}^H \mathbf{w}}} \sum_{m=1}^{N_c} \sum_{n=1}^{N_s} w_n x_{n,m} g\left(t - (m-1)T_c\right)$$

$$\times \exp\left[j2\pi\left(n - \frac{N_s+1}{2}\right)\frac{t}{T_c}\right]$$

$$= \frac{\exp\left[-j\pi(N_s-1)\frac{t}{T_c}\right]}{\sqrt{\mathbf{w}^H \mathbf{w}}} \sum_{m=1}^{N_c} g\left(t - (m-1)T_c\right)$$

$$\times \sum_{n=1}^{N_s} w_n x_{n,m} \exp\left[j2\pi(n-1)\frac{t}{T_c}\right]$$

(5.28)

where $\mathbf{w}$ is the waveform weight vector, w_n is the complex weight vector associated with the n-th subcarrier, $x_{n,m}$ is the m-th element of the sequence modulating carrier n ($|x_{n,m}| = 1$), and $g(t) \equiv 1$ for $0 \leq t < T_b$ (otherwise, $g(t) = 0$).

The subcarrier frequencies $f_n = \left(n - \frac{N_s+1}{2}\right)\frac{1}{T_c}$ are spaced in frequency by

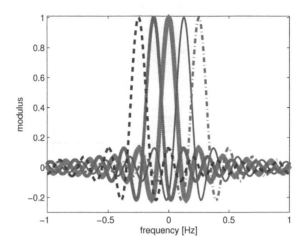

FIGURE 5.7: Spectra of an example OFDM pulse with five single-chip subcarriers.

$1/T_c$, so that in the chip duration the subcarriers are time orthogonal

$$\frac{1}{T_c} \int_0^{T_c} \left(w_{n_i} e^{j2\pi f_{n_i} t} \right)^* \left(w_{n_k} e^{j2\pi f_{n_k} t} \right) \, dt = \begin{cases} w_{n_i}^* w_{n_k}, & n_i = n_k, \\ 0, & n_i \neq n_k \end{cases} \qquad (5.29)$$

Correspondingly the spectrum of each subcarrier over the duration of a chip is

$$\text{rect}\left(\frac{t}{T_c} \right) e^{j2\pi f_n t} \Longleftrightarrow T_c \frac{\sin\left[\pi \left(f - \frac{n}{T_c} \right) T_c \right]}{\pi \left(f - \frac{n}{T_c} \right) T_c} = T_c \text{sinc}\left[\left(f - \frac{n}{T_c} \right) T_c \right]$$

$$(5.30)$$

Since the sinc $(fT_c - n)$ has a first null at the inverse of the signal duration T_b, the spectra of the subcarriers overlap, but the subcarrier signals are mutually orthogonal and the modulation symbol can be successfully recovered by a correlation processing algorithm. Figure 5.7 shows the spectra of an example OFDM pulse with five single-chip subcarriers.

OFDM signal can be easily implemented via the discrete Fourier transform

(DFT). First, rearrange Eq. (5.28)

$$
s_{\text{OFDM}}(t) = \frac{1}{\sqrt{\mathbf{w}^H \mathbf{w}}} \sum_{m=1}^{N_c} \sum_{n=1}^{N_s} w_n x_{n,m} g\left(t - (m-1)T_c\right)
$$

$$
\times \exp\left\{ j2\pi \left(n - \frac{N_s + 1}{2} \right) \frac{t}{T_c} \right\}
$$

$$
= N_s \exp\left\{ -j\pi(N_s - 1)\frac{t}{T_c} \right\}
$$

$$
\times \sum_{m=1}^{N_c} g\left(t - (m-1)T_c\right) \left\{ \frac{1}{N_s} \sum_{n=1}^{N_s} x_{n,m} \exp\left[j2\pi(n-1)\frac{t}{T_c} \right] \right\}
$$

$$(5.31)$$

Sampling $s_{\text{OFDM}}(t)$ in time at a rate of $\frac{N_s}{T_c}$ yields

$$
s_{\text{OFDM}}[k] = s_{\text{OFDM}}(t)\delta\left(t - \frac{(k-1)T_c}{N_s} \right)
$$

$$
= N_s \exp\left\{ -j\pi(N_s - 1)\frac{k-1}{N_s} \right\} \sum_{m=1}^{N_c} g\left(\frac{(k-1)T_c}{N_s} - (m-1)T_c \right)
$$

$$
\times \left[\frac{1}{N_s} \sum_{n=1}^{N_s} x_{n,m} \exp\left(j2\pi(n-1)\frac{k-1}{N_s} \right) \right]
$$

$$
= N_s \exp\left\{ -j\pi(N_s - 1)\frac{k-1}{N_s} \right\}
$$

$$
\times \sum_{m=1}^{N_c} g\left(\frac{(k-1)T_c}{N_s} - (m-1)T_c \right) \left[\mathbf{W}_{N_s}^{-1} \mathbf{x}_m \right]
$$

$$(5.32)$$

where the last term on the right represents the N_s-point inverse DFT (IDFT) of the $N_s \times 1$ column vector $\mathbf{x}_m$, which itself is the m-th column of the weighted code matrix $\mathbf{X}$. The DFT matrix is a complex-valued, symmetric $N_s \times N_c$ matrix defined as $\mathbf{W}_{N_s} = \left[W_{N_s}^{(n-1)(k-1)} \right]$ where $W_{N_s} = e^{-j2\pi/N_s}$. The matrix $\mathbf{W}_{N_s}^{-1} = \mathbf{W}_{N_s}^H / N_s$ is the IDFT.

To efficiently utilize the frequency diversity and effectively ensure the waveform orthogonality, we should allocate the subcarriers for the radar as spread out as possible and, further, hop the subcarriers every symbol time. The hop patterns should be as "apart" as possible for neighboring radars. In the following, we discuss the design of periodic hopping patterns that meet these broad design rules that repeat every N_s OFDM symbol intervals.

The periodic hopping pattern of the N_s subcarriers can be represented by a square matrix with entries from the set of virtual channels. Each pulse hops over different subcarriers at different OFDM symbol times. Each row of the hopping matrix corresponds to a subcarrier and each column represents an OFDM symbol time, and the entries represent the pulse that uses that subcarrier in different OFDM symbol times. In particular, the (i, j) entry of the matrix corresponds to the pulse number when the i-th subcarrier is taken at OFDM symbol time j. We require that every pulse hop over all the subcarriers in each period for maximal frequency diversity. Furthermore, in any OFDM symbol the pulses occupy different subcarriers. These two requirements correspond to the constraint that each row and column of the hopping matrix contains every pulse number exactly once. Such a matrix is called as Latin square [89]. Figure 5.8 shows the hopping patterns of the 5 pulses over the 5 OFDM symbol times (i.e., $N_s = 5$). The horizontal axis corresponds to the OFDM symbol times and the vertical axis denotes the 5 physical subcarriers, and the subcarriers used in the pulses are denoted by darkened squares. The corresponding hopping pattern matrix is

$$\begin{bmatrix} 0 & 1 & 2 & 3 & 4 \\ 2 & 3 & 4 & 0 & 1 \\ 4 & 0 & 1 & 2 & 3 \\ 1 & 2 & 3 & 4 & 0 \\ 3 & 4 & 0 & 1 & 2 \end{bmatrix} \tag{5.33}$$

The pulse 0 happens at the OFDM symbol time and subcarrier pairs $(0, 0)$, $(1, 2)$, $(2, 4)$, $(3, 1)$, $(4, 3)$. The number 0 appears at exactly these locations in the hopping matrix as well. Our design rule is to minimize the overlap between the OFDM pulses of neighboring radars.

5.5.2   Ambiguity Function Analysis

To optimally design OFDM waveforms, the radar ambiguity function can be employed to measure the waveform's range (time delay) resolution, Doppler frequency resolution, sidelobes in both range and Doppler dimensions. The use of the conventional ambiguity function implies that all signals are relatively narrow band such that the term Doppler shift is meaningful. This can be expressed as signal bandwidth B_r, which is much less than the RF carrier frequency f_{RF}. A rule of thumb is that B_r is less than or on the order of 10% of f_{RF} [237]. To overcome this problem, we use the wideband ambiguity function derived in [441]. The wideband ambiguity function of signal $s_0(t)$ is defined as

$$\chi_{wb}(\tau, \alpha) = \sqrt{|\alpha|} \int_{-T_p/2}^{T_p/2} s_0(t) \cdot s_0^*(\alpha(t - \tau)) \mathrm{d}t \tag{5.34}$$

where α is the time scaling factor of the received signal relative to the transmitted signal

$$\alpha = \frac{c_0 - v_s}{c_0 + v_s} \tag{5.35}$$

This wideband ambiguity function takes into account the Doppler shift of each subband individually whereas the narrowband ambiguity function would be

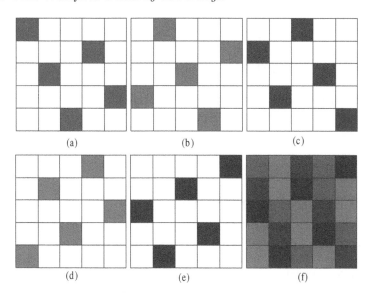

FIGURE 5.8: The OFDM pulse frequency hopping patterns for $N_s = 5$.

applied in only one subband causing all other subbands to be Doppler shifted incorrectly. Correspondingly, the wideband ambiguity function of an OFDM signal is

$$
\chi_{wb}(\tau, \alpha) = \sqrt{|\alpha|} \int_{-T_p/2}^{T_p/2} \sum_{k_1=1}^{N_s} x(k_1) e^{j2\pi k_1 \Delta f t} \sum_{k_2=1}^{N_s} x(k_2) e^{-j2\pi k_2 \Delta f \alpha (t-\tau)} \mathrm{d}t
$$

$$
= T_p \sqrt{|\alpha|} \sum_{k_1=1}^{N_s} \sum_{k_2=1}^{N_s} x(k_1) x(k_2) e^{j2\pi k_2 \Delta f \alpha \tau}
$$

$$
\cdot \frac{\sin[\pi \Delta f (k_1 - k_2 \alpha) T_p]}{\pi \Delta f (k_1 - k_2 \alpha) T_p}
$$

$$
\text{(5.36)}
$$

As an example, we suppose the following simulation parameters: $N_c = 256$, $T_b = 1 \times 10^{-8}s$, and the baseband bandwidth is $B_r = 1/(2T_b) = 50MHz$ divided by a factor of 2 ensures that we are sampling at the Nyquist rate to avoid aliasing. The OFDM pulse duration is $T_c = N_s \cdot T_b = 2.56 \times 10^{-6}s$ from which the subband spacing is calculated as $\Delta f = 0.39MHz$. Figure 5.9 shows the wideband ambiguity functions of example OFDM signals for different number of carrier N_s. It can be noticed that, when the bandwidth remains at a constant for any number of subcarriers, increasing the number of subcarriers improves the Doppler ambiguity performance without affects on the mainlobe delay ambiguity performance. Thus, for minimum ambiguity we should generate the OFDM waveform with the maximum allowable number of subcarriers; however, increasing the number of subcarriers will increase the OFDM pulse duration. The frequency content of a single OFDM pulse signal is shown in Figure 5. 10. It can be noticed that, although it is not uniform across the bandwidth like conventional LFM waveforms, its bandwidth is covered with no visible gaps. The frequency amplitude at $5MHz$ outside of the waveform's bandwidth is about $-25.3dB$, which can be further improved by weighting each subcarrier differently.

5.6   OFDM Chirp Waveform

As chirp (also called LFM) waveforms have been widely used in different radars due to their good waveform properties such as high range resolution, constant modulus, Doppler tolerance, implementation simplicity, and low ambiguity function sidelobes, from a practical point of view we think that MIMO SAR should use chirp-based waveforms [425], so as to reduce the tough requirement of high transmitting power peak. An adaptive LFM waveform diversity was proposed in [431], but only the simple up- and down-chirp waveform was used. Consequently, only two simultaneous transmissions are allowed in the SAR system. In this section, we extend it to OFDM chirp waveforms [418].

5.6.1   Chirp Diverse Waveform

We describe the OFDM chirp waveform from a general chirp signal, which can be represented by the starting frequency f_s, the chirp rate k_r, and the chirp duration T_p. Neglecting amplitude and carrier frequency terms, a chirp signal can be represented by

$$s(t) = \text{rect}\left[\frac{t}{T_p}\right] \cdot \exp\left\{j2\pi\left(f_s t + \frac{1}{2}k_r t^2\right)\right\} \qquad (5.37)$$

The correlation between any two chirp waveforms $s_i(t)$ and $s_j(t)$, where the subscripts i and j relate the quantities to one of the two chirp waveforms, can be easily derived as [426]

$$R_{s_i s_j}(\tau) = \int_{-\infty}^{+\infty} s_i^*(t) s_j(t+\tau) dt$$

$$= \int_{t_1}^{t_2} \exp\left\{j\frac{\pi}{\sqrt{2}}\left(\frac{f_{sj} + k_{rj}\tau - f_{si}}{\sqrt{k_{rj} - k_{ri}}} + \sqrt{2(k_{rj} - k_{ri})}t\right)^2\right\} dt$$

$$\times \exp\left\{j2\pi f_{sj} + j\pi k_{rj}\tau^2 - j\pi\frac{(f_{sj} + k_{rj}\tau - f_{si})^2}{k_{rj} - k_{ri}}\right\}$$

$$(5.38)$$

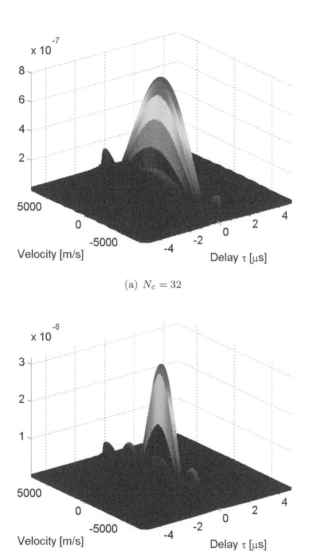

(a) $N_c = 32$

(b) $N_c = 64$

FIGURE 5.9: Wideband ambiguity functions of example OFDM pulse.

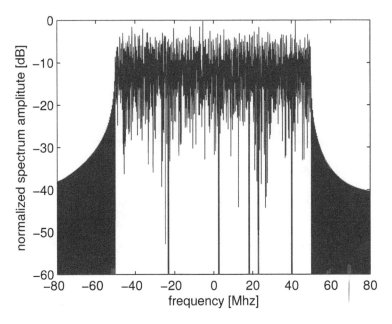

FIGURE 5.10: Spectrum of an example OFDM pulse transmitted by one antenna.

where k_{ri} and k_{rj} denote the chirp rates of the waveform i and j, respectively. The terms t_1 and t_2 denote the integration limits. Their values depend on the range of the variable τ as the following form

$$\text{if } 0 \leq \tau \leq T_p, \qquad \text{then, } t_1 = 0, \ t_2 = T_p - \tau, \qquad (5.39a)$$

$$\text{if } -T_p \leq \tau \leq 0, \qquad \text{then, } t_1 = -\tau, \ t_2 = T_p \qquad (5.39b)$$

Denoting

$$\gamma(t) = \frac{f_{sj} + k_j\tau - f_{si}}{\sqrt{k_{rj} - k_{ri}}} + \sqrt{2(k_{rj} - k_{ri})}t \qquad (5.40)$$

we can get

$$\mathrm{d}\gamma = \sqrt{2(k_{rj} - k_{ri})}\mathrm{d}t \qquad (5.41)$$

Accordingly, Eq. (5.38) can be further simplified to

$$R_{s_is_j}(\tau) = \frac{\exp\left\{j2\pi f_{sj} + j\pi k_{rj}\tau^2 - j\pi\frac{(f_{sj}+k_{rj}\tau-f_{si})^2}{k_{rj}-k_{ri}}\right\}}{\sqrt{2(k_{rj} - k_{ri})}}$$

$$\times \int_{\gamma(t_1)}^{\gamma(t_2)} \exp\left(j\frac{\sqrt{2}\pi}{2}\gamma^2\right) \mathrm{d}\gamma \qquad (5.42)$$

with

$$\int_{\gamma(t_1)}^{\gamma(t_2)} \exp\left(j\frac{\sqrt{2}\pi}{2}\gamma^2\right)\mathrm{d}\gamma$$
$$= C(\gamma(t_2)) + jS(\gamma(t_2)) - C(\gamma(t_1)) - jS(\gamma(t_1)) \tag{5.43}$$

where $C(\gamma)$ and $S(\gamma)$ denote the Fresnel integrals defined as

$$C(\gamma) = \int_0^\gamma \cos\left(\frac{\pi\nu^2}{2}\right)\mathrm{d}\nu \approx \frac{1}{2} + \frac{1}{\pi\gamma}\sin\left(\frac{\pi}{2}\gamma^2\right) \tag{5.44a}$$

$$S(\gamma) = \int_0^\gamma \sin\left(\frac{\pi\nu^2}{2}\right)\mathrm{d}\nu \approx \frac{1}{2} - \frac{1}{\pi\gamma}\cos\left(\frac{\pi}{2}\gamma^2\right) \tag{5.44b}$$

$$C(\gamma) + jS(\gamma) = \int_0^\gamma \exp\left(+j\frac{\pi}{2}\gamma^2\right)\mathrm{d}\gamma \tag{5.44c}$$

Equation (5.42) can then be used to determine the correlation characteristics between any two chirp waveforms. From a practical point of view, suppose the chirp waveforms have equal frequency bandwidth and inverse or equal chirp rate (i.e., $|k_{ri}| = |k_{rj}|$), there are four different combinations between these chirp waveforms, as shown in Figure 5.11. As an example, Figure 5.12 shows the correlation characteristics of the different chirp combinations shown in Figure 5.11. Note that the parameters of $B_r = 10MHz$, $f_{s_2} = 0Hz$ and $T_p = 10 \cdot 10^{-6}s$ are assumed in the simulations. A quantity of importance is the relative level between the correlation of identical chirp waveforms and the correlation of different chirp waveforms. Hence, the chirp waveforms with high cross-correlational suppression are desired. The results of different starting frequencies with an equal chirp rate are shown in Figure 5.12(a) and (b). It is obvious that the performance of cross-correlation suppression will improve with the increase of the separation between two starting frequencies (i.e. $f_{si} - f_{sj}$). The results of different starting frequencies with inverse chirp rate are shown in Figure 5.12(c) and (d). It is also obvious that the maximum and occupied time of the correlation will decrease with the increase of the separation between two starting frequencies.

From Figure 5.12 it is deduced that the chirp waveforms with adjacent starting frequency and inverse chirp rate can provide a good cross-correlation suppression; hence, it is possible to use an arbitrary number of chirp waveforms that having good cross-correlation performance, provided that they occupy adjacent or non-overlapping frequency bands. Note that, if the echo is Doppler shifted, this is also true because Doppler effect brings an equivalent frequency shift. However, using an adjacent starting frequency means wider total RF bandwidth and more complexity for the SAR hardware system. From a practical point of view, we consider that practical SAR system should use the chirp waveforms with equal chirp duration and equal absolute of chirp rate $|k_{ri}| = |k_{rj}| = |k_r|$ (i and j denote any two chirp waveforms), so as to reduce the complexity of hardware design and make the imaging performance being independent of the chirp parameters used for the actual chirp signal.

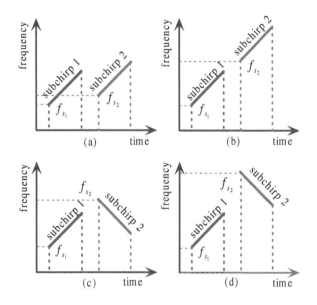

FIGURE 5.11: Different combinations between two subchirp waveforms: (a) $k_{r1} = k_{r2}$, (b) $k_{r1} = k_{r2}$, (c) $k_{r1} = -k_{r2}$, (d) $k_{r1} = -k_{r2}$.

We consider three typical chirp waveforms, two up-chirp signals $s_a(t)$ and $s_b(t)$, and one down-chirp signal $s_c(t)$. From Eq.(5.42) we can get

$$R_{ab}(\tau) = R_{ba}(\tau) = \text{rect}\left[\frac{|\tau|}{T_p}\right] \frac{\sin[\pi(f_{sb} - f_{sa} + k_r\tau)T_p]}{\pi(f_{sb} - f_{sa} + k_r\tau)}$$
$$\times \exp\left\{j\pi[(f_{sb} - f_{sa})T_p + (f_{sb} + f_{sa})\tau + 2k_rT_p\tau]\right\} \tag{5.45}$$

$$R_{ac}(\tau) = R_{ca}(\tau) = \frac{\exp\left[j2\pi f_{sc} + j\pi k_r\tau^2 - j\pi\frac{(f_{sc} - f_{sa} + k_r\tau)^2}{k_r}\right]}{2\sqrt{k_r}}$$
$$\times [C(\gamma(t_j)) + jS(\gamma(t_j)) - C(\gamma(t_i)) - jS(\gamma(t_i))] \tag{5.46}$$

$$R_{bc}(\tau) = R_{cb}(\tau) = \frac{\exp\left[j2\pi f_{sc} + j\pi k_r\tau^2 - j\pi\frac{(f_{sc} - f_{sb} + k_r\tau)^2}{k_r}\right]}{2\sqrt{k_r}}$$
$$\times [C(\gamma(t_j)) + jS(\gamma(t_j)) - C(\gamma(t_i)) - jS(\gamma(t_i))] \tag{5.47}$$

$$R_{bb}(\tau) = \text{rect}\left[\frac{|\tau|}{T_p}\right] \frac{\sin[\pi k_r\tau(T_p - |\tau|)]}{\pi k_r\tau} \exp\left\{j\pi(2f_{sa} + k_rT_p)\tau\right\} \tag{5.48}$$

As an example, assuming the following parameters $k_r = 5 \cdot 10^{11}$ Hz/s, $B_a = B_b = B_c = 500$ MHz, $f_{sa} = -250$ MHz, $f_{sb} = 0$ Hz and $f_{sc} = 250$ MHz, the corresponding cross-correlation between any two chirp waveforms

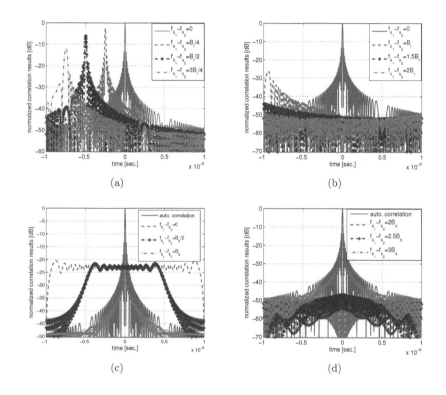

FIGURE 5.12: Correlation characteristics of the different chirp combinations: (a) and (b) have equal chirp rates, (c) and (d) have inverse chirp rates.

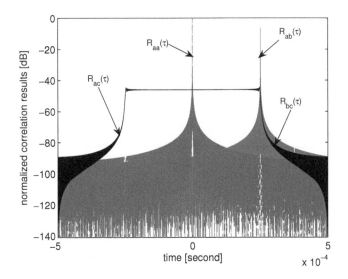

FIGURE 5.13: Correlation results for different combination of chirp waveforms.

is shown in Figure 5.13. It is obvious that the auto-correlation $R_{aa}(\tau)$ shows a higher peak than the cross-correlation peaks. The results show that, if the chirp waveforms with $|k_{ri}| = |k_{rj}|$ and different starting frequencies are used, in SAR imaging applications their echoes can be separated during subsequent signal processing.

Thereafter, matched filtering can be applied successfully. To describe this process, we consider a transmitted down-chirp signal

$$s_{dw}(t) = \text{rect}\left[\frac{t}{T_p}\right] \exp\left(-j\pi k_r t^2\right) \tag{5.49}$$

Matched filtering this down-chirp signal with its reference function

$$G_{dw}(f) = \exp\left(-j\pi \frac{f^2}{k_r}\right) \tag{5.50}$$

yields

$$r_{\text{unamb}}(t) = \exp\left(-\frac{j\pi}{4}\right) \sqrt{k_r T_p^2} \text{sinc}\left[\pi k_r T_p t\right] \tag{5.51}$$

But, for the up-chirp signal

$$s_{dw}(t) = \text{rect}\left[\frac{t}{T_p}\right] \exp\left(-j\pi k_r t^2\right) \tag{5.52}$$

matched filtering it with the reference function $G_{dw}(f)$ will yield

$$r_{\text{amb}}(t) = \frac{1}{2}\text{rect}\left[\frac{t}{2T_p}\right] \exp\left(-j\pi \frac{k}{2}t^2\right) \tag{5.53}$$

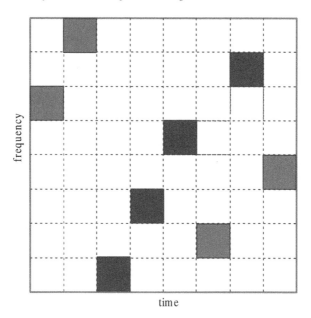

FIGURE 5.14: Example chirp diverse waveforms with 8 subchirp signals.

Comparing Eqs. (5.51) and (5.53) we can see the chirp diverse waveforms can be separated with an appropriate filter. Therefore, although the total imaged scatterers contribute to the received SAR returns, the chirp signals can be separated and fused by utilizing their different cross- and auto-correlation characteristics. That is to say, the specific chirp signal can be extracted by its auto-correlation processing.

Suppose that 8 orthogonal chirp diverse waveforms shown in Figure 5.14 are transmitted simultaneously by 8 antennas, Figure 5.15(a) − (h) show the spectra transmitted by the 8 antennas, respectively. The total transmit spectra is shown in Figure 5.15(i). Equivalently, a wide-band chirp waveform is transmitted. Each receiving antenna will receive the whole echoes associated with the chirp diverse waveform. As the transmitted chirp diverse waveform signals are orthogonal, their matched filtering can be obtained separately with the responding reference functions. Figure 5.16(b) − (i) show example matched filtering results for a single point target.

If the subchirp spectra synthesis (transmit beamforming) is employed, the final matched filtering output resolution can be approximately improved by a factor of 8 (i.e., the number of the subchirp signals), as shown in Figure 5.17. However, note that continuous frequency coverage is assumed in the above simulation. If there are frequency overlap or frequency separation between two subchirp waveforms, unwanted sidelobes will be generated in the multi-

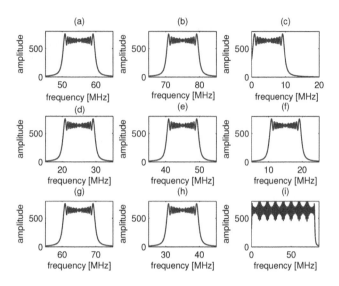

FIGURE 5.15: The spectra transmitted by the MIMO SAR subantennas: (a)-(h) denote the spectra transmitted by the subantenna (a)-(h), respectively, (i) is the equivalent wideband transmission.

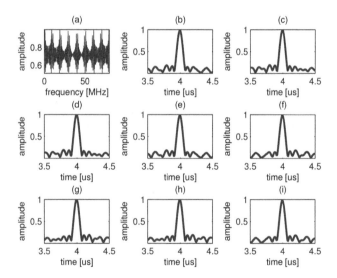

FIGURE 5.16: An example of matched filtering results of a subantenna for a single point target: (a) is the received spectra, (b)-(i) denote the separate matched filtering result of the chirp waveform (a)-(h), respectively.

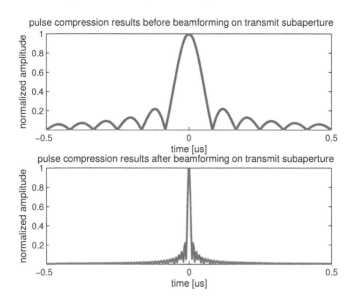

FIGURE 5.17: The comparative matched filtering output resolution: no transmit beamforming is employed in the above one and transmit beamforming is employed in the below one.

beam former, as shown in Figure 5.18. This information is useful for practical waveform design.

5.6.2   OFDM Chirp Diverse Waveform

To efficiently utilize the frequency diversity, we further extend the chirp diverse waveform to OFDM chirp diverse waveform. The baseband OFDM chirp waveform with P subcarriers and N_c temporal chips each with duration T_c and satisfying $N_c \cdot T_c \ll T_p$ can be expressed as

$$
s_k(t) = \sum_{p=0}^{P-1} \sum_{m=0}^{N_c-1} \hat{a}_{k,p,m} u(t - m \cdot T_c)
$$
$$
\times \exp\left[j2\pi\Delta f_{k,p,m}(t - m \cdot T_c)\right] \cdot \exp\left[j\pi\gamma_{k,p,m}(t - m \cdot T_c)^2\right]
$$
(5.54)

where $\Delta f_{k,p,m}$ and $\gamma_{k,p,m}$ denote respectively the corresponding baseband frequency and chirp rate for the subcarriers, and $u(t) = 1, 0 \leq t \leq T_c$. Each temporal chip is modulated by a complex weight

$$
\hat{a}_{k,p,m} = A_{k,p,m} \cdot \exp(j\phi_{k,p,m})
$$
(5.55)

The complex weighter's amplitude is limited to $A_{k,p,m} \in [0, A_{max}]$ where A_{max} is the largest allowable amplitude and the weighter's phase is limited

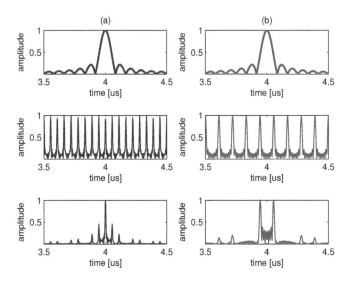

FIGURE 5.18: Impacts of the frequency overlap and frequency separation: (a) there is a frequency overlap of $0.5B_r$, (b) there is a separation of $0.5B_r$.

to $\phi_{k,p,m} \in [0, 2\pi)$. The correlation characteristics of the OFDM chirp diverse pulses can be evaluated by the correlation function

$$
\begin{aligned}
C_{k,k'}(\tau) =& \int_{-\infty}^{\infty} \sum_{p=0}^{P-1} \sum_{m=0}^{N_c-1} \sum_{p'=0}^{P-1} \sum_{m'=0}^{N_c-1} \hat{a}_{k,p,m} \hat{a}_{k',p',m'}^* \\
& \times u(t - m \cdot T_c) u(t - \tau - m' \cdot T_c) \\
& \times \exp\left[j2\pi \Delta f_{k,p,m}(t - m \cdot T_c)\right] \\
& \times \exp\left[-j2\pi \Delta f_{k',p',m'}(t - \tau - m' \cdot T_c)\right] \mathrm{dt} \\
& \times \exp\left[j\pi \gamma_{k,p,m}(t - m \cdot T_c)^2\right] \\
& \times \exp\left[-j\pi \gamma'_{k',p',m'}(t - \tau - m' \cdot T_c)^2\right] \qquad (5.56) \\
=& \sum_{p=0}^{P-1} \sum_{m=0}^{N_c-1} \sum_{p'=0}^{P-1} \sum_{m'=0}^{N_c-1} \hat{a}_{k,p,m} \hat{a}_{k',p',m'}^* \\
& \times \exp\left[j2\pi \frac{(\Delta f_{k,p,m} - \Delta f_{k',p',m'} T_c)}{2} \tau\right] \\
& \times \operatorname{sinc}\left[j2\pi \frac{(\Delta f_{k,p,m} - \Delta f_{k',p',m'} T_c)}{2} \tau\right]
\end{aligned}
$$

To separate the K OFDM chirp diverse pulses at receiver and minimize the interferences between subcarriers, the subcarrier frequency separation should be optimally designed. As discussed previously, if the subcarrier sig-

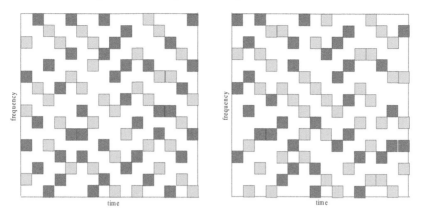

FIGURE 5.19: Two example pulses of the designed OFDM chirp diverse waveform.

nals with $|\Delta f_p - \Delta f_{p'}| \geq B$ where $B = 1/T_c$ is the chip bandwidth, the cross-correlation interferences can be ignored. Certainly, the subchirp signals with $|\Delta f_p - \Delta f_{p'}| \geq 2B$ offer an additional suppression of the cross-correlation components. However, given a required range resolution (determined by the $1/(PB)$), this also means a wider transmit/receive bandwidth requirement for the RF hardware. As a compromise, the subchirp signals which occupy adjacent starting frequency or $|\Delta f_p - \Delta f_{p'}| = 2B$ with inverse chirp rate are used for the OFDM chirp diverse waveforms.

We designed one kind of OFDM chirp diverse waveform with the following parameters: 16 subcarriers, 16 chips, the subcarrier bandwidth is $10MHz$, and the chip duration is $3\mu s$. Figure 5.19 gives two illustration pulses of the designed OFDM chirp diverse waveform. The possibilities of processing their returns can be investigated by analyzing the correlation performance. Considering the two OFDM chirp diverse pulses given in Figure 5.19, their comparative auto-correlation and cross-correlation are shown in Figure 5.20. Note that, to obtain an effective comparison, the shown cross-correlation amplitude has been normalized by the amplitude of the auto-correlation. It can be noticed that the OFDM chirp diverse waveform has perfectly correlation characteristics, a sharp auto-correlation peak and noise-like cross-correlation. Therefore, although the total imaged swath contribute to the received signal, the returns corresponding to different OFDM diverse chirp pulses can be separated and cooperatively processed by utilizing the good correlation performance.

To further investigate the Doppler tolerance of the OFDM chirp diverse waveform, Figure 5.21 shows the calculated ambiguity function of the designed OFDM chirp diverse waveform. As the value of $\chi(0,0)$ represents the matched filtering output without any mismatch, the sharper the function $|\chi(\tau, f_d)|$, the better range resolution and azimuth (Doppler) resolution can be obtained for the radar system. It can be easily noticed that the designed OFDM chirp

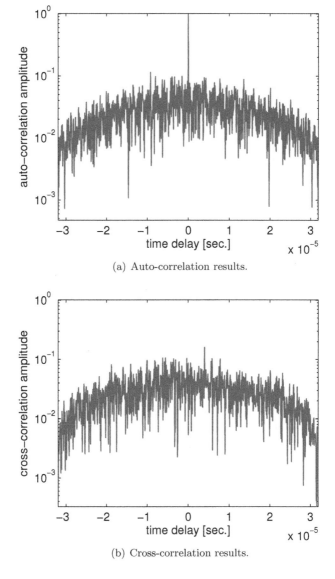

(a) Auto-correlation results.

(b) Cross-correlation results.

FIGURE 5.20: Comparative auto-correlation and cross-correlation of the OFDM chirp diverse waveform.

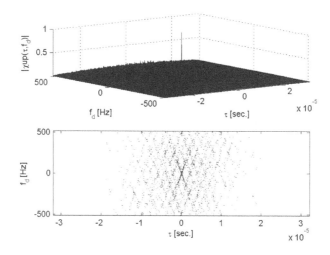

FIGURE 5.21: Ambiguity function of the designed OFDM chirp diverse waveform.

diverse waveform has a satisfactory ambiguity performance in range resolution and Doppler frequency resolution.

Figure 5.22 shows the amplitude and spectra of the designed OFDM chirp diverse waveform. It can be noticed that it has a good peak-average performance which is desired for the RF hardware transmitter and an almost constant modulus. This is rather different from conventional OFDM waveforms which are lack of a constant modulus. Although the spectra is not uniform across the bandwidth like conventional chirp waveforms, the bandwidth is covered with no visible gaps. Therefore, the designed OFDM chirp diverse waveform has not only a good ambiguity function but also a large time-bandwidth product.

5.6.3 Waveform Synthesis and Generation

Existing waveform synthesizer can be classified into: direct digital synthesizer (DDS), phased-locked loop (PLL) synthesizer, and DDS and PLL combined ones. DDS generates high-linearity and agile-frequency signals, but the transmitted RF LFM signals cannot be directly synthesized because the maximum usable clock frequency of state-of-art commercial DDS devices is limited up to $1GHz$. PLL synthesizers cannot achieve fast frequency switching speed while an arrow frequency step is required. Moreover, low phase noise and low spur level are also required for SAR waveforms [401, 421, 477].

To synthesize high-performance waveforms, we proposed a parallel configuration of full-coherent PLL synthesizer driven by DDSs, as shown in Figure 5.23. The DDS output frequency is determined by its clock frequency f_{clk}

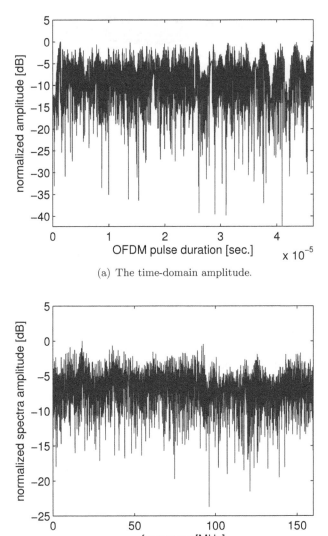

(a) The time-domain amplitude.

(b) The frequency-domain spectra magnitude.

FIGURE 5.22: The amplitude and spectra of the designed OFDM chirp diverse waveform.

and an D-bit tuning word $2^d(d \in [1, D])$ written to its register, where D is the register length. The value of 2^d is added to an accumulator at each clock update, and the resulting digital wave is converted into an analog wave in the D/A converter. The DDS output frequency is determined by

$$f_{\text{DDS}} = \frac{f_{\text{clk}} \cdot 2^d}{2^D}, d \in [1, 2, 3, \ldots, D-1] \tag{5.57}$$

By continuously increasing the input DDS phase increment, a baseband LFM signal can then be synthesized.

The baseband frequency signal is converted to the PLL reference frequency by a double-balance mixer and a variable frequency divider (VFD). In this way, the even harmonics, especially second harmonics can be cancelled out. To achieve the absence of spurs in the PLL's bandwidth B_L, it requires that [367]

$$\frac{f_{\text{clk}} + 10B_L}{l+1} < f_{\text{DDS}} < \frac{f_{\text{clk}} \cdot 2^d - 10B_L}{l}, l = 1, 2, 3, \ldots \tag{5.58}$$

This equation can be used to determine whether the high level spurs fall in the PLL loop bandwidth or not. If it is so, we appropriately tune the frequency setting word of the variable frequency divider division ratios N_{1a}/N_{1b} and N_{2a}/N_{2b} using the triple-tuned algorithm to output desired frequency signal with low spurious emissions. Note that the triple-tuned algorithm is a novel low spurious frequency setting technique, which was first proposed in [367]. Unlike the classic triple-tuned algorithm where only one frequency configuration is saved, here we save the whole desired frequency sets, so as to choose the best one.

Finally, the output LFM signals can be further up-converted by the wideband frequency multiplier M_{2a}/M_{2b}

$$\begin{cases} f_a = M_{2a} \left[M_{1a} f_{clk} - \frac{N_{2a}}{N_{1a}} (f_{clk} + f_{DDSa}) \right] \\ f_b = M_{2b} \left[M_{1b} f_{clk} - \frac{N_{2b}}{N_{1b}} (f_{clk} + f_{DDSb}) \right] \end{cases} \tag{5.59}$$

This waveform generator provides several significant advantages such as fast frequency switching speed, fine frequency resolution, good spectral purity. More importantly, this approach enables synthesizing multiple orthogonal waveforms simultaneously with consistent performance.

5.7 Constant-Envelope OFDM Waveform

While OFDM pulses coded with mutually orthogonal complementary sets [237, 382] would have good spectral containment and be quasi-orthogonal, the time-varying envelope of an OFDM signas requires linear RF power amplifiers

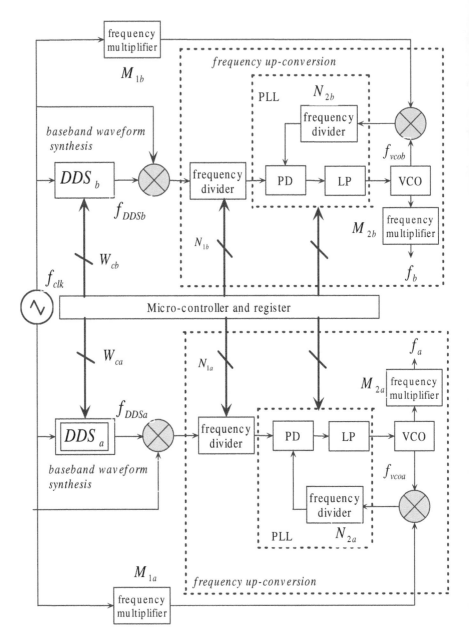

FIGURE 5.23: Block diagram of the parallel DDS-based waveform generator. Here only two orthogonal LFM waveforms are shown.

which are typically power inefficient. A solution to OFDM's peak-to-average power ratio (PAPR) problem called constant-envelope OFDM (CE-OFDM) was proposed for digital communication systems [375]. It is based on using a real-valued baseband composite OFDM signal to phase modulate a RF carrier resulting in a single-carrier, polyphase-coded RF signal with a constant amplitude envelope. Another benefit of CE-OFDM is that the amplitude of the real-valued OFDM phase modulation signal can be made relatively smooth, such that resulting single-carrier waveform does not have drastic phase discontinuities. This characteristic will produce a more spectrally-efficient signal than a generic polyphase-coded signal.

5.7.1 Peak-to-Average Power Ratio

Since a baseband OFDM signal is the coherent superposition of a large number of modulated subcarriers, the magnitude of the composite signal is time-varying. When the subcarriers happen to add in phase a high instantaneous signal magnitude relative to the average level will result. The time-varying baseband signal magnitude is an issue for the RF power amplifier after the baseband signal has been upconverted to the RF carrier frequency in the OFDM transmitter. The pertinent metric is the PAPR of the real-valued RF signal.

High PAPR OFDM signals lead to the requirement for high dynamic range RF power amplifiers. Furthermore, when the signal changes from a low instantaneous power level to a high instantaneous power level large signal amplitude swings are encountered. Unless the transmitter's power amplifier is linear across this range of signal levels, high out-of-band harmonic distortion will result. Practical amplifiers exhibit a finite amplitude range, in which they can be considered almost linear. In order to prevent severe clipping of the high OFDM signal peaks, the power amplifier must not be driven to saturation. Unfortunately, conventional power amplifiers trade power efficiency for linearity. In digital communications a significant amount of research has been done to reduce the PAPR of OFDM signals so that lower-cost, power-efficiency RF power amplifiers can be utilized in the OFDM transmitter [317].

The PAPR problem is also a key issue when considering OFDM as a radar modulation. RF power amplifiers in radar transmitters are usually operated in saturation when they are enabled during pulse transmission. This precludes the use of intrapulse amplitude modulation techniques for pulse compression and requires the modulated pulse to have a constant envelope. As a result, generally only single-carrier phase and/or frequency modulations are used for pulse compression. However, the multicarrier nature of OFDM provides unique advantages over conventional single-carrier radar modulations. References [237, 292] discussed two radar-specific techniques in detail: Schroeder subcarrier phasing of an identical sequence and consecutive ordered cyclic shifts of an ideal chirp-like sequence. They can reduce PAPR to the order of 1.5 to 2.

5.7.2   Constant-Envelope OFDM Pulse

CE-OFDM pulse achieves a constant envelope by phase modulating the RF carrier frequency with a real-valued baseband OFDM signal. To generate a real-valued baseband OFDM signal, a real-valued $N_s \times N_c$ M-ary code matrix $\mathbf{X}$ is modulated with a set of N_s real-valued sinusoids as opposed to complex sinusoids. The cosine-based subcarriers are expressed as

$$p_{n,m}(t) = g(t - (m-1)T_c) \cos\left(\pi \frac{(n - \frac{1}{2})}{T_c} t \right) \tag{5.60}$$

where $g(t) = \text{rect}\left(\frac{t}{T_c} - \frac{1}{2} \right)$, $n = 1, 2, \ldots, N_s$, and $m = 1, 2, \ldots, N_c$. The subcarrier frequencies are separated by $\frac{1}{2T_c}$ to achieve time orthogonality [357]. The phase-modulated CE-OFDM signal can then be expressed as [375]

$$s_{\text{CE-OFDM}}(t) = \text{rect}\left[\frac{t}{M_c T_c} - \frac{1}{2} \right] \exp\left(j \sum_{m=1}^{N_c} \phi_m(t) \right) \tag{5.61}$$

where the phase for the m-th chip is

$$\phi_m(t) = \left(\theta_m + 2\pi\gamma C_N \sum_{n=1}^{N_s} \mathbf{X}_{n,m} p_{n,m}(t) \right) \text{rect}\left(\frac{t}{T_c} - m + \frac{1}{2} \right) \tag{5.62}$$

The term γ is the modulation index, C_N is a normalized constant, and θ_m is a phase memory term.

The term $C_N = \sqrt{2/(N_s \sigma_{\mathbf{X}}^2)}$ with $\sigma_{\mathbf{X}}^2$ being the variance of the code element $x_{n,m}$ normalizes the real-valued OFDM signal variance to unity. For a random M-ary code matrix, where the elements $x_{n,m}$ have equal likelihood of being any of the elements that make up the set $\{\pm 1, \pm 3, \ldots, \pm(M-1)\}$, the variance is [375]

$$\sigma_{\mathbf{X}}^2 = \frac{1}{M} \sum_{k=1}^{M} (2k - 1 - M)^2 = \frac{M^2 - 1}{3} \tag{5.63}$$

The phase memory term θ_m is used to provide continuous phase transitions between chips to improve spectral efficiency.

CE-OFDM signal can be implemented in a manner like the generic OFDM waveforms. Stochastic optimization techniques can be employed to design CE-OFDM pulse modulation sets [196].

5.8   Conclusion

Waveform diversity and design has received much attention in MIMO radar society, but little work can be found in MIMO SAR imaging related literature. In this chapter, we introduced several representative waveform diversity and design methods that may be suitable for MIMO SAR high-resolution imaging. Particularly, we present the OFDM chirp waveform, which has a large time-bandwidth product. Although the spectra is not uniform across the bandwidth like conventional LFM waveforms, the bandwidth is covered with no visible gaps. Moreover, to avoid the high peak-to-average power ratio problem, constant-envelope waveforms are desired. Further work should be carried out on this topic.

6

MIMO SAR in High-Resolution Wide-Swath Imaging

Efficient SAR imaging techniques should provide high-resolution imagery over a wide area of surveillance, but it is a contradiction due to the minimum antenna area constraint. The illuminated area must be restricted, so as to avoid ambiguous returns. Conventional methods use either a small portion of a large Tx/Rx antenna array or a separate transmit aperture to illuminate a large footprint on the ground; however, these methods cannot provide efficiently high-resolution wide-swath (HRWS) remote sensing capability. MIMO SAR provides potential solutions to resolving the disadvantages of conventional single-antenna SAR in HRWS remote sensing. Consider a MIMO SAR system with a transmitting array equipped with M colocated antennas and a receiving arrays equipped with N colocated antennas. There will be a total of $M \times N$ independent returns. Compared to phased-array SAR where only a total of N coefficients can be obtained for the matched filtering, MIMO SAR gives more coefficients and therefore provides more degrees-of-freedom. More importantly, MIMO SAR enables a flexible and reconfigurable SAR system design. In this chapter, we discuss mainly MIMO SAR potential and challenges in HRWS remote sensing.

This chapter is organized as follows. Section 6.1 makes a critical overview of SAR HRWS imaging. It is not intended to cite all work performed in the related areas, but the important accomplishments relevant to the development of the MIMO SAR HRWS imaging are summarized. Section 6.2 introduces the advantages of MIMO SAR in HRWS imaging. Section 6.3 discusses the multidimensional waveform encoding SAR HRWS imaging. The main procedures are described in detail, and a mathematical derivation of the scheme is given. Next, Section 6.4 presents the MIMO SAR with multiple antennas placed only in the elevation dimension for HRWS imaging. This approach employs multiple-antenna configuration in elevation dimension, which is divided into multiple subapertures. In this way, multiple pairs of transmit-receive virtual beams directing to different subswathes are formed simultaneously. Equivalently a large swath can be synthesized. A range-Doppler-based image formation algorithm is also derived. Section 6.5 presents the space-time encoding MIMO SAR for HRWS imaging. This approach employs MIMO configuration in the elevation dimension and the Alamouti space-time coding scheme in the azimuth dimension.

6.1   Introduction

Spotlight SAR can acquire remote sensing imagery with spatial resolutions in the meter or even decimeter range [47, 290, 355], and ScanSAR can frequently monitor large areas; however, it is not possible to combine the benefits from spotlight SAR and ScanSAR imaging in one and the same data take. The rising demand for HRWS imaging motivated research toward the development of new radar techniques that allow for the acquisition of high-resolution SAR imagery without the classical coverage limitations imposed by range and azimuth ambiguities.

Since current single-antenna SAR techniques cannot provide efficiently HRWS imaging, several multichannel- or multiaperture-based suggestions [249, 361, 408, 453] have been proposed to alleviate the requirements imposed on the minimum antenna area constraint. All of them use either a small portion of a large Tx/Rx antenna array or a sparate transmitting aperture to illuminate a large footprint on the ground. The scattered signal is then received by multiple independent subaperture elements that are connected to individual recording channels. The combination of the subaperture signals enables the formation of an arbitrary number of narrow Rx beams a *posteriori* to the digital signal recording.

The multibeam SAR proposed in [187] employs a squinted imaging geometry. In contrast to a conventional side-looking SAR, the antenna is rotated about its vertical axis and uses multiple horizontally displaced aperture elements with increased vertical extension for signal reception. Such an antenna arrangement enables the formation of multiple narrow beams that limit the Rx antenna footprints in both the along-track and cross-track directions to a small area on the Earth's surface. This allows for unambiguous SAR imaging narrow swathes, and by combining the images from multiple Rx beams, one will be able to map a wide swath at nearly constant incident angle. A drawback of this solution is impaired spatial resolution and performance associated with high squint angles, and further problems arise for subsequent SAR processing due to the large range walk results from the linear component of the range cell migration [216].

The displaced phase center antenna (DPCA) is another represent HRWS imaging technique. The basic idea is to divide the receiving antenna in the along-track direction into multiple subapertures. In this way, the DPCA SAR benefits from the whole antenna length regarding azimuth ambiguity suppression, while the azimuth resolution is determined by the length of a single subaperture, thus decoupling the restrictions on high-resolution and wide-swath. However, the DPCA technique does not allow for an increase of the unambiguous swath width, but it is only well suited to improve the azimuth resolution. Moreover, the relation between platform speed and along-track subchannel offset has to be adjusted in order to obtain a signal that is equivalently sampled as a single-channel signal of the same effective sampling rate [146, 214, 262].

Another potential approach is the quad-element rectangular array SAR [45], which combines the advantages of gathering additional samples in azimuth to suppress azimuth ambiguities and simultaneously enabling an enlarged swath for a fixed pulse repetition frequency (PRF). This system employs, in addition to the DPCA technique in the azimuth dimension, multiaperture rows in the elevation dimension. An inherent disadvantage is that the resulting swath is no longer contiguous because the receiver must be switched off during transmission. This leads to blind ranges and, therefore, gaps in the cross-track direction of the acquired scene.

An extension of the quad-element rectangular array SAR was proposed in [121, 176, 361]. This system combines a separate transmitting antenna with a large receiving array. The small transmitting antenna illuminates a wide swath on the ground, and the large receiving aperture compensates the Tx gain loss by a real-time digital beamforming process called scanning on receive. Each azimuth panel of the receiving antenna is again divided into multiple vertically arranged subapertures. The signals from the individual subapertures are then combined in real-time to form a narrow elevation beam that follows the radar pulse as it travels on the ground. In this way, we can collect the radar echoes from a wide swath despite using a receiving aperture with large vertical extension.

Digital beamforming on the receiver was further suggested in [147, 217, 466, 467], where the receiving antenna is split into multiple subapertures. In contrast to analog beamforming, the received signals of each subaperture element are separately amplified, down-converted and digitized. This enables an *a posteriori* combination of the recorded subaperture signals to form multiple beams with adaptive shapes. As noted in [216], the additional information about the direction of the scattered radar echoes can then be used to:

- suppress spatially ambiguous signal returns from the ground;
- increase the receiving antenna gain without a reduction of the imaged area;
- suppress spatially localized interference;
- gain additional information about the dynamic behavior of the scatterers and their surroundings.

Several proposals have been made to use digital beamforming on the receiver for SAR HRWS imaging [466]. They employ a small fixed aperture antenna to illuminate a wide swath. The scattered radar echoes are then recorded by a large antenna array, which enables the suppression of SAR ambiguities and compensates the gain loss that arises from a wider illumination. As discussed in Chapter 4, however, when the sensing matrix is ill-conditioned the performance of the beamforming on the receiver will be significantly degraded.

The last but not the least approach is to suppress the range ambiguities, so that the azimuth resolution can be improved without degrading the range-ambiguity-to-signal ratio (RASR) performance. One principal suppression of range ambiguities is accomplished by designing the appropriate antenna elevation pattern [104]; however, range ambiguities will appear if the antenna directivity in the direction of one or more of the ambiguous ranges is not sufficiently low. Narrow elevation beamwidth favors range ambiguity suppression, but consequently, the swath width must be decreased. Another common approach to resolve range ambiguity is to use several coherent processing intervals (CPIs) within a single dwell, where each CPI has a different PRF. Resolving the true range and Doppler is performed noncoherently [445]. This is a rather inefficient use of the transmit energy. To overcome this disadvantage, an alternative approach based on inter-pulse binary coding is proposed in [236]. The suggested inter-pulse coding is based on periodic Ipatov binary sequences, which exhibit ideal cross-correlation. Since the correlation receiver needs to perform linear processing, the transmitter/receiver isolation must be large enough. The depth-of-focus approach proposed in [44] can defocus range ambiguities and thereby reduce their peak levels. However, the energy introduced by widening the elevation beam will result in degraded integration sidelobe performance. Moreover, due to the energy of a near ambiguous range zone is always greater than that of the focal range zone, a pixel value is determined more by the scatterers outside the resolution cell than the scatterers inside the resolution cell. Range ambiguity suppression by transmitting alternatively up and down wideband chirp signals was investigated in [283]. This approach allows only a very limited range ambiguity suppression performance and cannot be used for high PRF radars [426].

Range ambiguities can also be suppressed by the azimuth phase coding technique [81]. This approach is conceived for conventional single-antenna SAR systems and is based on three main steps [29]: i) azimuth phase modulation on transmission; ii) azimuth phase demodulation on reception; and iii) filtering of the SAR azimuth signal over the processing bandwidth. The modulation and demodulation produce a frequency displacement between the spectrum of the useful signal and that of the range ambiguity. Then, in the presence of Doppler oversampling, the final step allows to filter out the ambiguous signal power, which is located outside the bandwidth of interest. In [72], the advantages offered by the azimuth phase coding are explained by considering conventional systems and multichannel systems with multiple independent receivers, i.e., imaging multiple subswaths, each according to a conventional method.

Modulating successively transmitted pulses by orthogonal codes can significantly reduce the range ambiguity peaks. The basic scheme is illustrated in Figure 6.1, where different OFDM pulses are transmitted in different pulse repetition interval (PRI). In the K adjacent PRIs, we transmit K different OFDM pulses, $s_1(t)$, $s_2(t)$, ..., $s_K(t)$, which meet with the following condition

$$\int_{T_p} s_m(t)s_n^*(t)\mathrm{d}t = \delta(t), (m, n) \in [1, 2, 3, \ldots, K], \qquad (6.1)$$

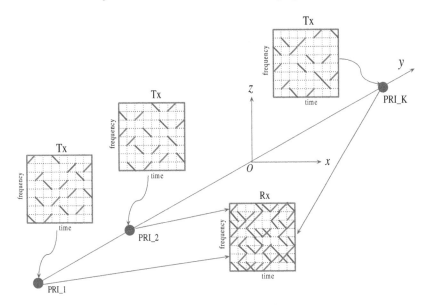

FIGURE 6.1: Illustration of the range ambiguity suppression with OFDM chirp diverse waveform. Different OFDM chirp diverse pulses are transmitted in different PRI.

where t is the time index within the radar pulse. The overlapped radar returns can be separated by separately matched filtering.

$$y_k(t) \doteq \int_{T_p} x(t)s_k^*(t)\mathrm{d}t, k \in [1, 2, 3, \ldots, K]. \tag{6.2}$$

where $y_k(t)$ and $x(t)$ denote the matched filtering output for the k-th pulse signal and the received overlapping returns, respectively.

As an illustration example, three different OFDM chirp diverse pulses are transmitted in one PRI. Figure 6.2 shows the corresponding transmission and reception relations, where the horizontal axis is the fast time in the range dimension and the vertical axis is the slow time in the azimuth dimension. Taking Figure 6.2(a) as an example, the pulse a is transmitted at PRI_1 (see Figure 6.1); at PRI_2, the pulse b is transmitted, so there are pulses b and a in the propagation channel. Similarly, at PRI_3, the pulse c is transmitted, so there are pulses c, b, and a in the propagation channel. The three PRIs (PRI_1, PRI_2, and PRI_3) form one waveform repetition interval (WRI), which is equivalent to one conventional PRI used in classic SAR systems. Suppose the three OFDM chirp diverse pulses are perfectly orthogonal; after separately matched filtering, we can obtain three virtual PRIs. Equivalently, the effective PRF can be improved by the virtual PRIs with a factor of three. This approach aims to eliminate the range ambiguities, instead of just suppressing them like other techniques.

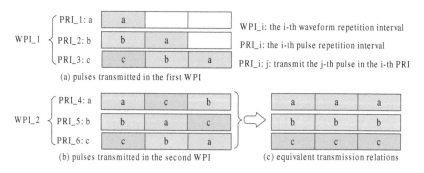

(a) pulses transmitted in the first WPI

(b) pulses transmitted in the second WPI   (c) equivalent transmission relations

FIGURE 6.2: Illustration of the transmission and reception relations in the case of three subpulses, where the horizontal axis denotes three subswaths.

6.2 MIMO SAR System Scheme

Similar to MIMO wireless communications, the idea of MIMO radar has also drawn considerable attention in recent years [170, 244, 245]. The fundamental difference between MIMO radar and other radars is that unlike beamforming, array radar or space-time adaptive processing (STAP), which presuppose a high correlation between signals either transmitted or received by an array, MIMO radar exploits the independence between signals at the array elements [123]. The performance of conventional radar systems is limited by target scintillations. Both experimental measurements and theoretical results demonstrate that scintillations of 10 to 15 dB in the reflected energy can occur if the target aspect is changed as little as one milliradian [349]. The application of MIMO techniques to radar can greatly improve the detection and estimation performance due to reduced target fades, and offer many potential advantages, including improved resolution and sensitivity. For example, with adequately separated transmitting/receiving array elements, a MIMO radar can probe a fluctuating target from different aspect angles, which offers a detection diversity gain [123, 170]. MIMO radar can provide higher spatial resolution and degrees-of-freedom [28], better parameter identifiability [246], and allow direct application of adaptive array techniques [455].

In summary, the gains available in MIMO radar include [404]:

- *Spatial Multiplexing Gain*—offers a linear (in the number of transmit-receive antenna pairs) capability increase, compared to systems with a single-antenna at one or both sides of the link, at no additional power or bandwidth expenditure [374]. The corresponding gain is available if the propagation channel exhibits rich scattering and can be realized by simultaneous transmission of independent data streams in the same frequency band. The receiver exploits differences in the spatial signatures induced by the MIMO channel onto the multiplexed data streams to separate the different signals, thereby realizing a capability gain.

. *Array Gain*—refers to the average increase in the signal-to-noise ratio (SNR) at the receiver that arises for the coherent combining effect of multiple antennas at the receiver or transmitter or both.

. *Diversity Gain*—leads to improved link reliability by rendering the channel "less fading" and by increasing the robustness to co-channel interference. Diversity gain is obtained by transmitting the signal over multiple independently fading dimensions in time, frequency, and space and by performing proper combination in the receiver. Spatial diversity is particularly attractive when compared to time and frequency diversity, as it does not incur an expenditure in transmission time or bandwidth.

More details can be found in [245] and references therein. In this chapter, we consider only MIMO SAR imaging, which combines MIMO radar and SAR.

6.2.1   Signal Models

There are two kinds of general MIMO SAR configuration (see Figure 6.3), single phase center multibeam (SPCM) and multiple phase center multibeam (MPCM). In the SPCM MIMO SAR, a distinct channel is associated with each of the receiving beams and, hence, the data are split according to azimuth angular position or, equivalently, instantaneous Doppler frequency center in the azimuth direction. By this, given knowledge of the relative squint angles of each beam (hence the Doppler center frequency for each beam) and assuming suitable isolation between the beams, each channel can be sampled at a Nyquist rate appropriate to the bandwidth covered by each narrow beam, instead of that covered by the full beamwidth. This arrangement enables correct sampling of the azimuth spectrum with a PRF fitting the total antenna azimuth length, which is significantly smaller than the general PRF requirement. The MPCM MIMO SAR also synthesizes multiple receiving beams in the azimuth direction; however, the operating mode of this system is quite different from that of the previous one. In this case, the system transmits multiple broad beams and receives the radar returns in multiple beams which are displaced in the along-track direction. The motivation is that multiple independent sets of target returns are obtained for each transmitted pulse if the distance between phase centers is suitably set. This method basically implies that we may broaden the azimuth beam from diffraction-limited width, giving rise to improved resolution, without having to increase the system operating PRF.

One of the main advantages of MIMO SAR is that the degrees-of-freedom can be greatly increased by the concept of virtual array provided by the multiple antennas. Consider a MIMO SAR system equipped with M colocated transmitting antennas and N colocated receiving antennas (see Figure 6.4).

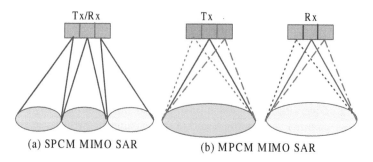

FIGURE 6.3: Geometry mode of the MIMO SAR antennas.

Suppose both the transmitting and receiving arrays are close to each other in space (possibly the same array) so that they see targets at the same directions. The received signal at each receiving antenna is the weighted summation of all the transmitted waveform

$$r_n(t) = \sum_{m=1}^{M} a_{n,m} s_m(t), m \in [1, 2, \ldots, M], n \in [1, 2, \ldots, N] \qquad (6.3)$$

where $r_n(t)$ is the received signal at the n-th antenna, $s_m(t)$ is the transmitted waveform at the m-th antenna, and $a_{n,m}$ is the channel coefficient with the m-th antenna as input and the n-th antenna as output. When the transmitted waveforms are designed to be orthogonal

$$\int s_m(t) s_n^*(t) \mathrm{d}t = \delta_{m,n} \qquad (6.4)$$

At each receiving antenna, these orthogonal waveforms can then be extracted by M matched filters. There are a total of $M \times N$ extracted signals. Compared to traditional phased-array SAR, where the same waveform is used at all the transmitting antennas and a total of N coefficients are obtained for the matched filtering, MIMO SAR gives more coefficients and therefore provides more degrees-of-freedom.

Suppose there are P point targets, the received MIMO SAR signals can be written in a vector form

$$\mathbf{x}(t) = \sum_{p=1}^{P} \sigma_p \left[\mathbf{a}^T(\theta_p) \mathbf{s}(t) \right] \mathbf{b}(\theta_p) + \boldsymbol{\varepsilon}(t) \qquad (6.5)$$

where θ_p is the target direction, σ_p is the complex-valued reflection coefficient of the focal point θ_p for the p-th point target, $\boldsymbol{\varepsilon}(t)$ is the noise vector, and $\mathbf{a}(\theta_p)$ and $\mathbf{b}(\theta_p)$ are the actual transmitter and actual receiver steering vectors associated with the direction θ_p. Without loss of generality, we ignore the noise in the following discussions. The SAR returns due to the m-th transmitted

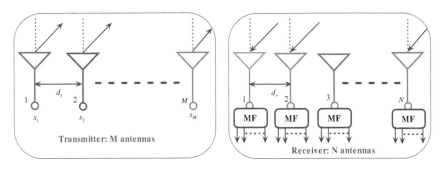

FIGURE 6.4: Illustration of an example MIMO SAR system.

waveform can be extracted by matched filtering the received signal to each of the waveforms $s_m(t)$

$$\mathbf{y}_m = \int \mathbf{x}(t)s^*(t)\mathrm{d}t \tag{6.6}$$

The $MN \times 1$ virtual target signal vector can then be written as

$$\mathbf{y} = \sigma_s \mathbf{a}(\theta_s) \otimes \mathbf{b}(\theta_s) \tag{6.7}$$

where $\otimes$ and θ_s denote the Kronker product and the target direction, respectively. Note that here perfect waveform orthogonality is assumed. This equation can be represented by

$$\mathbf{y} = \sigma_s \mathbf{v}(\theta_s) \tag{6.8}$$

where

$$\mathbf{v}(\theta_s) = \mathbf{a}(\theta_s) \otimes \mathbf{b}(\theta_s) \tag{6.9}$$

is the $MN \times 1$ steering vector associated with an virtual array of $M \times N$ sensors.

Suppose the transmitter has M antennas whereas the receiver has N antennas, Eq. (6.9) means that an virtual antenna array with utmost $M \times N$ non-overlapped virtual transmitting/receiving elements may be obtained to take full advantages of the MIMO antenna array. Different antenna array configurations have different spatial sampling characteristics and signal processing complexity. Therefore, MIMO SAR antenna array configuration should be optimally designed for specified applications.

6.2.2 Equivalent Phase Center

Consider a linear transmitting array with M antenna elements and a linear receiving array with N antenna elements. Without loss of any generality, suppose the transmitting and receiving arrays are parallel and colocated. The

m-th transmitting antenna is located at $x_{T,m} = \frac{\lambda}{2}u_m$ and the n-th receiving antenna is located at $x_{R,n} = \frac{\lambda}{2}v_n$. Consider a far-field point target, the transmitter and receiver steering vectors can be represented, respectively, by

$$\mathbf{a}(\theta_s) = [e^{ju_1\pi\sin\theta_s}, e^{ju_2\pi\sin\theta_s}, \ldots, e^{ju_M\pi\sin\theta_s}]^T \tag{6.10}$$

$$\mathbf{b}(\theta_s) = [e^{jv_1\pi\sin\theta_s}, e^{jv_2\pi\sin\theta_s}, \ldots, e^{jv_N\pi\sin\theta_s}]^T \tag{6.11}$$

From Eq. (6.9) we can get

$$\mathbf{v}(\theta_s) = \begin{bmatrix} e^{j(v_1+u_1)\pi\sin\theta_s} & e^{j(v_1+u_2)\pi\sin\theta_s} & \cdots & e^{j(v_1+u_M)\pi\sin\theta_s} \\ e^{j(v_2+u_1)\pi\sin\theta_s} & e^{j(v_2+u_2)\pi\sin\theta_s} & \cdots & e^{j(v_2+u_M)\pi\sin\theta_s} \\ \vdots & \vdots & \cdots & \vdots \\ e^{j(v_N+u_1)\pi\sin\theta_s} & e^{j(v_N+u_2)\pi\sin\theta_s} & \cdots & e^{j(v_N+u_M)\pi\sin\theta_s} \end{bmatrix} \tag{6.12}$$

Note that the amplitude of the signal reflected by the target has been normalized to unity. That is, the target response in the m-th matched filtering output of the n-th receiving antenna is expressed as

$$v_{m,n}(\theta_s) = e^{j(v_n+u_m)\pi\sin\theta_s} \tag{6.13}$$

It can be noticed that the phase differences are created by both the transmitting antenna locations and the receiving antenna locations. The target response expressed in Eq. (6.13) is the same as the target response received by a receiving array with $M \times N$ antenna elements located at

$$\{x_{T,m} + x_{R,m}\}, \quad m \in [1,2,\ldots,M], n \in [1,2,\ldots,N] \tag{6.14}$$

The phase differences are created by both the transmitting and receiving antenna locations. This MN-element array is just the virtual antenna array. An utmost of $M \times N$-element virtual array can be obtained by using only $M + N$ physical antenna elements. It is as if we have a receiving array of $M \times N$ elements. The virtual antenna array can be seen as a way to sample the electromagnetic wave in the spatial domain. The degrees-of-freedom can greatly increase the design flexibility of the MIMO SAR systems.

To investigate the effective phase center caused by the virtual antenna array, we consider several typical linear array configurations of MIMO SAR systems.

Case A: Transmitter is Same to Receiver, $M = N = L$

If the transmitting array and the receiving array are uniform linear arrays, we assume that the first element of $\mathbf{a}(\theta_s)$ and $\mathbf{b}(\theta_s)$, respectively, is the reference element. From Eqs. (6.10) and (6.11) we have

$$\mathbf{a}(\theta_s) = \mathbf{b}(\theta_s) = [1, e^{j\pi\sin\theta_s}, e^{j2\pi\sin\theta_s}, \ldots, e^{j(M-1)\pi\sin\theta_s}]^T \tag{6.15}$$

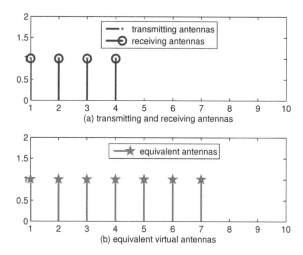

FIGURE 6.5: Virtual phase center of the uniform linear array.

Eq. (6.13) can then be reexpressed as

$$\mathbf{v}(\theta_s) = \begin{bmatrix} 1 & e^{j\pi \sin \theta_s} & \cdots & e^{j(L-1)\pi \sin \theta_s} \\ e^{j)\pi \sin \theta_s} & e^{j2\pi \sin \theta_s} & \cdots & e^{jL\pi \sin \theta_s} \\ \vdots & \vdots & \cdots & \vdots \\ e^{j(L-1)\pi \sin \theta_s} & e^{jM\pi \sin \theta_s} & \cdots & e^{j(2L-3)\pi \sin \theta_s} \end{bmatrix} \tag{6.16}$$

In this case, the number of effective virtual phase center is $2L - 1$ with the biggest virtual aperture length of $2L - 2$. Suppose $M = N = 4$, Figure 6.5 shows the corresponding virtual arrays.

If the transmitting array and receiving array are nonuniform linear array, we can express the steering vector as

$$\mathbf{a}(\theta_s) = \mathbf{b}(\theta_s) = [e^{ju_1 \pi \sin \theta_s}, e^{ju_2 \pi \sin \theta_s}, \dots, e^{ju_M \pi \sin \theta_s}]^T \tag{6.17}$$

In this case, the Eq. (6.13) can then be reexpressed as

$$\mathbf{v}(\theta_s) = \begin{bmatrix} e^{j(2u_1)\pi \sin \theta_s} & e^{j(u_1+u_2)\pi \sin \theta_s} & \cdots & e^{j(u_1+u_M)\pi \sin \theta_s} \\ e^{j(u_2+u_1)\pi \sin \theta_s} & e^{j(u_2+u_2)\pi \sin \theta_s} & \cdots & e^{j(u_2+u_M)\pi \sin \theta_s} \\ \vdots & \vdots & \cdots & \vdots \\ e^{j(u_M+u_1)\pi \sin \theta_s} & e^{j(u_M+u_2)\pi \sin \theta_s} & \cdots & e^{j(2u_M)\pi \sin \theta_s} \end{bmatrix} \tag{6.18}$$

It can be proved that the utmost number of effective virtual phase centers is $\frac{L(L+1)}{2}$. Suppose also $M = N = 4$, Figure 6.6 shows the corresponding virtual arrays.

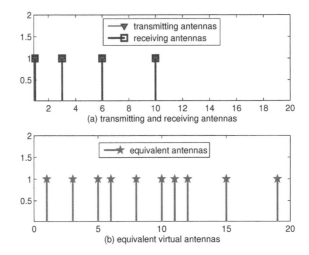

FIGURE 6.6: Virtual phase centers of the nonuniform linear array.

Case B: Transmitter and Receiver Have No Overlapped Elements

In this case, it can be easily concluded from Eq. (6.12) that the utmost number effective virtual phase centers is $M \cdot N$. Suppose $M + N = L$, the utmost number of effective virtual phase centers can be determined by $L_v = N(L - N) \le L^2/4$. When there is $M = 3, N = 4$, Figure 6.7 shows two typical virtual arrays, one is uniform array and the other is nonuniform array.

6.3   Multidimensional Waveform Encoding SAR HRWS Imaging

The multidimensional waveform encoding SAR concept, which indicates that the transmitted waveform should be encoded spatiotemporally and hence become a joint function of space and time variables, was proposed in [215]. The multidimensional waveform encoding affords waveform diversity on the transmitter, and thus better system performance can be obtained, provided that appropriate processing techniques are applied on the receiver to exploit the waveform diversity effectively.

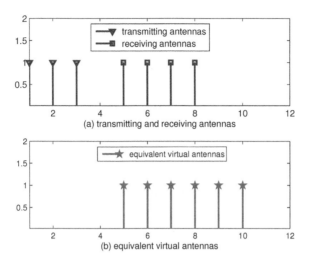

(a) Uniform transmitting and/or receiving array

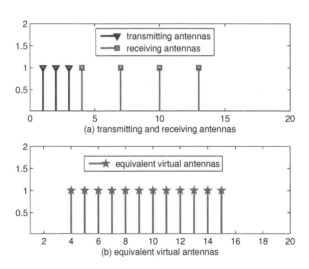

(b) Nonuniform transmitting and/or receiving array

FIGURE 6.7: Two typical virtual arrays for case B.

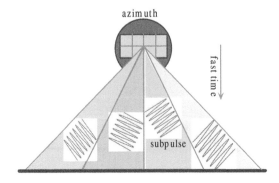

FIGURE 6.8: An example multidimensional encoding for the transmitted radar pulse.

6.3.1   Multidimensional Encoding Radar Pulses

The multidimensional waveform encoding is characterized by [216]

$$w\left(\tau, \theta_{el}, \theta_{az}\right) \neq h(\tau) \cdot a(\theta_{el}) \cdot b(\theta_{az}) \qquad (6.19)$$

where $h(\tau)$ is the temporal structure of the modulated RF radar pulse, θ_{el} is the weighting function from the antenna pattern in the elevation dimension, and θ_{az} is the weighting function for the antenna pattern in the azimuth dimension. A very simple means to obtain a multidimensional waveform is the division of a long transmitting pulse into multiple subpulses, as shown in Figure 6.8. Each subpulse is then associated with a different antenna beam via switching the phase coefficients in the Tx/Rx modules between successive subpulses.

The multidimensional waveform encoding can be extended to an arbitrary spatiotemporal radar illumination, where each direction has its own temporal transmit signal with different power, duration, and phase code. The overall transmitted range frequency spectrum can be decomposed into multiple subbands. Each subband is then associated with a different subaperture and/or beam. Such a frequency decomposition of the transmitted range pulse may be combined with intrapulse aperture switching and/or beam-steering in the azimuth dimension. This enables the illumination of a wide swath on the ground notwithstanding a large extension of the total Tx antenna array in the elevation dimension and the simultaneous suppression of azimuth ambiguities. The spatiotemporal excitation coefficients in the transmitter could even be adaptively selected by evaluating the recorded samples from previous radar echoes [216]. As an example, we consider the automatic compensation of angular differences in the received Rx signal power arising from variations in the length and attenuation of the propagation path and/or spatial inhomogeneities in the first-order scattering statistics of the imaged scene. The latter may be due to changing incident angles, land-water or water-ice transitions, as well as varying types of land cover. A spatiotemporal adaptation of the transmitted signal power is hence well-suited to provide a more homogenous image quality.

Consider a multidimensional waveform encoding SAR using the nonseparable radar pulse, which consists of K temporal adjacent linear FM subpulses with the same carrier frequency, FM rate, and bandwidth. Suppose each of these subpulses can be considered as a time-shifted version of another one, and the transmitting interval between them equals the duration of the subpulse. The transmitted signal can then be expressed as [118]

$$
\begin{aligned}
s(t) =\ &\mathrm{rect}\left[\frac{t+T}{T}\right] \exp\left\{j2\pi f_c(t+T) + j\pi k_r\,(t+T)^2\right\} \\
&+ \mathrm{rect}\left[\frac{t}{T}\right] \exp\left\{j2\pi f_c t + j\pi k_r t^2\right\} \\
&+ \mathrm{rect}\left[\frac{t-T}{T_s}\right] \exp\left\{j2\pi f_c(t-T) + j\pi k_r(t-T)^2\right\}
\end{aligned}
\tag{6.20}
$$

where t is the range fast time, T is the subpulse duration, f_c is the carrier frequency, and k_r is the chirp rate. Note that, for simplicity and without loss of generality, only three subpulses are assumed.

As shown in Figure 6.8, three contiguous azimuth beams can be formed and switched in the transmitter from subpulse to subpluse. Each subbeam covers a portion of the full illuminated azimuth footprint. During the first subpulse, all phase shifters are steered toward the forward beam position, and the entire antenna array is used to coherently transmit the first subpulse in that direction. During the second subpulse, the phase shifters are reset to point toward the middle beam position and the second subpulse is transmitted. The third subpulse is transmitted by the third subbeam that points backward. The azimuth antenna beamwidth for each subbeam can be approximated, respectively, as

$$
\theta_1 \approx \frac{\lambda}{L_a} \tag{6.21a}
$$

$$
\theta_2 \approx 0 \tag{6.21b}
$$

$$
\theta_3 \approx -\frac{\lambda}{L_a} \tag{6.21c}
$$

where L_a is the overall antenna length. Different from the azimuth dimension, in the elevation dimension all the subbeams illuminate the same region.

The radar echoes associated with different subpulses, which arrive at the receiver from different elevation angles simultaneously, should be separated by digital beamforming on the elevation dimension. Next, these echoes are passed through appropriate azimuth beamformers to remove their first-order azimuth ambiguity components. Finally, the azimuth subband signals are coherently combined to achieve the full Doppler bandwidth for a high azimuth resolution [216].

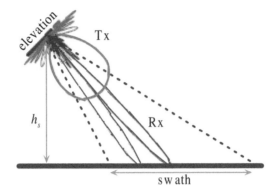

FIGURE 6.9: Illustration of the intrapulse beamsteering in the elevation dimension.

6.3.2   Intrapulse Beamsteering in the Elevation Dimension

Intrapulse beamsteering in the elevation dimension enables an illumination of a wide swath with a sequence of narrow and high-gain antenna beams, as shown in Figure 6.9. Such a staggered illumination is similar to traditional ScanSAR mode, with the major difference that each transmitted pulse now illuminates not only one but all subswaths simultaneously. The illumination sequence within the Tx pulse can be arranged in any order. If we start from a far-range illumination and subsequently proceed to the near range, the overall receiving window can be shortened, thereby reducing the amount of data to be recorded [216]. The short receiving window will result in increased time available for the transmission of multiple subpulses. This reduces the RF peak power requirements of the transmitter and gives further margin to switch between the subpulses, thereby simplifying the electrical system design. Another advantage is the reduced gain loss at the border of the swath compared to a conventional radar illuminator. The staggered illumination allows for a flexible distribution of the signal energy. For example, longer transmitting pulses can be used for the subbeams with higher incident angles. This illumination strategy is well-suited to compensate the SNR loss due to both the typical decrease of the backscattering with increasing incident angles and the additional free space loss from a larger range.

Two solutions to implement the intrapulse beamsteering in the elevation dimension are described in [216]. The first solution uses different azimuth parts of the large antenna array for each beam, thereby ensuring a low duty cycle for each Tx/Rx module notwithstanding the long transmitting pulse. In this way, both azimuth beam broadening and multiaperture high-resolution imaging can be achieved simultaneously. The second solution uses the full antenna array for the total pulse-length, which has the advantage of further reducing the peak power requirements for each Tx/Rx module. The illumination is then achieved either by switching between multiple moderately phase tapered beams or even via continuous beamsteering from far range to near range. The latter requires an appropriate temporal modulation of the transmitted RF signal such that each scatter receives a sufficient bandwidth for high range resolution.

According to the intrapulse beamsteering operation scheme, the radar echoes from different subswathes will overlap in the receiver. The temporal overlapped radar echoes from different subswathes should be resolved in the spatial domain by digital beamforming on the receiver, which exploits the relation between time delay and elevation angle in the side-looking radar imaging geometry [412]. Since subpulse within one complete radar pulse are transmitted in sequence, the scattered signals from different subpulses will then arrive from different elevation angles. It is hence possible to separate radar echoes from adjacent subpulses spatially by forming multiple scanning beams, each of which is related to one subpulse during the echo-receiving period.

In order to effectively separate the overlapped echoes, the elevation antenna pattern of each formed subbeam in the elevation dimension should have maximum gain in the direction along which echoes of corresponding subpulse arrive and simultaneously deep nulls at the positions with echoes from other subpulses. This problem can be formulated as [118]

$$\mathbf{w}_k^* \mathbf{a}_{k'} = \begin{cases} 1, k = k' \\ 0, k \neq k' \end{cases} \tag{6.22}$$

where $\mathbf{w}_k$ is the desired beamformer weighting vector for the k-th subbeam, and $\mathbf{a}_{k'}$ is the receiver steering vector related to the k'-th subpulse

$$\mathbf{a}_{k'} = \begin{bmatrix} 1 & \exp\left\{j\frac{2\pi}{\lambda}d_{se}\sin(\alpha_{k'})\right\} & \cdots & \exp\left\{j\frac{2\pi}{\lambda}(K-1)d_{se}\sin(\alpha_{k'})\right\} \end{bmatrix}^T \tag{6.23}$$

The d_{se} is the space between two adjacent subapertures in the elevation dimension, K is the total number of subapertures in the elevation dimension, $\mathbf{a}_{k'}$ is the off-boresight angles along the echoes arriving from the k'-th subpulse.

Taking three subswaths as an example, the three off-boresight angles can be derived, respectively, as [299, 362]

$$\alpha_1(t) = \arccos\left(\frac{4R_{\text{orbit}}^2 - 4R_e^2 + (t+T_s)^2 c_0^2}{4R_{\text{orbit}}(t+T_s)c_0}\right) - \eta \tag{6.24a}$$

$$\alpha_2(t) = \arccos\left(\frac{4R_{\text{orbit}}^2 - 4R_e^2 + t^2 c_0^2}{4R_{\text{orbit}}tc_0}\right) - \eta \tag{6.24b}$$

$$\alpha_1(t) = \arccos\left(\frac{4R_{\text{orbit}}^2 - 4R_e^2 + (t-T_s)^2 c_0^2}{4R_{\text{orbit}}(t-T_s)^2 c_0}\right) - \eta \tag{6.24c}$$

where R_{orbit} is the radar orbit radius, R_e is the Earth radius, and η is the antenna boresight off-nadir angle. Eq. (6.23) can then be represented by

$$\mathbf{v}_k^* \mathbf{A} = \mathbf{e}_k^T \tag{6.25}$$

where $\mathbf{e}_k$ is the k-th column of an identity matrix and $\mathbf{A}$ is the receiving array matrix with columns being the receiver steering vectors, $\mathbf{A} = \begin{bmatrix} \mathbf{a}_1 & \mathbf{a}_2 & \mathbf{a}_3 \end{bmatrix}$ [118]. We then have

$$\mathbf{v}_k = \mathbf{A}(\mathbf{A}^*\mathbf{A})^{-1}\mathbf{e}_k \tag{6.26}$$

It can be noticed that the $\mathbf{v}_k$ is a time-varying beamformer weighting vector due to the variation of the off-boresight angles as the subpuses run over the ground surface.

6.3.3 Digital Beamforming in Azimuth

Multiple apertures are also employed in the azimuth dimension. Specific azimuth beamformers should steer nulls in the resulting joint azimuth antenna pattern to the angles corresponding to the ambiguous Doppler frequencies. Each subpulse employs a single narrow transmitting beam and multiple broad receiving beams. Different azimuth beamformer weighting vectors should be applied to different subpulse echoes, since their azimuth spectrum characteristics are largely different. The joint spatiotemporal azimuth processing algorithm proposed in [145, 214] can be employed to unambiguously recover the original broad azimuth spectrum according to their respective space-frequency spectrum.

For the k-th subpulse echo, the corresponding azimuth beamformer weighting vector $\mathbf{p}_k$, which needs to steer nulls in the directions where its ambiguity components are coming from, can be obtained by solving the following matrix equation

$$\mathbf{H}_k\mathbf{p}_k = \begin{bmatrix} 0 & 1 & 0 \end{bmatrix}^T \tag{6.27}$$

where $\mathbf{H}_k$ is the receiving array manifold matrix with rows being the steering vectors associated with signal and ambiguity components. For three azimuth subapertures, it is expressed as [118]

$$\mathbf{H}_k = \begin{bmatrix} \exp\left\{j\frac{2\pi}{\lambda}d_{sa}\sin\left(\theta_{\text{k-amb-l}}\right)\right\} & 1 & \exp\left\{-j\frac{2\pi}{\lambda}d_{sa}\sin\left(\theta_{\text{k-amb-l}}\right)\right\} \\ \exp\left\{j\frac{2\pi}{\lambda}d_{sa}\sin\left(\theta_{\text{k-s}}\right)\right\} & 1 & \exp\left\{-j\frac{2\pi}{\lambda}d_{sa}\sin\left(\theta_{\text{k-s}}\right)\right\} \\ \exp\left\{j\frac{2\pi}{\lambda}d_{sa}\sin\left(\theta_{\text{k-amb-r}}\right)\right\} & 1 & \exp\left\{-j\frac{2\pi}{\lambda}d_{sa}\sin\left(\theta_{\text{k-amb-r}}\right)\right\} \end{bmatrix}^T \tag{6.28}$$

where $\theta_{\text{k-s}}$ is the view angle of the useful signal, and $\theta_{\text{k-amb-l}}$ and $\theta_{\text{k-amb-r}}$ denote the view angles of first-order azimuth ambiguities on the left and right sides of the useful signal, respectively.

Using the relation between azimuth squint angle θ and Doppler frequency

$$f_d = \frac{v_s}{\lambda}\sin(\theta) \tag{6.29}$$

with the knowledge about frequency intervals where ambiguities and echo signals are located for different subpulses, we can replace the azimuth angle variable θ in Eq. (6.28) with the frequency variable f_a. The azimuth beam-former weighting vectors $\mathbf{p}_1$, $\mathbf{p}_2$ and $\mathbf{p}_3$ can then be derived, respectively, as [118, 424]

$$\mathbf{p}_1 = \left[\begin{array}{ccc} \dfrac{\exp\{-j\frac{\pi}{v_s}d_{sa}(f_a+f_{dc})\}}{2[1-\cos(\frac{\pi}{v_s}d_{sa}f_p)]} & \dfrac{-\cos(\frac{\pi}{v_s}d_{sa}f_p)}{1-\cos(\frac{\pi}{v_s}d_{sa}f_p)} & \dfrac{\exp\{j\frac{\pi}{v_s}d_{sa}(f_a+f_{dc})\}}{2[1-\cos(\frac{\pi}{v_s}d_{sa}f_p)]} \end{array} \right]^T \tag{6.30}$$

$$\mathbf{p}_2 = \left[\begin{array}{ccc} \dfrac{\exp\{-j\frac{\pi}{v_s}d_{sa}(f_a)\}}{2[1-\cos(\frac{\pi}{v_s}d_{sa}f_p)]} & \dfrac{-\cos(\frac{\pi}{v_s}d_{sa}f_p)}{1-\cos(\frac{\pi}{v_s}d_{sa}f_p)} & \dfrac{\exp\{j\frac{\pi}{v_s}d_{sa}(f_a)\}}{2[1-\cos(\frac{\pi}{v_s}d_{sa}f_p)]} \end{array} \right]^T \tag{6.31}$$

$$\mathbf{p}_3 = \left[\begin{array}{ccc} \dfrac{\exp\{-j\frac{\pi}{v_s}d_{sa}(f_a-f_{dc})\}}{2[1-\cos(\frac{\pi}{v_s}d_{sa}f_p)]} & \dfrac{-\cos(\frac{\pi}{v_s}d_{sa}f_p)}{1-\cos(\frac{\pi}{v_s}d_{sa}f_p)} & \dfrac{\exp\{j\frac{\pi}{v_s}d_{sa}(f_a-f_{dc})\}}{2[1-\cos(\frac{\pi}{v_s}d_{sa}f_p)]} \end{array} \right]^T \tag{6.32}$$

where f_{dc} and f_p denote the Doppler frequency centroid associated with the squint transmitting subbeams and the value of the PRF, respectively. A co-herent combination of these beamformer outputs can then recover the full Doppler spectrum for high-resolution SAR imaging. This obtained Doppler spectrum no longer suffers from an aliasing problem.

6.3.4 Range Ambiguity to Signal Ratio Analysis

Consider also three azimuth subapertures, the received signal power associated with each subpulse can be represented, respectively, by [118]

$$E_{\text{signal-1}} = \int_{f_{dc}-B_{sd}/2}^{f_{dc}+B_{sd}/2} |A_1(f_a)|^2 \, \mathrm{d}f_a \tag{6.33}$$

$$E_{\text{signal-2}} = \int_{-B_{sd}/2}^{B_{sd}/2} |A_2(f_a)|^2 \, \mathrm{d}f_a \tag{6.34}$$

$$E_{\text{signal-3}} = \int_{-f_{dc}-B_{sd}/2}^{-f_{dc}+B_{sd}/2} |A_3(f_a)|^2 \, \mathrm{d}f_a \tag{6.35}$$

with B_{sd} being the Doppler bandwidth for each subbeam. The two-way antenna patterns associated with the three subpulses are given by

$$A_1(f_a) = \text{sinc}\left[\frac{L_a}{2v_s}(f_a - f_{dc})\right] \cdot \text{sinc}\left[\frac{d_{sa}}{2v_s}f_a\right] \tag{6.36a}$$

$$A_2(f_a) = \text{sinc}\left[\frac{L_a}{2v_s}(f_a)\right] \cdot \text{sinc}\left[\frac{d_{sa}}{2v_s}f_a\right] \tag{6.36b}$$

$$A_3(f_a) = \text{sinc}\left[\frac{L_a}{2v_s}(f_a + f_{dc})\right] \cdot \text{sinc}\left[\frac{d_{sa}}{2v_s}f_a\right] \tag{6.36c}$$

The corresponding total signal power is

$$E_{\text{signal}} = E_{\text{signal-1}} + E_{\text{signal-2}} + E_{\text{signal-3}} \tag{6.37}$$

Similarly the respective ambiguity power associated with the three sub-pulses are [118]

$$
\begin{aligned}
E_{\text{amb-1}} = \sum_{k=1}^{\infty} \Bigg\{ & \int_{-B_{sd}/2}^{B_{sd}/2} \left| \mathbf{p}_1^T \mathbf{h}_{\text{azi}}(f_a + f_{dc} + k f_p) A_1(f_a + f_{dc} + k f_p) \right|^2 \mathrm{d}f_a \\
& + \int_{-B_{sd}/2}^{B_{sd}/2} \left| \mathbf{p}_1^T \mathbf{h}_{\text{azi}}(f_a + f_{dc} - k f_p) A_1(f_a + f_{dc} - k f_p) \right|^2 \mathrm{d}f_a \Bigg\}
\end{aligned}
$$

$$(6.38)$$

$$
E_{\text{amb-2}} = 2 \sum_{k=1}^{\infty} \int_{-B_{sd}/2}^{B_{sd}/2} \left| \mathbf{p}_2^T \mathbf{h}_{\text{azi}}(f_a + k f_p) A_2(f_a + k f_p) \right|^2 \mathrm{d}f_a \tag{6.39}
$$

$$
\begin{aligned}
E_{\text{amb-3}} = \sum_{k=1}^{\infty} \Bigg\{ & \int_{-B_{sd}/2}^{B_{sd}/2} \left| \mathbf{p}_1^T \mathbf{h}_{\text{azi}}(f_a - f_{dc} + k f_p) A_1(f_a - f_{dc} + k f_p) \right|^2 \mathrm{d}f_a \\
& + \int_{-B_{sd}/2}^{B_{sd}/2} \left| \mathbf{p}_1^T \mathbf{h}_{\text{azi}}(f_a - f_{dc} - k f_p) A_1(f_a - f_{dc} - k f_p) \right|^2 \mathrm{d}f_a \Bigg\}
\end{aligned}
$$

$$(6.40)$$

The corresponding total ambiguity power is

$$
E_{\text{ambiguity}} = E_{\text{amb-1}} + E_{\text{amb-2}} + E_{\text{amb-2}} \tag{6.41}
$$

The azimuth ambiguity to signal ratio (RASR) can then be computed as

$$
\text{RASR} = \frac{E_{\text{ambiguity}}}{E_{\text{signal}}} \tag{6.42}
$$

Different from traditional single-antenna SAR, in the multidimensional waveform encoding SAR the total range ambiguity has extra range ambiguities due to the mutual interferences between the echoes corresponding to adjacent subpulses transmitted in one PRI. More details can be found in [118].

6.4   MIMO SAR HRWS Imaging

In this section, we present an alternative solution to HRWS imaging with MIMO antennas. The basic idea is to form multiple transmitting and receiving subapertures, which are steered toward different subswaths. The multiple antennas can be arranged in the along-track (azimuth, flight direction), cross-track (elevation, perpendicular to along-track), or in both dimensions. But in this section the multiple antennas are placed only in the cross-track dimension. Figure 6.10 illustrates an example MIMO SAR with multiple elevation antennas. These subapertures can be disjointed or overlapped in space. Multiple pairs of virtual transmitting and receiving beams can then be formed simultaneously in the direction of a desired subswath.

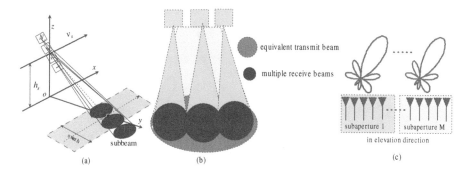

FIGURE 6.10: Illustration of the MIMO SAR with multiple elevation antennas.

6.4.1   Transmit Subaperturing MIMO Technique

In recent years, several attempts to exploit jointly the benefits of phased-array and MIMO arrays have been reported [138, 175, 454]. The general idea is to divide the transmitting array into multiple subarrays that are allowed to overlap. Each subarray is used to coherently transmit a waveform which is orthogonal to the waveforms transmitted by other subarrays. Coherent processing gain can be achieved by designing the weighting vector of each subarray to form a beam towards a certain direction in space. The subarrays are combined jointly to form a MIMO radar resulting in higher angular resolution capabilities. The advantages of this transmit subaperturing MIMO array over traditional phased-array and MIMO arrays are summarized as [175, 243]

i)   enjoys all the advantages of the MIMO array, i.e., it enables improving angular resolution, detecting a higher number of targets, improving parameter identifiability, and extending the array aperture.

ii)   enables the use of existing beamforming techniques at both the transmitting and receiving ends.

iii)   provides the means for designing the overall beampattern of the virtual array.

iv)   offers a tradeoff between angular resolution and robustness against beam-shape loss.

v)   offers improved robustness against strong interference.

Figure 6.11 illustrates an example transmit subaperturing MIMO array with two subarrays. In general, suppose the transmitting array is divided into K subarrays ($1 \leq K \leq M$ with M being the number of the transmitting array elements). Each transmitting subarray can be composed in any number of array elements ranging from 1 to M such that no subarray is exactly the same to another subarray. All elements of the k-th subarray are used to

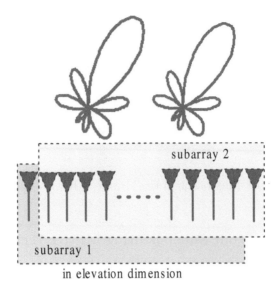

FIGURE 6.11: Illustration of an example transmit subaperturing array. Only two subarrays are shown, it is for illustration only.

coherently emit the signal $\phi_k(t)$, so that a beam is formed towards a certain direction in space. The beamforming weighting vector can be properly designed to maximize the coherent processing gain. Moreover, different waveforms are transmitted by different subarrays.

Suppose the k-th subarray consists of $N_k < M$ elements, the equivalent baseband signal can be modeled as

$$s_k(t) = \sqrt{\frac{N_t}{K}}\phi_k(t)\widetilde{\mathbf{w}}_k^*, \qquad k = 1, 2, \dots, K \tag{6.43}$$

with

$$\int_{T_p} \phi_k(t)\phi_{k'}^H(t)\mathrm{d}t = \delta(k - k') \tag{6.44}$$

where $\widetilde{\mathbf{w}}_k$ is the $M \times 1$ unit-norm complex vectror which consists of N_k beamforming weights corresponding to the active elements of the k-th subarray. The energy of $s_k(t)$ within one radar pulse is given by

$$E_k = \int_{T_p} s_k^H(t)s_k(t)\mathrm{d}t = \frac{M}{K} \tag{6.45}$$

which means that the total transmitted energy for the transmit subaperturing array system within one radar pulse is equal to M.

The signal reflected by a hypothetical target located at direction θ in the far-field is modeled as [138]

$$x(t,\theta) = \sqrt{\frac{M}{K}}\beta(\theta)\sum_{k=1}^{K}\mathbf{w}_k^H\mathbf{a}_k(\theta)e^{-j\boldsymbol{\tau}_k(\theta)}\phi_k(t) \tag{6.46}$$

where $\beta(\theta)$ is the reflection coefficient of the hypothetical target, and $\mathbf{w}_k$ and $\mathbf{a}_k(\theta)$ are the $N_k \times 1$ beamforming vector and steering vector, respectively, which contain only the elements corresponding to the active elements of the k-th subarrays, and $\boldsymbol{\tau}_k(\theta)$ is the time required for the signal to travel across the spatial displacement between the first element of the transmitting subarray and the first element of the k-th subarray. Note that the first element of $\mathbf{a}_k$ is taken as the reference element.

Denoting

$$\mathbf{c}(\theta) \doteq \left[\mathbf{w}_1^H\mathbf{a}_1(\theta), \mathbf{w}_2^H\mathbf{a}_2(\theta), \ldots, \mathbf{w}_K^H\mathbf{a}_K(\theta)\right]^T \tag{6.47a}$$

$$\mathbf{d}(\theta) \doteq \left[e^{-j\boldsymbol{\tau}_1(\theta)}, e^{-j\boldsymbol{\tau}_2(\theta)}, \ldots, e^{-j\boldsymbol{\tau}_K(\theta)}\right]^T \tag{6.47b}$$

Equation (6.46) can be rewritten as

$$x(t,\theta) = \sqrt{\frac{M}{K}}\beta(\theta)\left[\mathbf{c}(\theta)\odot\mathbf{d}(\theta)\right]^T\phi_K(t) \tag{6.48}$$

where $\phi_K(t) \doteq [\phi_1(t), \phi_2(t), \ldots, \phi_K(t)]$ is the $K \times 1$ vector of waveforms. Suppose the target of interest is observed in the background of D interferences with reflection coefficients $\{\beta_i\}_{i=1}^{D}$ and locations $\{\theta_i\}_{i=1}^{D}$, the $N \times 1$ received complex vector can be written as

$$\mathbf{y} = \sqrt{\frac{M}{K}}\beta_s\mathbf{u}(\theta_s) + \sum_{i=1}^{D}\sqrt{\frac{M}{K}}\beta_i\mathbf{u}(\theta_i) + \mathbf{n} \tag{6.49}$$

where the $KN \times 1$ vector $\mathbf{u}(\theta)$ is the virtual steering vector associated with direction θ

$$\mathbf{u}(\theta) \doteq \left[\mathbf{c}(\theta)\odot\mathbf{d}(\theta)\right]\otimes\mathbf{b}(\theta) \tag{6.50}$$

with $\mathbf{b}(\theta)$ being the receiver steering vector, $\mathbf{n}$ is the $KN \times 1$ noise term.

Specifically, if the transmitting array is designed as one subarray and only one waveform is emitted, i.e, $K = 1$, the virtual steering vector simplifies to

$$\mathbf{u}(\theta) = \left[\mathbf{w}^H\mathbf{a}(\theta)\right]\mathbf{b}(\theta) \tag{6.51}$$

where $\mathbf{w}^H\mathbf{a}(\theta)$ is just the uplink coherent processing gain of conventional phased-array radar towards the direction θ, and $\mathbf{w}$ is the phased-array transmit beamformer weight vector. In contrast, if $K = M$ is chosen, the virtual steering vector is just for a general MIMO array

$$\mathbf{u}(\theta) = \mathbf{a}(\theta)\otimes\mathbf{b}(\theta) \tag{6.52}$$

This explains why MIMO radar has the largest degrees-of-freedom at the price of having no coherent processing at the transmitter.

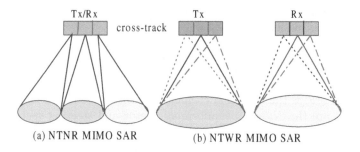

(a) NTNR MIMO SAR   (b) NTWR MIMO SAR

FIGURE 6.12: Two different MIMO SAR operation modes. Only three beams are shown, it is for illustration only.

6.4.2 Transmit Subaperturing for HRWS Imaging

When multiple antennas are placed only in the cross-track dimension, there are two different configurations, as shown in Figure 6.12. We refer them as narrow-beam transmission narrow-beam reception (NTNR) MIMO SAR and narrow-beam transmission wide-beam reception (NTWR) MIMO SAR, respectively. Note that these two operation modes are different from the single-phase center multibeam (SPCM) SAR and multiple-phase center multibeam (MPCM) SAR in classic DPCA technique [77], where the multiple antennas are placed in the along-track dimension (but they are placed only in the cross-track dimension in this section). In the following, we discuss them separately.

6.4.2.1 NTNR Operation Mode

The NTNR MIMO SAR operation mode shown in Figure 6.12(a) is to transmit orthogonal signal pulses by multiple narrow beams directing toward different subswaths and also receive the corresponding returns by narrow contiguous beams that span the whole transmitting beams. A distinct channel is associated with each receiving beam, and hence, the returns can be separated by utilizing the waveform orthogonality. As a compromise between computation complexity and imaging performance, the number of transmitting beams or subswaths is determined by

$$M = \left\lceil \frac{2R_{\max}\mathrm{PRF}}{c_0} \right\rceil - \left\lfloor \frac{2R_{\min}\mathrm{PRF}}{c_0} \right\rfloor \tag{6.53}$$

where $\lceil \cdot \rceil$ and $\lfloor \cdot \rfloor$ denote, respectively, the maximum and minimum integer, and $R_{\max}$ and $R_{\min}$ denote, respectively, the maximum and minimum slant range within the imaged swath.

Although narrow-beams are employed in the receiving antennas (see Figure 6.13), the desired receiving channel will receive the other channels' returns from its antenna sidelobe. Therefore, the MIMO SAR received signals at each

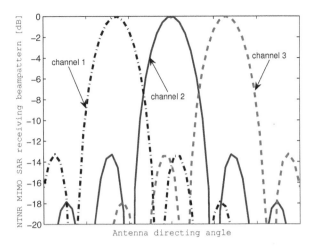

FIGURE 6.13: NTNR MIMO SAR receiving beampattern distribution.

receiving channel can be expressed as

$$x_n(t) = \sum_{m=1}^{M} \alpha_{m,n} s_m(t - \tau_{m,n}) \tag{6.54}$$

where $m \in [1, 2, \ldots, M]$ and $n \in [1, 2, \ldots, N]$, $x_n(t)$ is the received signal at the n-th channel, $\alpha_{m,n}$ is the channel coefficient for the m-th transmitting beam and the n-th receiving channel, and $\tau_{m,n}$ is the time for the signal propagating from the m-th transmitting beam to the n-th receiving channel.

If the transmitted signals are orthogonal, the echo in the n-th receiving channel corresponding to the m-th transmitted signal can be extracted separately by M matched filtering

$$
\begin{aligned}
y_n(t) &= x_n(t) \otimes s_m^*(-t) \\
&= \left[\sum_{m=1}^{M} \alpha_{m,n} s_m(t - \tau_{m,n}) \right] \otimes s_m^*(-t)
\end{aligned}
\tag{6.55}
$$

Thus, M ($M = N$ is assumed) independent transmit-receive beams directing toward different elevation angles can then be formed simultaneously. This provides a potential solution to HRWS imaging. It is well-known that, for a fixed PRF f_p that is appropriate to the desired azimuth resolution requirement, the swath for conventional single-antenna SAR is determined by

$$\frac{c_0}{2}\left(\frac{i}{f_p} + T_p\right) < R < \frac{c_0}{2}\left(\frac{i+1}{f_p} - T_p\right) \tag{6.56}$$

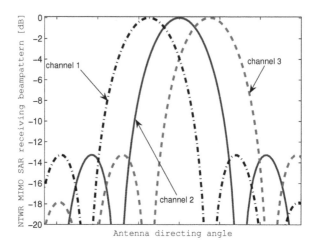

FIGURE 6.14: NTWR MIMO SAR receiving beampattern distribution.

In contrast, the swath for the NTNR MIMO SAR is determined by

$$\frac{c_0}{2}\left(\frac{i}{f_p}+T_p\right) < R < \frac{c_0}{2}\left(\frac{i+M}{f_p}-T_p\right) \tag{6.57}$$

This means that the NTNR MIMO SAR can obtain a wider (M times) swath width than that implied for conventional single-antenna SAR without degrading the RASR performance. Therefore, this operation mode can be seen as multiple (equal to the number of the transmit beams) independently operating single-antenna SAR systems.

6.4.2.2 NTWR Operation Mode

The NTWR MIMO SAR operation mode shown in Figure 6.12(b) also employs multiple narrow transmitting beams that contiguously span the desired swath, but it is quite different from the NTNR operation mode. In this case, the MIMO SAR employs multiple (N) wide-beams, which are displaced in the cross-track direction, as shown in Figure 6.14.

The signals arriving at the n-th receiving channel can be represented by

$$x_n'(t) = \sum_{m=1}^{M} \beta_{m,n}s_m(t-\tau_{m,n}) \tag{6.58}$$

Note that, here, the channel coefficients $\beta_{m,n}$ are different from the $\alpha_{m,n}$ given in Eq. (6.54) because they have different receiving beampatterns (see Figures 6.13 and 6.14). Suppose M transmitting antennas and N receiving antennas,

TABLE 6.1: The comparisons between conventional single-antenna SAR (case a), NTNR MIMO SAR (case b) and NTWR MIMO SAR (case c) in case of three transmitting beams and three receiving channels.

parameters	case a	case b	case c
equivalent phase centers	1	3	5
degrees-of-freedom	1	3	9
swath width	w_s	$3\,w_s$	$3\,w_s$
output SNR	SNR_0	$3\,\text{SNR}_0$	$3\,\text{SNR}_0$
AASR	AASR_0	AASR_0	AASR_0
system complexity	simple	middle	complex
transmitted signals	1	3	3
received data rate	dr_0	3dr_0	3dr_0

there will be $M \times N$ independent transmitting-receiving beams

$$y_n'(t) = x_n'(t) \otimes s_m^*(-t)$$
$$= \left[\sum_{m=1}^{M} \alpha_{m,n} s_m(t - \tau_{m,n}) \right] \otimes s_m^*(-t) \tag{6.59}$$

Similarly, for a fixed PRF that is appropriate for the desired azimuth resolution requirement, a wider (M times) swath than that implied Eq. (6.56) can be obtained without degrading the RASR performance.

Figure 6.15 compares the equivalent phase centers of the two MIMO SAR operation modes. It can be noticed that the NTWR mode can obtain more equivalent phase centers than the NTNR, which means that more degrees-of-freedom can be obtained for the NTWR mode. More detailed comparisons are given in Table 6.1 (AASR denotes the azimuth ambiguity to signal ratio), where a fixed PRF, range ambiguity to signal ratio (RASR) and transmitted power requirements are assumed. Note that perfect waveform orthogonality is assumed in the system performance analysis. It can be noticed that, both NTNR and NTWR modes can obtain wider swath, better SNR and more degrees-of-freedom than conventional single-antenna mode at the cost of increased system complexity and data processing. An equal swath width can be obtained for the NTNR MIMO SAR and NTWR MIMO SAR, but the NTWR MIMO SAR can obtain more degrees-of-freedom and better SNR performance at the cost of a bigger system complexity.

6.4.3 Range Ambiguity to Signal Ratio Analysis

The MIMO SAR beampattern, specially its sidelobe characteristics, will influence the imaging performance. Ambiguity noise is thus an important consideration. Since the multiple antennas are placed only in the elevation dimension, we analyze only the range ambiguities.

Range ambiguities result from preceding and succeeding pulse echoes ar-

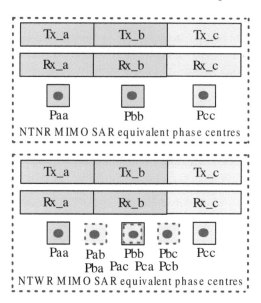

FIGURE 6.15: The equivalent phase centers of the two MIMO SAR operation modes, where P_{ij} ($i \in [a, b, b]$, $j \in [a, b, b]$) denotes the equivalent phase center from the i-th transmitting beam to the j-th receiving channel.

riving at the antenna simultaneously with the desired return. At a given time t_i within the data record window, ambiguous signals arrive from the ranges of

$$R_{ij} = \frac{c_0}{2} \left(t_i + \frac{j}{\mathrm{PRF}} \right), \quad j = \pm 1, \pm 2, \ldots, \pm n_h \tag{6.60}$$

where j, the pulse number ($j = 0$ for the desired pulse), is positive for preceding interfering pulses and negative for succeeding ones. The value $j = n_h$ is the number of pulses to the horizon. The range ambiguity is evaluated by the RASR function.

Consider the geometry of range ambiguity shown in Figure 6.16, the RASR for conventional single-antenna SAR is given by [76]

$$\mathrm{RASR}_{sa} = \frac{\displaystyle\sum_{i=1}^{N_i} \sum_{j=-n_h, j \neq 0}^{n_h} \frac{\sigma_{ij} G_{ij}^2}{R_{ij}^2 \sin(\eta_{ij})}}{\displaystyle\sum_{i=1}^{N_i} \frac{\sigma_{ij} G_{ij}^2}{R_{ij}^3 \sin(\eta_{ij})}} \tag{6.61}$$

where σ_{ij} is the normalized backscatter coefficient at a given incidence angle η_{ij} and G_{ij} is the cross-track antenna pattern at a given antenna boresight angle γ_{ij}. The η_{ij} and γ_{ij} can be calculated, respectively, by

$$\eta_{ij} = \sin^{-1} \left(\frac{R_e + h_s}{R_e} \sin(\gamma_{ij}) \right) \tag{6.62}$$

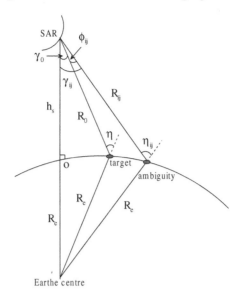

FIGURE 6.16: The geometry of range ambiguity.

$$\gamma_{ij} = \cos^{-1}\left(\frac{R_{ij}^2 + (R_e + h_s)^2 - R_e^2}{2R_{ij}(R_e + h_s)}\right) \tag{6.63}$$

where R_e and h_s denote the Earth radius and radar platform altitude, respectively. For a uniformly illuminated antenna, the far-field antenna pattern is given by

$$G_{ij} = \text{sinc}^2\left(\frac{\pi W_r}{\lambda}\sin(\phi_{ij})\right) \tag{6.64}$$

where W_r is the antenna width in the elevation dimension and ϕ_{ij} is the beam off-boresight elevation angle.

The range ambiguity of the NTNR MIMO SAR is slightly different from that of the conventional single-antenna SAR. Consider the illustration of NTNR MIMO SAR range ambiguities shown in Figure 6.17, the RASR for the three channels can be expressed as

$$\text{RASR}_{\text{ntnr}_k} = \frac{\displaystyle\sum_{i=1}^{N_i}\sum_{j=-n_h, j\neq 0}^{n_h}\frac{\sigma_{nk_{ij}}G_{nk_{ij}}^2}{R_{nk_{ij}}^2\sin(\eta_{nk_{ij}})}}{\displaystyle\sum_{i=1}^{N_i}\frac{\sigma_{nk_{ij}}G_{nk_{ij}}^2}{R_{nk_{ij}}^3\sin(\eta_{nk_{ij}})}}, \quad k=1,2,3 \tag{6.65}$$

where

$$G_{nk_{ij}}^2 = \text{sinc}^2\left(\frac{\pi W_r}{\lambda}\sin(\phi_{nkij})\right) \tag{6.66}$$

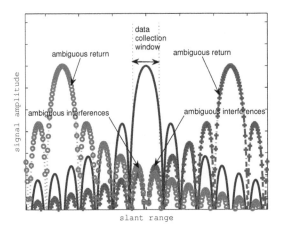

FIGURE 6.17: Illustration of the NTNR MIMO SAR range ambiguities.

with ϕ_{nkij} being the corresponding off-boresight elevation angle for the k-th channel. The other variables are defined in the same manner as that of Eq. (6.61) (but they may have different values). Note that perfect waveform orthogonality is assumed in the above equations. Since each transmitting-receiving beam is narrower than the conventional single-antenna SAR, the NTNR MIMO SAR RASR is quite small as compared with conventional single-antenna SAR. The NTNR MIMO SAR RASR can be evaluated by

$$\text{RASR}_{\text{ntnr}} \approx \frac{1}{3} \sum_{k=1}^{3} \text{RASR}_{\text{ntnr}_k} \tag{6.67}$$

Different from the NTNR MIMO SAR, the NTWR MIMO SAR has complex range ambiguities because a larger receiving beam is employed in each channel, as shown in Figure 6.18. The far-field antenna pattern from the m-th transmitting beam to the n-th receiving channel is

$$G^2_{w_{mn}} = \text{sinc}^2\left(\frac{\pi W_r}{\lambda}\sin(\phi_m)\right) \cdot \text{sinc}^2\left(\frac{\pi W_r}{\lambda}\sin(\phi_m + |m-n|\phi_s)\right),$$
$$m = 1, 2, 3; n = 1, 2, 3 \tag{6.68}$$

where $\phi_s \approx \lambda/W_r$. The corresponding RASR can be derived as

$$\text{RASR}_{\text{ntwr}_{mn}} = \frac{\displaystyle\sum_{i=1}^{N_i} \sum_{j=-n_h, j\neq 0}^{n_h} \frac{\sigma_{mn_{ij}} G^2_{w_{mn}}}{R^2_{mn_{ij}} \sin(\eta_{mn_{ij}})}}{\displaystyle\sum_{i=1}^{N_i} \frac{\sigma_{mn_{ij}} G^2_{w_{mn}}}{R^3_{mn_{ij}} \sin(\eta_{mn_{ij}})}} \tag{6.69}$$

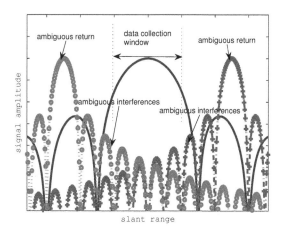

FIGURE 6.18: Illustration of the NTWR MIMO SAR range ambiguities.

where $m = 1, 2, 3; n = 1, 2, 3$. The variables are defined in the same manner as that of Eq. (6.65) (but they may have different values). Note also that perfect waveform orthogonality is assumed in the above equations. The averaged NTWR MIMO SAR RASR can then be evaluated by

$$\text{RASR}_{\text{ntwr}} \approx \frac{1}{9} \sum_{m=1}^{3} \sum_{n=1}^{3} \text{RASR}_{\text{ntnr}_{mn}} \qquad (6.70)$$

6.4.4 Image Formation Algorithms

To focus the MIMO SAR raw data, we introduce a range-Doppler-based algorithm. Like conventional single-antenna SAR processor, the task of the MIMO SAR processor is to implement a filter, matched in both the range and azimuth dimensions, to focus the raw data.

Suppose the signal transmitted by the m-th beam is

$$s_m(t) = c_m(t) \cdot \exp(j2\pi f_c t) \qquad (6.71)$$

where $c_m(t)$ is a wideband encoded signal in baseband used for the m-th transmitting beam. Note that the same carrier frequency f_c is assumed for all the beams. The returns (after being demodulated) received by the n-th channel is

$$s_{mn}(t, \tau) = A_0 c_m \left(t - \frac{R_{mn}(\tau)}{c_0} \right) \cdot \exp \left(-j2\pi f_c \frac{R_{mn}(\tau)}{c_0} \right) \qquad (6.72)$$

where $R_{mn}(\tau)$ is the slant range for the m-th transmitting beam to the n-th receiving channel with τ being the azimuth slow time, and A_0 denotes the amplitude.

After separately matched filtering, we can get

$$s_{mn_{rc}}(t, \tau) = A_0 p_{r_m}\left(t - \frac{R_{mn}(\tau)}{c_0}\right) \cdot \exp\left(-j2\pi f_c \frac{R_{mn}(\tau)}{c_0}\right) \qquad (6.73)$$

where $p_{r_m}(t)$ is the corresponding range compression of the range signal $s_m(t)$, which can be approximately seen as a sinc function. As the NTNR operation mode can be seen as multiple independent narrow-beam SARs and they can be easily processed like conventional single-antenna SAR, in the following we discuss only the NTWR operation mode.

Consider the MIMO SAR geometry with three transmitting beams and three receiving channels in the cross-track, as shown in Figure 6.19 (shows only the NTWR operation mode). Suppose the radar platform moves at a constant speed v_s and the closest slant range for the first beam is R_{1_0}. At the azimuth time τ, the range histories for the first receiving beam can then be derived as

$$R_{11}(\tau) = \sqrt{R_{l_0}^2 + (v_s\tau)^2} + \sqrt{R_{l_0}^2 + (v_s\tau)^2} \approx 2R_{1_0} + \frac{v_s^2\tau^2}{R_{1_0}} \qquad (6.74)$$

$$\begin{aligned} R_{21}(\tau) &= \sqrt{R_{l_0}^2 + (v_s\tau)^2} + \sqrt{(R_{l_0} + d\sin(\theta))^2 + (v_s\tau)^2} \\ &\approx R_{l_0} + \frac{v_s^2\tau^2}{2R_{1_0}} + R_{1_0} + d\sin(\theta) + \frac{v_s^2\tau^2}{2(R_{1_0} + d\sin(\theta))} \qquad (6.75) \\ &\approx 2R_{1_0} + d\sin(\theta) + \frac{v_s^2\tau^2}{R_{1_0}} \end{aligned}$$

$$\begin{aligned} R_{31}(\tau) &= \sqrt{R_{l_0}^2 + (v_s\tau)^2} + \sqrt{(R_{l_0} + 2d\sin(\theta))^2 + (v_s\tau)^2} \\ &\approx R_{l_0} + \frac{v_s^2\tau^2}{2R_{1_0}} + R_{1_0} + d\sin(\theta) + \frac{v_s^2\tau^2}{2(R_{1_0} + d\sin(\theta))} \qquad (6.76) \\ &\approx 2R_{1_0} + 2d\sin(\theta) + \frac{v_s^2\tau^2}{R_{1_0}} \end{aligned}$$

where $d\sin(\theta)$ with d being the antenna spacing is the range difference due to the displaced antenna distance in the elevation dimension. The above three range histories have the following relation

$$R_{31}(\tau) - R_{21}(\tau) \approx R_{21}(\tau) - R_{11}(\tau) \approx d\sin(\theta) \qquad (6.77)$$

Therefore, after being registered with the range difference given in Eqs. (6.74)-(6.77), for the first receiving channel we have

$$\begin{aligned} s_{1_{rc}}(t, \tau) = {}&3A_0 p_{r_1}\left(t - \frac{2R_{11}(\tau)}{c_0}\right) \\ &\times \exp\left(-j4\pi f_c \frac{R_{1_0}}{c_0}\right) \cdot \exp\left(-j\pi \frac{2v_s^2}{\lambda R_{1_0}}\tau^2\right) \end{aligned} \qquad (6.78)$$

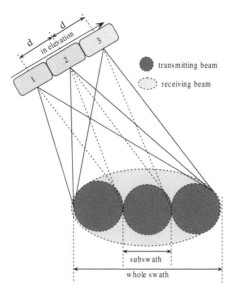

FIGURE 6.19: The geometry of range ambiguity.

where the 3 means that the signal amplitude can be increased by a factor of 3 (equal to the number of the transmitting beams). Fourier transforming with the variable τ yields

$$
\begin{aligned}
s_{1_{rc}}(t, f_a) =& 3A_0 p_{r_1}\left(t - \frac{2R_{1_{rd}}(f_a)}{c_0}\right) \\
& \times \exp\left(-j4\pi f_c \frac{R_{1_0}}{c_0}\right) \cdot \exp\left(j\pi \frac{f_a^2}{k_{a1}}\right)
\end{aligned}
\tag{6.79}
$$

where k_{a1} and $R_{1_{rd}}(f_a)$ denote, respectively, the azimuth chirp rate and range cell migration

$$
k_{a1} \approx \frac{2v_s^2}{\lambda R_{1_0}}
\tag{6.80}
$$

$$
R_{1_{rd}}(f_a) = R_{1_0} + \frac{\lambda R_{1_0} f_a^2}{8v_s^2}
\tag{6.81}
$$

After range cell migration correction, we then have

$$
\begin{aligned}
s_{1_{rc}}(t, f_a) =& 3A_0 p_{r_1}\left(t - \frac{2R_{1_0}(f_a)}{c_0}\right) \\
& \times \exp\left(-j4\pi f_c \frac{R_{1_0}}{c_0}\right) \cdot \exp\left(j\pi \frac{f_a^2}{k_{a1}}\right)
\end{aligned}
\tag{6.82}
$$

Using the azimuth reference function

$$
H_{1_a}(f_a) = \exp\left(-j\pi \frac{f_a^2}{k_{a1}}\right)
\tag{6.83}
$$

azimuth compression processing yields

$$s_{1_{ac}}(t, f_a) = s_{1_{rc}}(t, f_a) \cdot H_{1_a}(f_a)$$
$$= 3A_0 p_{r_1}\left(t - \frac{2R_{1_0}(f_a)}{c_0}\right) \cdot \exp\left(-j4\pi f_c \frac{R_{1_0}}{c_0}\right) \quad (6.84)$$

The focused image for the first subswath can then be obtained by inverse Fourier transform

$$s_{1_{ac}}(t, \tau) = 3A_0 p_{r_1}\left(t - \frac{2R_{1_0}(f_a)}{c_0}\right) p_{a_1}(\tau) \exp\left(-j4\pi f_c \frac{R_{1_0}}{c_0}\right) \quad (6.85)$$

where $p_{a_1}(\tau)$ determines the azimuth resolution, which can be approximately seen as a sinc function.

The images for the other two subswaths can be obtained in the same manner

$$s_{2_{ac}}(t, \tau) = 3A_0 p_{r_2}\left(t - \frac{2R_{2_0}(f_a)}{c_0}\right) p_{a_2}(\tau) \exp\left(-j4\pi f_c \frac{R_{2_0}}{c_0}\right) \quad (6.86)$$

$$s_{3_{ac}}(t, \tau) = 3A_0 p_{r_3}\left(t - \frac{2R_{3_0}(f_a)}{c_0}\right) p_{a_3}(\tau) \exp\left(-j4\pi f_c \frac{R_{3_0}}{c_0}\right) \quad (6.87)$$

Finally, the whole swath MIMO SAR imagery can be obtained by synthesizing the three subswaths

$$s_{ws}(t, \tau) = s_{1_{ac}}(t, \tau) + s_{2_{ac}}(t, \tau) + s_{3_{ac}}(t, \tau) \quad (6.88)$$

The detailed processing steps of this image formation algorithm in case of three transmitting beams and three receiving channels are illustrated in Figure 6.20. This algorithm can be extended to the MIMO SAR with over three transmitting beams or three receiving channels in the same manner.

6.4.5   Numerical Simulation Results

To evaluate the quantitative performance of the MIMO SAR, we consider an example system. Table 6.2 gives the corresponding system parameters. Note that, as a compromise between computation complexity and system performance, two subswathes are designed in the simulations. A far-field, flat-Earth, free-space and single-polarization mode is assumed. It is also assumed that the radar platform moves at a constant speed and operates in stripmap mode. On transmit, activating all elements gives two beams illuminating the whole swath. On receive, the energy returned from each subswath is focused by one channel for the NTNR MIMO SAR operation mode or cooperatively processed by two channels for the NTWR MIMO SAR operation mode.

Note that, without loss of generality, to avoid the problem of "out of memory" in the PC (personal computer)-based MATLAB numerical simulations,

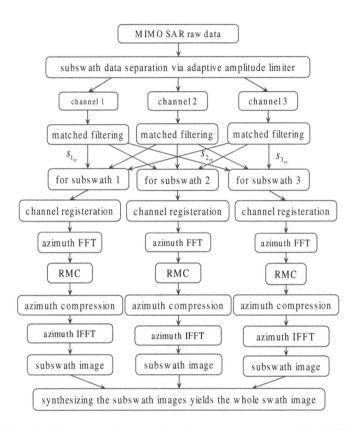

FIGURE 6.20: The range-Doppler-based processing steps of the MIMO SAR imaging algorithm in the case of three transmitting beams and three receiving channels, where RMC denotes the range migration correction.

TABLE 6.2: The system simulation parameters for the NTNR and NTWR MIMO SAR systems.

Parameters	Values	Units
carrier frequency	5.3	GHz
platform velocity	500	m/s
platform altitude	60	km
antenna length in elevation	0.06	m
antenna length in azimuth	8.6	m
earth radius	6370	km
minimum slant range	58.6	km
maximum slant range	175.8	km
PRF	2000	Hz
pulse duration time	2	μs
pulse bandwidth	100	MHz
number of transmitting beams	2	$--$
number of subswaths	2	$--$
number of receiving channels	2	$--$

a relative narrow bandwidth is employed for the transmitted signals. It can be easily derived that there will be range ambiguities; consequently, the whole swath data cannot be successfully focused by conventional single-antenna SAR systems. Suppose there are two point targets locating at (70km, 100m) and (138.594km, 9.8046m) (the two terms in the brackets denote, respectively, the slant range position and azimuth distance). Their range ambiguous signals will be generated at the slant range positions of 128.594 km and 80 km. There will be ambiguous signals for the conventional single-antenna SAR due to range ambiguities. If the azimuth reference function is matched to the range ambiguous signal, they will be ghost images. Figure 6.21 gives the computer simulation results. The whole swath data are processed by the imaging algorithm one by one.

It is also necessary to compare the NTNR and NTWR MIMO SAR RASR performance to conventional single-antenna SAR. Using also the system parameters listed in Table 6.2, Figure 6.22 gives the comparative RASR performances between the two MIMO SARs and conventional single-antenna SAR as a function of slant range. Note that perfect waveform orthogonality is assumed in the simulation. It can be noticed that the averaged RASR for the single-antenna SAR, NRNT MIMO SAR and NTWR MIMO SAR are about -35 dB, -43 dB and -39 dB, respectively. These results clearly show that a significant RASR performance improvement is obtained for the MIMO SAR, especially the NTNR MIMO SAR due to its relative narrow receiving

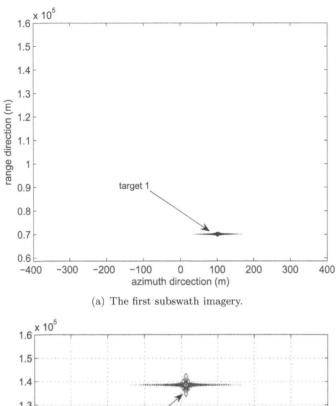

(a) The first subswath imagery.

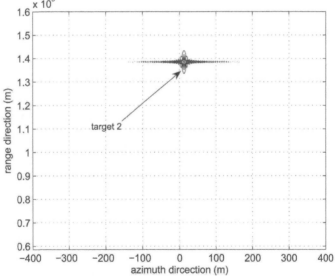

(b) The second subswath imagery.

FIGURE 6.21: The MIMO SAR wide-swath imaging simulation results.

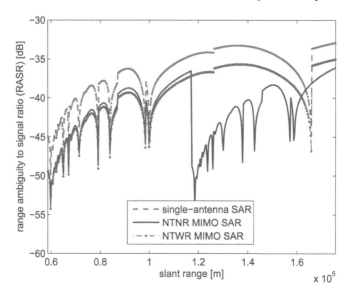

FIGURE 6.22: Comparative RASR performance.

beamwidth. This means that, when compared to conventional single-antenna SAR, a wider swath can be obtained for the MIMO SAR without decreasing the operating PRF, which means that a higher azimuth resolution can be achieved.

6.5   Space-Time Coding MIMO SAR HRWS Imaging

Like space-time coding MIMO wireless communications [313], space-time coding MIMO SAR is also a possible technique [202]. In the following, we introduce a space-time coding MIMO SAR for HRWS imaging [411]. This approach employs MIMO configuration in the elevation dimension and space-time coding in the azimuth dimension, along with waveform diversity and DPCA techniques. The basic idea is to divide the total transmitting antenna elements into multiple groups, with each forming a subbeam, thus offering several benefits, including improved ambiguity suppression for HRWS imaging, improved SNR performance, and flexible operational configuration.

6.5.1   Space-Time Block Coding

Alamouti invented the simplest space-time block coding (STBC) in 1998 [4], although he did not coin the term "space-time block code". Space-time block coding, introduced in [372], generalizes the transmission scheme discovered by Alamouti to an arbitrary number of transmitting antennas and is able to achieve the full diversity promised by the transmitting and receiving antennas. These codes retain the property of having a very simple maximum likelihood decoding algorithm based only on linear processing at the receiver.

The space-time block code proposed by Alamouti [4] uses the complex orthogonal design

$$\begin{bmatrix} s_1 & s_2 \\ -s_2^* & s_1^* \end{bmatrix} \tag{6.89}$$

At a given symbol period, two signals are simultaneously transmitted from the two antennas. The signal transmitted from the first antenna is denoted by s_1 and from the second antenna by s_2. During the next symbol period signal $(-s_2^*)$ is transmitted from the first antenna, and signal s_1^* is transmitted from the second antenna.

Alamouti's scheme is the only STBC that can achieve its full diversity gain without needing to sacrifice its data rate [373]. Strictly, this is only true for complex modulation symbols. The significance of Alamouti's proposal is that it was the first demonstration of a method of encoding which enables full diversity with linear processing at the receiver. Earlier proposals for transmit diversity required processing schemes which scaled exponentially with the number of transmitting antennas. Furthermore, it was the first open-loop transmit diversity technique which had this capability. Subsequent generalizations of Alamouti's concept have led to a tremendous impact on wireless communications.

One particularly attractive feature of the STBC is that maximum likelihood decoding can be achieved at the receiver with only linear processing. Consider the signal received at the n-th antenna

$$x_n(t) = \sum_{m=1}^{nT_s} \alpha_{mn} s_m(t) + N_n(t) \tag{6.90}$$

where $N_n(t)$ is a sample of additive white Gaussian noise. Maximum-likelihood detection amounts to minimizing the decision statistic

$$\sum_{n=1}^{N} \left(|x_n(1) - \alpha_{1n} s_1 - \alpha_{2n} s_2|^2 + |x_n(2) + \alpha_{1n} s_2^* - \alpha_{2n} s_1^*|^2 \right) \tag{6.91}$$

over all possible values of s_1 and s_2. The minimal values are the receiver estimates of s_1 and s_2, respectively. This is equivalent to minimize the decision statistic

$$\left| \left[\sum_{n=1}^{N} (x_n(1)\alpha_{1n}^* + (x_n(2))^* \alpha_{2n}) \right] - s_1 \right|^2 + \left(-1 + \sum_{n=1}^{N} \sum_{m=1}^{2} |\alpha_{mn}|^2 \right) |s_1|^2 \tag{6.92}$$

for decoding s_1 and the decision metric

$$\left|\left[\sum_{n=1}^{N}(x_n(1)\alpha_{2n}^* + (x_n(2))^*\alpha_{1n})\right] - s_2\right|^2 + \left(-1 + \sum_{n=1}^{N}\sum_{m=1}^{2}|\alpha_{mn}|^2\right)|s_2|^2$$

(6.93)

for decoding s_2.

Tarokh et al. [372, 373] designed a set of higher-order STBCs. Two examples STBCs used respectively for three and four transmitting antennas are

$$\begin{bmatrix} s_1 & s_2 & s_3 \\ -s_2 & s_1 & -s_4 \\ -s_3 & s_4 & s_1 \\ -s_4 & -s_3 & s_2 \\ s_1^* & s_2^* & s_3^* \\ -s_2^* & s_1^* & -s_4^* \\ -s_3^* & s_4^* & s_1^* \\ -s_4^* & -s_3^* & s_2^* \end{bmatrix}$$

(6.94)

$$\begin{bmatrix} s_1 & s_2 & s_3 & s_4 \\ -s_2 & s_1 & -s_4 & s_3 \\ -s_3 & s_4 & s_1 & -s_2 \\ -s_4 & -s_3 & s_2 & s_1 \\ s_1^* & s_2^* & s_3^* & s_4^* \\ -s_2^* & s_1^* & -s_4^* & s_3^* \\ -s_3^* & s_4^* & s_1^* & -s_2^* \\ -s_4^* & -s_3^* & s_2^* & s_1^* \end{bmatrix}$$

(6.95)

which have equal power from all antennas in all time-slots. There exist also quasi-orthogonal STBCs that achieve higher data rates at the cost of intersymbol interference. An example quasi-orthogonal STBC is [185]

$$\begin{bmatrix} s_1 & s_2 & s_3 & s_4 \\ -s_2^* & s_1^* & -s_4^* & s_3^* \\ -s_3^* & -s_4^* & s_1^* & s_2^* \\ s_4 & -s_3 & -s_2 & s_1 \end{bmatrix}$$

(6.96)

The orthogonality criterion only holds for columns (1 and 2), (1 and 3), (2 and 4) and (3 and 4).

6.5.2   Space-Time Coding MIMO SAR Scheme

Figure 6.23 shows a typical side-looking space-time coding MIMO SAR with space-time coding transmission in the azimuth dimension and MIMO configuration in the elevation dimension. The waveforms used in any two subarrays are orthogonal, but the antenna elements within one subarray transmit the same waveform. Each antenna receives all the echoes, which are coherently

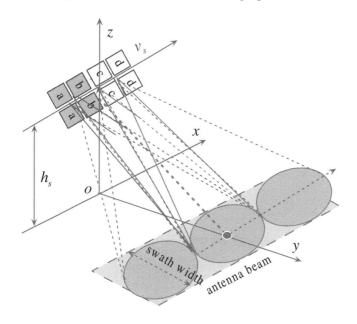

FIGURE 6.23: Side-looking space-time coding MIMO SAR.

combined to form a single SAR image. Moreover, the DPCA in the azimuth dimension is applied to reduce the required operating PRF for wide-swath imaging.

6.5.2.1 Space-Time Coding Transmission in Azimuth

The simplest Alamouti code is suitable for the space-time coding MIMO SAR system with two transmitting antennas [201, 202, 203]. That is to say, two different signals, s_1 and s_2, are transmitted simultaneously from the antennas 1 and 2,respectively, during the first signal period, following which the signals s_2^* and $-s_1^*$ are transmitted from antennas 1 and 2, respectively. We assume that the target's radar cross section (RCS) remains constant over consecutive signal periods. The channel response for the n-th receiving antenna can then be represented by

$$\mathbf{H} = \begin{bmatrix} h_{1,n} & h_{2,n} \end{bmatrix} \tag{6.97}$$

where $h_{1,n}$ and $h_{2,n}$ denote the channel response of the two transmitted signals, respectively.

The signals received by the n-th receiving antenna over consecutive signal periods are

$$\begin{cases} r_{1n} = s_1 \otimes h_{1,n} + s_2 \otimes h_{2,n} + n_{1,n}, \\ r_{2n} = s_2^* \otimes h_{1,n} - s_1^* \otimes h_{2,n} + n_{2,n}, \end{cases} \tag{6.98}$$

where $n_{1,n}$ and $n_{2,n}$ denote individual additive white Gaussian noise. This signal model is formulated in time domain. Transforming them into frequency domain yields

$$\begin{bmatrix} R_{1n} \\ R_{2n} \end{bmatrix} = \begin{bmatrix} S_1 & S_2 \\ S_2^* & -S_1^* \end{bmatrix} \cdot \begin{bmatrix} H_{1,n} \\ H_{2,n} \end{bmatrix} + \begin{bmatrix} N_{1,n} \\ N_{2,n} \end{bmatrix} \tag{6.99}$$

where $R_{1n}(R_{2n})$, $S_1(S_2)$, $S_2^*(S_1^*)$, $H_{1,n}(H_{2,n})$ and $N_{1,n}(N_{2,n})$ denote the Fourier transforming representations of $r_{1n}(r_{2n})$, $s_1(s_2)$, $s_2^*(s_1^*)$, $h_{1,n}(h_{2,n})$ and $n_{1,n}(n_{2,n})$, respectively.

Next, the received radar echoes should be separated by decoding processing. As the transmitted signal matrix is known for both the transmitter and the receiver, the decoding matrix can be easily constructed as

$$\mathbf{D} = \begin{bmatrix} S_1 & S_2^* \\ S_2 & -S_1^* \end{bmatrix} \tag{6.100}$$

Since S_1 and S_2 are orthogonal, the decoded signals can be represented by

$$\begin{aligned} \begin{bmatrix} R_{1n}' \\ R_{2n}' \end{bmatrix} &= \begin{bmatrix} S_1 & S_2^* \\ S_2 & -S_1^* \end{bmatrix} \cdot \begin{bmatrix} R_{1n} \\ R_{2n} \end{bmatrix} \\ &= \begin{bmatrix} (|S_1|^2 + |S_2|^2)H_{1,n} \\ (|S_1|^2 + |S_2|^2)H_{2,n} \end{bmatrix} + \begin{bmatrix} N_{1,n} - N_{2,n} \\ N_{1,n} + N_{2,n} \end{bmatrix} \end{aligned} \tag{6.101}$$

It is noticed from Eq. (6.101) that, after Alamouti decoding, the received two signals have the same transmit information ($|S_1|^2 + |S_2|^2$) and dependent channel response. It can also be concluded that the Alamouti scheme extracts a diversity order of 2 if two orthogonal signals are transmitted with the same power which means $|S_1|^2 = |S_2|^2$. After matched filtering, the reconstructed R_{1n}' and R_{2n}' achieve a gain of $12dB$ while the noise level will increase $6dB$. Thus, the equivalent diversity gain is $6dB$. This gain is attractive for spaceborne SAR systems.

6.5.2.2   MIMO Configuration in Elevation

In the elevation direction, the total antenna array elements are divided into multiple subarrays. These subarrays can be disjoint or overlapped in space, but disjoint subarrays are assumed in this chapter. Orthogonal waveforms are transmitted from different subarrays to illuminate a wide swath (see Figure 6.23). Each receiving subarray can receive all the reflected signals. Suppose the subarrays are pointing in the same direction and the backscattering from the target are in the far field. If M transmitting subarrays and N receive subarrays are employed, there will be $M \times N$ different returns for the receiver. For the m-th ($m = 1, 2, \ldots, M$) transmitting subarray and n-th ($n = 1, 2, \ldots, N$)) receiving subarray, the radar response or scattering function can be approximated as a realization of one random process denoted by $a_{n,m}$ ($m = 1, 2, \ldots, M; n = 1, 2, \ldots, N$). The random process is such that over the time period of the transmit signal duration, the measured amplitude should be the same, i.e., $|a_{n,m}|$ is in fact a constant (for one set of simultaneous measurements employing different codes).

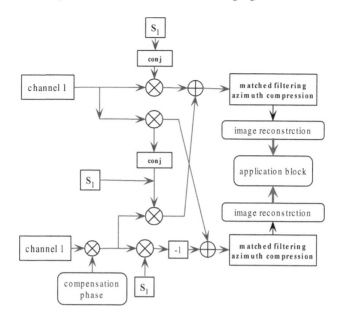

FIGURE 6.24: The STBC MIMO SAR with Alamouti decoder.

The received signal at the n-th receive subarray due to the m-th transmit waveform can be represented by

$$r_{n,m}(t) = s_m(t,n)a_{n,m} + n_{n,m}(t) \tag{6.102}$$

The term $s_m(t,n)$ denotes the signals transmitted by m-th subarray and received by the nth subarray. The term $n_{n,m}(t)$ is an additive noise process independent of the response function $a_{n,m}$. As there are M transmitted signals, the received signals at the n-th receive subarray are the linear combination of all such signals

$$\mathbf{r}_n = \sum_{m=1}^{M} r_{n,m}(t) = \sum_{m=1}^{M} s_m(t,n)a_{n,m} + \sum_{m=1}^{M} n_{n,m}(t) = \mathbf{s}(n)\mathbf{a}_n + \mathbf{n}_n \tag{6.103}$$

The terms $\mathbf{s}(n)$, $\mathbf{a}_n$ and $\mathbf{n}_n$ are the matrixes whose columns are the $s_m(t,n)$, $a_{n,m}$ and $n_{n,m}(t)$, respectively. As there are total of N receiving subarrays, the data from each subarray can be composed into a single vector

$$\mathbf{R} = \left[\mathbf{r}_1{}^T, \mathbf{r}_2{}^T, \ldots, \mathbf{r}_N{}^T\right]^T$$
$$= \mathbf{S} \cdot \mathbf{A} + \mathbf{N} \tag{6.104}$$

where $\mathbf{S} = \text{diag}[\mathbf{s}(1), \mathbf{s}(2), \ldots, \mathbf{s}(N)]$, $\mathbf{A} = [\mathbf{a}_1, \mathbf{a}_2, \ldots, \mathbf{a}_N]^T$, $\mathbf{N} = [\mathbf{n}_1, \mathbf{n}_2, \ldots, \mathbf{n}_N]$ and T denotes the transpose. Equation (6.104) is just the

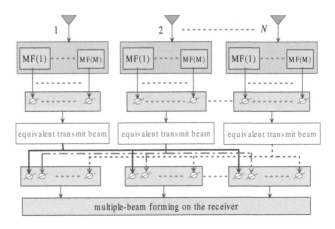

FIGURE 6.25: Separately matched filtering and multibeam forming on receiver in elevation for the STBC MIMO SAR.

signal model used in this chapter to describe the received MIMO SAR data in the elevation direction.

Each of the receiving subarray's signals are mixed, digitized and stored. A posteriori, digital beamforming on receiver can then be carried out by a joint spatiotemporal processing of the collected data [148, 467].

6.5.3 Digital Beamforming in Elevation

In the elevation direction each receiving channel's signals are mixed, digitized, and stored. Next, multibeam forming should be carried out for subsequent azimuth signal processing and image formation processing. The output of the beamforming filter can be represented by [92]

$$\mathbf{w}^T\mathbf{s} + \mathbf{w}^T\mathbf{v} = \sum_{n=1}^{N}\sum_{m=1}^{M} w_{mn}(s_{mn} + v_{mn}) \qquad (6.105)$$

where $\mathbf{w}$ denotes the beamforming filter coefficients, $\mathbf{s}$ and $\mathbf{v}$ are the target and noise vectors, respectively. The terms w_{mn}, s_{mn} and v_{mn} are the filter weighting coefficient, target signal and white noise, respectively, for the output of the matched filter at subarray n and matched to the orthogonal waveform transmitted by the subarray m. Figure 6.25 shows the processing steps of separately matched filtering and multibeam forming on receiver for the STBC MIMO SAR system.

Suppose the transmitting array is a uniform linear array and M equal-sized transmitting subarrays with an equal element size of L are employed. An identical waveform is transmitted from each antenna element within the same subarray, but the transmitted waveforms between any two subarrays are orthogonal. In this way, beamforming is possible at the subarray level, and the equivalent wideband transmission and improved range resolution can be obtained by separately matched filtering and cooperative processing of the multiple echoes. Without loss of generality, the noise and antenna gain are ignored in the following discussion.

The transmitted signals arriving at a specific target can be represented by

$$\mathbf{s}(t) = \sum_{m=1}^{M} \mathbf{s}_m \left(t - \frac{\tau}{2}\right) F_m(\theta_{m,l}, \phi_t) \exp\left[-jM(m-1)\frac{2\pi d \sin(\phi_t)}{\lambda}\right] \quad (6.106)$$

where $F_m(\theta_{m,l}, \phi_t)$ is the m-th transmitting subarray antenna pattern with direction angle $\theta_{m,l}$ (l denotes the l-th antenna element in each subarray) and target angle ϕ_t. Note that the transmitting antenna gain is ignored. Suppose $\phi_t = \theta_{m,l}$ where $l = 1, 2, \ldots, L$, we then have

$$F_m(\theta_{m,l}, \phi_t) = \sum_{l=1}^{L} \exp[j(i-1)2\pi d(\sin(\theta_{m,l}) - \sin(\phi_t))] = L \quad (6.107)$$

In this case, Eq. (6.106) can be transformed into

$$\begin{aligned}
\mathbf{s}(t) &= \sum_{m=1}^{M} \mathbf{s}_m \left(t - \frac{\tau}{2}\right) L \exp\left[-jM(m-1)\frac{2\pi d \sin(\phi_t)}{\lambda}\right] \\
&= L\mathbf{v}_t^T(\phi_t)\mathbf{s}\left(t - \frac{\tau}{2}\right)
\end{aligned} \quad (6.108)$$

where $\mathbf{s}(t) = [\mathbf{s}_1(t), \mathbf{s}_2(t), \ldots, \mathbf{s}_M(t)]^T$ is the transmitted signal vector, $\mathbf{v}_t(\phi_t) = [1, e^{-j2\pi[(d_r \sin \phi_t)/\lambda]}, \ldots, e^{-j2\pi(M-1)[(d_r \sin \phi_t)/\lambda]}]^T$ is the transmitter steering vector. Note that the frequency synchronization and separately frequency demodulation have been assumed in the system.

The received signals can be written as a vector $\mathbf{x}(t) = [\mathbf{x}_1(t), \mathbf{x}_2(t), \ldots, \mathbf{x}_N(t)]^T$

$$\mathbf{x}(t) = L\mathbf{v}_r(\phi_r)\mathbf{v}_t^T(\phi_t)\mathbf{s}(t - \tau) \quad (6.109)$$

where $\mathbf{v}_r(\phi_r) = [1, e^{-j2\pi[(d \sin \phi_r)/\lambda]}, \ldots, e^{-j2\pi(N-1)[(d \sin \phi_r)/\lambda]}]^T$ is the receiver steering vector. Note that the Doppler frequency shift is ignored, i.e., the principle of stop-and-go is assumed. The matched filtering of each receiving channel can be represented by

$$\begin{aligned}
z_{nm}(t) &= x_n(t) \otimes s_m^*(t) = F^{-1}\{F\{x_n(t)\} \cdot F\{s_m^*(t)\}\}, \\
&\quad m \in [1, 2, \ldots, M], \, n \in [1, 2, \ldots, N]
\end{aligned} \quad (6.110)$$

where the F^{-1} and F denote the inverse Fourier transform operation and Fourier transform operation, respectively. Note that each receiver within the N receiving subarray use M different reference functions for its matched filtering. The term $z_{nm}(t)$ is the output of the matched filter at the receiving subarray n and matched to the orthogonal waveform transmitted by the transmit subarray m. The total matched filtering results can then be represented in a vector form as

$$\mathbf{z}_r(t) = LA_a \frac{\sin[\pi B_s(t-\tau)]}{\pi B_s(t-\tau)} \mathbf{v}_r(\phi_r)\mathbf{v}_t^T(\phi_t)$$

$$\times \begin{bmatrix} e^{j2\pi f_1(t-\tau)} & 0 & \cdots & 0 \\ 0 & e^{j2\pi f_2(t-\tau)} & \cdots & 0 \\ \vdots & \vdots & \ddots & \vdots \\ 0 & 0 & 0 & e^{j2\pi f_M(t-\tau)} \end{bmatrix} \quad (6.111)$$

where A_a denotes the amplitude term, f_i $(i = 1, 2, \ldots, M)$ is the starting frequency of each signal.

Suppose the receiving antenna direction is synchronized with the transmitting antenna direction which is easily implemented for colocated MIMO SAR, we then have [253]

$$\mathbf{z}_{in}(t) = \mathbf{z}_r(t) \cdot \mathbf{v}_t(\phi_t)$$
$$= LA_a \frac{\sin[\pi B_s(t-\tau)]}{\pi B_s(t-\tau)} \mathbf{v}_r(\phi_r) e^{j2\pi f_{\min}(t-\tau)} \frac{\sin[MB_s\pi(t-\tau)]}{\sin[B_s\pi(t-\tau)]} \quad (6.112)$$

where $f_{\min}$ is the smallest frequency center among the M orthogonal subchirp waveforms. Then, from the receiving array response vector $\mathbf{v}_r(\phi_r)$ we can get

$$\mathbf{z}_{out}(t) = \mathbf{v}_r^T(\phi_r) \cdot \mathbf{z}_{in}(t)$$
$$= LNA_a \frac{\sin[\pi B_s(t-\tau)]}{\pi B_s(t-\tau)} \cdot \frac{\sin[MB_s\pi(t-\tau)]}{\sin[B_s\pi(t-\tau)]} \cdot e^{j2\pi f_{\min}(t-\tau)} \quad (6.113)$$

Then

$$|\mathbf{z}_{out}(t)| = \left| LNA_a \frac{\sin[\pi B_s(t-\tau)]}{\pi B_s(t-\tau)} \cdot \frac{\sin[MB_s\pi(t-\tau)]}{\sin[B_s\pi(t-\tau)]} \right| \quad (6.114)$$

The equivalent range resolution is

$$\rho_r = \frac{c_0}{2MB_s} \quad (6.115)$$

Note that unoverlapped frequency spectrum is assumed for the transmitted subpulses.

Therefore, if M orthogonal random LFM waveforms without overlapped frequency spectrum (each waveform has a frequency bandwidth of B_s) are employed, the range resolution of the MIMO SAR can be improved by M times of the general SAR using a frequency bandwidth of B_s. In this way, high range resolution imaging can be obtained.

6.5.4 Azimuth Signal Processing

In the azimuth direction, the space-time coding scheme results in equivalent phase centers when the radar moves from one position to another. This enables a coherent combination of the subsampled signals with the DPCA technique, which synthesizes multiple receiving beams that are displaced in the along-track direction [262]. It implies that we can broaden the azimuth beam from the diffraction-limited width, giving rise to an improved resolution without having to increase the operating PRF of the system.

For the signals transmitted at azimuth time τ_1, the equivalent beam two-way phase centers between the equivalent transmitting beam d and the equivalent receiving beam a can be represented by

$$R_{da}(\tau_1) = \sqrt{\left[\frac{3L_a}{2} - R_c \tan(\alpha(\tau_1))\right]^2 + R_c^2} \\ + \sqrt{\left[\frac{3L_a}{2} + R_c \tan(\alpha(\tau_1))\right]^2 + R_c^2} \tag{6.116}$$

where L_a is the distance between subarrays in the azimuth direction. Similarly, for signals transmitted by the equivalent beam b at azimuth time τ_2 and received by the equivalent beam c we have

$$R_{bc}(\tau_2) = \sqrt{\left[\frac{L_a}{2} - R_c \tan(\alpha(\tau_2))\right]^2 + R_c^2} \\ + \sqrt{\left[\frac{L_a}{2} + R_c \tan(\alpha(\tau_2))\right]^2 + R_c^2} \tag{6.117}$$

The corresponding equivalent phase difference between $R_{da}(\tau_1)$ and $R_{bc}(\tau_2)$ is

$$\Delta\Phi(\tau_1, \tau_2) = \frac{4\pi}{\lambda}[R_{da}(\tau_1) - R_{bc}(\tau_2)] \tag{6.118}$$

As the equivalent receiving beam is assumed to be coincident in the far-field region and is of the same width as the equivalent transmitting beam, for two successive pulse repeated interval (PRI=1/PRF) there is

$$\alpha(\tau_1) \approx \alpha(\tau_2) \tag{6.119}$$

As an example, assuming an X-band MIMO SAR system with the following parameters: $\lambda = 0.03m$, $R_c = 700km$ and $L_a = 4m$, Figure 6.26 shows the corresponding equivalent phase difference $\Delta\Phi(\tau_1, \tau_2)$ as a function of the instantaneous squint angle α. We notice that the phase difference is small and may be neglected during subsequent image formation processing. More importantly, in this case the channel responses $H_{1,n}$ and $H_{2,n}$ at azimuth

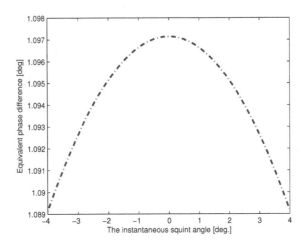

FIGURE 6.26: Equivalent phase difference $\Delta\Phi(\eta_1, \eta_2)$ between two successive PRI transmission as a function of the instantaneous squint angle.

time τ_1 and τ_2 can be seen as being equivalent. Equation (6.101) can then be simplified to

$$\begin{bmatrix} R_{1n}' \\ R_{2n}' \end{bmatrix} = \begin{bmatrix} (|S_1|^2 + |S_2|^2)H_{1,n} \\ (|S_1|^2 + |S_2|^2)H_{1,n} \end{bmatrix} + \begin{bmatrix} N_{1,n} - N_{2,n} \\ N_{1,n} + N_{2,n} \end{bmatrix} \qquad (6.120)$$

Therefore, the signals received by the antenna n can be recombined, so as to further improve its SNR performance.

The range from the equivalent MIMO SAR phase center to an arbitrary point target can then be represented by

$$R(\tau) = \sqrt{R_0^2 + v_s^2(\tau - \tau_0)^2 - 2v_s(\tau - \tau_0)R_0 \cos(\theta)} \qquad (6.121)$$

where R_0 is the slant-range to broadside of antenna center, τ_0 is the time at target broadside, and θ is the angle of the target off broadside. In SAR applications, cubic and higher-order terms may be ignored in the azimuth phase history of targets. In this case, the above equation can be simplified into

$$R(\tau) = R_0 - v_s \cos(\theta)(\tau - \tau_0) + \frac{v_s^2 \sin^2(\theta)(\tau - \tau_0)^2}{2R_0} \qquad (6.122)$$

Therefore, for the case illustrated in the bottom of the Figure 6.26, the Doppler phase histories of the four received subarrays can be represented, respectively, by

$$\psi_{aa}(\tau) = \frac{1}{\lambda} \left[R\left(\tau - \frac{1.5L_a}{v_s}\right) + R\left(\tau - \frac{1.5L_a}{v_s}\right) \right] \tag{6.123a}$$

$$\psi_{ab}(\tau) = \frac{1}{\lambda} \left[R\left(\tau - \frac{1.5L_a}{v_s}\right) + R\left(\tau - \frac{0.5L_a}{v_s}\right) \right] \tag{6.123b}$$

$$\psi_{ac}(\tau) = \frac{1}{\lambda} \left[R\left(\tau - \frac{1.5L_a}{v_s}\right) + R\left(\tau + \frac{0.5L_a}{v_s}\right) \right] \tag{6.123c}$$

$$\psi_{ad}(\tau) = \frac{1}{\lambda} \left[R\left(\tau - \frac{1.5L_a}{v_s}\right) + R\left(\tau + \frac{1.5L_a}{v_s}\right) \right] \tag{6.123d}$$

$$\psi_{ba}(\tau) = \frac{1}{\lambda} \left[R\left(\tau - \frac{0.5L_a}{v_s}\right) + R\left(\tau - \frac{1.5L_a}{v_s}\right) \right] \tag{6.123e}$$

$$\psi_{bb}(\tau) = \frac{1}{\lambda} \left[R\left(\tau - \frac{0.5L_a}{v_s}\right) + R\left(\tau - \frac{0.5L_a}{v_s}\right) \right] \tag{6.123f}$$

$$\psi_{bc}(\tau) = \frac{1}{\lambda} \left[R\left(\tau - \frac{0.5L_a}{v_s}\right) + R\left(\tau + \frac{0.5L_a}{v_s}\right) \right] \tag{6.123g}$$

$$\psi_{bd}(\tau) = \frac{1}{\lambda} \left[R\left(\tau - \frac{0.5L_a}{v_s}\right) + R\left(\tau + \frac{1.5L_a}{v_s}\right) \right] \tag{6.123h}$$

where $\psi_{mn}(\tau)$ denotes the phase of the signal transmitted from the equivalent subarray m and received by the equivalent subarray n. For example, Eq. (6.123c) can be extended into

$$\psi_{ac}(\tau) = \frac{1}{\lambda} \left[2R_0 - v_s \cos(\theta) \left(\tau - \frac{0.5L_a}{v_s} - \tau_0\right) \right.$$
$$\left. + \frac{v_s^2 \sin^2(\theta)(\theta - \frac{0.5L_a}{v_s} - \theta_0)^2}{R_0} \right] + \frac{L_a^2 \sin^2(\theta)}{R_0 \lambda} \tag{6.124}$$

For spaceborne SAR, L_a and R_0 and λ are typically on the orders of 1 m and 700 km, respectively. In this case, the last term $L_a^2 \sin^2(\theta)/R_0\lambda$ is small and can be ignored. We then have

$$\psi_{ac}(\tau) = \frac{2}{\lambda} R\left(\tau - \frac{0.5L_a}{v_s}\right) = \psi_{bb}(\tau) \tag{6.125}$$

In a like manner, we can get

$$\psi_{ad}(\tau) = \psi_{bc}(\tau) = \frac{2}{\lambda} R(\tau) \tag{6.126}$$

and

$$\psi_{aa}(\tau) = \frac{2}{\lambda} R \left(\tau - \frac{1.5 L_a}{v_s} \right) \tag{6.127a}$$

$$\psi_{bd}(\tau) = \frac{2}{\lambda} R \left(\tau + \frac{0.5 L_a}{v_s} \right) \tag{6.127b}$$

$$\psi_{ab}(\tau) = \psi_{ba}(\tau) = \frac{2}{\lambda} R \left(\tau - \frac{L_a}{v_s} \right) \tag{6.127c}$$

To construct one synthetic aperture, the system must operate with a PRF, which leads, after combination of data streams, to a properly sampled synthetic aperture appropriate to the beamwidth of the system. Once the DPCA condition is met, the operating PRF of the system is always equal to the Nyquist sampling rate for the diffraction-limited beamwidth of the antennas. In this way, for a fixed PRF value, the antenna length of the MIMO SAR can remain roughly constant with increasing $N_{ea}/2$ (N_{ea} is the number of equivalent beams in azimuth, and 2 results from the Alamouti coding scheme) the achievable azimuth resolution. Alternatively, for a fixed azimuth resolution, an increase in swath coverage can be obtained. If the DCPA condition is not matched, the gathered azimuth samples will be spaced non-uniformly. This problem can be solved using the reconstruction filter algorithms detailed in [146].

6.6 Conclusion

High-resolution wide-swath imaging is an important application of MIMO SAR. In this chapter, we discussed two basic MIMO operation modes, NTNR and NTWR, for high-resolution wide-swath imaging. The basic idea is to form multiple transmitting and receiving subapertures, which are steered toward different subswaths. They can implement a wider swath without decreasing the operating PRF, which means that a higher azimuth resolution can be obtained. Furthermore, we discussed a space-time coding MIMO SAR for high-resolution wide-swath imaging. This employs MIMO configuration in elevation and space-time coding in azimuth. It is necessary to note that since MIMO SAR antennas can be arranged in the along-track, cross-track, or in both dimensions, different configurations call for different image formation algorithms and thus further research work is required.

7

MIMO SAR in Moving Target Indication

SAR was originally developed to image stationary scenes. SAR imaging of stationary scenes from both spaceborne and airborne platforms has been well understood; however, with the advancement of sophisticated SAR signal processing and imaging method more specialized remote sensing problems are being studied, including the detection, parameter estimation and imaging of the ground moving targets in SAR scenes. When conventional imaging algorithms are applied to SAR data containing moving targets, the targets will be defocused and displaced from their true location in the image. It is necessary to combine both SAR and ground moving target indication (GMTI) capabilities to form images of the terrain and to detect, estimate motion parameters and focus ground moving targets. GMTI using MIMO SAR is interesting because the target indication performance can be improved by MIMO SAR's larger virtual aperture. The minimum detectable speed of a target can be improved by the larger virtual aperture and longer coherent processing interval.

This chapter is organized as follows. Section 7.1 offers a critical overview of SAR ground moving target imaging. Section 7.2 introduces the MIMO SAR with multiple antennas in the azimuth dimension. Section 7.3 introduces an adaptive matched filtering algorithm. Next, Section 7.4 presents the ground moving target imaging via azimuth three-antenna MIMO SAR. After cancelling the clutter and estimating the along-track velocity of the moving targets with the double-interferometry technique and estimating the Doppler parameters of the moving targets, the moving targets are focused with a uniform chirp scaling (CS) image formation algorithm. Furthermore, Section 7.5 presents the ground moving target imaging via azimuth two-channel MIMO SAR, which cooperatively utilizes the displaced phase center antenna (DPCA) and along-track interferometry (ATI) processing techniques. The system scheme, motion parameters estimation and moving targets focusing algorithms are derived.

7.1 Introduction

GMTI is of great important for surveillance and reconnaissance. Despite numerous SAR GMTI algorithms being reported [315, 451, 457, 478, 479], significant challenges in detecting and measuring target motion information exist. The main challenge is to detect moving targets, estimate their motion parameters, and then focus and insert them into the imaged scene [259].

Several single-channel SAR GMTI methods have been proposed in the literature [242, 275, 475]. A basic limitation of single-channel SAR GMTI is that the moving targets' Doppler frequency shift must be greater than the clutter Doppler spectrum width, such that a portion of the target signal falls outside of the clutter background and may be detected. This is usually achieved by using a high pulse repetition frequency (PRF) and a narrow azimuth antenna beamwidth. Even so, it is difficult to detect the moving targets with small across-track velocity components since the target spectra will be superimposed upon the clutter spectra. Moreover, a small azimuth bandwidth places a limitation on the azimuth resolution and a high PRF reduces the unambiguous swath width. To detect slowly moving targets in which the Doppler shift is not separated from the clutter, multiple-channel or multiple-antenna SAR GMTI has received much attention in recent years [157, 158, 319]. Multi-channel SAR provides additional degrees-of-freedom with which clutter may be suppressed.

In SAR imaging, the moving target with a cross-track speed will generate a differential phase shift. This phase shift could be detected by interferometric combination of the signals from a two-channel ATI SAR [39, 55, 166, 329]. However, the complex signals of the moving targets will be contaminated by the clutter that lies near the zero-phase line in the complex plane. If the clutter contribution is not negligible when compared to the signal power, the estimation of target radial velocity from the contaminated interferometric phase may lead to erroneous results. This effect will become more serious for slowly moving targets, and the moving targets will be indistinguishable from the clutter. Moreover, in this case the targets' impulse response is not a normal delta function, particularly for the moving targets that are poorly focused due to unmatched azimuth compression filter. This leads to a point target's response overlaps with several neighboring resolution cells. For this reason, ATI SAR is usually regarded as a clutter-limited GMTI detector [63].

Another representative SAR GMTI is the DPCA clutter cancellation technique, which provides sub-clutter visibility of dim, slowly moving targets [84, 193, 199]. Ideally, stationary clutter can be suppressed to the noise floor, remaining only the signals reflected from moving targets. However, stationary clutter may contribute as phase noise and consequently complicate the process of extracting moving targets. The DPCA is thus regarded as a noise-limited GMTI detector [63]. Moreover, in two-channel DPCA SARs there will be no additional degrees-of-freedom that can be used to estimate the moving targets' position information.

To overcome the aforementioned weakness of the ATI SAR GMTI and DPCA SAR GMTI and utilize their advantages, MIMO SAR-based GMTI approaches are introduced in this chapter. GMTI is an interesting application for MIMO SAR. Target angle-estimation performance is dependent upon the antenna array size, and is therefore improved by MIMO's larger virtual aperture. The minimum detectable velocity of a target is improved by both the larger virtual aperture size and the longer coherent processing interval [129]. Because the MIMO virtual array enables the use of sparse real arrays, a larger effective aperture can be employed. The virtual array has a large, filled aperture with many more virtual degrees of freedom than the physical array. Clutter suppression techniques take advantage of angle-Doppler coupling of the stationary clutter. The larger MIMO virtual array increases the distance, in terms of angular resolution cells, between the clutter and the target within a given Doppler bin, making clutter suppression easier. The minimum detectable velocity of a target depends also on the length of the coherent processing interval. The increase in coherent processing interval length of the MIMO radar reduces the size of a Doppler cell. The smaller Doppler cell both reduces the amount of clutter in a bin and increases the distance, in the terms of Doppler resolution cells, from clutter to the target, making clutter suppression easier.

7.2 MIMO SAR with Multiple Antennas in Azimuth

Figure 7.1 shows the geometry of a MIMO SAR with multiple antennas in the azimuth dimension. The SAR platform is moving along x-axis at a constant speed of v_s and at an altitude of h_s. Suppose the transmitting array has M colocated antennas whereas the receiving array has N colocated antennas. The waveforms used in any two transmitting antennas are orthogonal, so that their echoes can be separated in subsequent signal processing. For each of the M transmitted signals, there are N coherent returns; hence, there will be $M \times N$ different returns, each is weighted by different amplitude and phase. Each receiving antenna signal is separately demodulated, digitized and stored. Next, multiple pairs of equivalent transmitter-to-receiver phase centers can be formed by an jointly spatiotemporal processing. This provides a potential solution to ground moving target imaging.

Suppose there are three along-track antennas. The range history from the central antenna to a given moving target locating at (x_0, y_0) with a velocity of (v_x, v_y) (v_x and v_y denote the velocity elements in x-axis and y-axis respectively) is represented by

$$R_2(\tau) = \sqrt{[x_0 + (v_x - v_s)\tau]^2 + (y_0 + v_y\tau)^2 + h_s^2}$$
$$\approx R_c + v_y\tau + \frac{[x_0 + (v_s - v_x)\tau]^2 + (v_y\tau)^2}{2R_c} \tag{7.1}$$

where τ is the azimuth slow time. The reference range R_c is

$$R_c = \sqrt{y_0^2 + h_s^2} \tag{7.2}$$

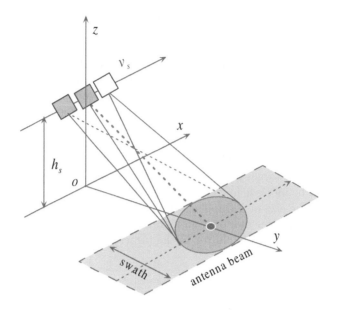

FIGURE 7.1: Geometry of MIMO SAR with multiple antennas in the azimuth dimension.

Similarly, the range histories from the target to the left antenna and right antenna are represented, respectively, by

$$R_1(\tau) \approx R_c + v_y\tau + \frac{[x_0 + d_a + (v_s - v_x)\tau]^2 + (v_y\tau)^2}{2R_c} \tag{7.3}$$

$$R_3(\tau) \approx R_c + v_y\tau + \frac{[x_0 - d_a + (v_s - v_x)\tau]^2 + (v_y\tau)^2}{2R_c} \tag{7.4}$$

where d_a is the separation distance between two neighboring antennas in azimuth. Without loss of generality, amplitude terms are ignored in the following discussions. Suppose $s_m(t)$ is the signal transmitted by the m-th ($m \in (1,2,3)$) transmitting antenna with carrier frequency $f_{c,m}$, the echoes received by the n-th ($n \in (1,2,3)$) receiving antenna can then be expressed as

$$
\begin{aligned}
s_{r_{m,n}}(t,\tau) = & s_m\left(t - \frac{R_m(\tau) + R_n(\tau)}{c_0}\right) \\
& \times \exp\left\{j2\pi f_c\left[t - \frac{R_m(\tau) + R_n(\tau)}{c_0}\right]\right\}
\end{aligned}
\tag{7.5}
$$

where t and f_c denote the range fast time and radar carrier frequency, respectively. As the transmitted waveforms are orthogonal, after separately range pulse compression, we have

$$s_{n,m}(t,\tau) = \exp\left(-j2\pi f_{c,m}\tau_{n,m}\right) \cdot \text{sinc}[\pi B_r(t - \tau_{n,m})] \tag{7.6}$$

where $\tau_{n,m}$ is the signal propagation time from the m-th transmitting antenna to the n-th receiving antenna

$$\tau_{n,m} = \frac{R_m(\tau) + R_n(\tau)}{c_0} \tag{7.7}$$

The term $s_{n,m}(t, \tau)$ denotes the matched filtering output at the n-th receiving antenna and matched to the m-th transmitting antenna waveform. This is just the signal model of the MIMO SAR with multi-antenna in azimuth, which means that there are a total of 3×3 matched filter outputs (or sub-images) resulting from a spatial convolution of the transmitting and receiving arrays.

As the multiple antennas are displaced in azimuth direction, it is effective to define "two-way" phase center as the mid-point between the transmitting and receiving phase centers. This provides a potential clutter suppression, like the DPCA technique. The clutter cancellation can be performed by subtracting the samples of the returns received by two-way phase centers located in the same spatial position, which are temporally displaced. The returns corresponding to stationary objects like the clutter coming from natural scene are cancelled, while the returns backscattered from moving targets have a different phase center in the two acquisitions and remain uncancelled. Therefore, all static clutter are cancelled, leaving only moving targets.

7.3 Adaptive Matched Filtering

The mixed MIMO SAR returns should be extracted separately for subsequent image formation algorithm. The traditional separately matched filtering used in Chapter 6 may be impacted by the cross-correlation interferences due to imperfect waveform orthogonality. For simplicity and without loss of generality, three transmitting beams using three orthogonal waveforms (maybe imperfect orthogonal) are assumed in the following discussions.

Taking the signals arriving at the first receiving channel as an example, we have

$$x_1(t) = \sum_{m=1}^{3} \alpha_{m,1} s_m(t - \tau_{m,1}) \tag{7.8}$$

where $\alpha_{m,1}$ is the channel coefficient for the m-th transmitting beam and the first receiving channel, and $\tau_{m,1}$ is the corresponding signal propagation time. Using the first transmitting beam signal $s_1(t)$ as the reference function, we can get

$$y_1(t) = \left[\sum_{m=1}^{3} \alpha_{m,1} s_m(t - \tau_{m,1})\right] \otimes s_1^*(-t)$$

$$= \alpha_{1,1} s_1(t - \tau_{1,1}) \otimes s_1^*(-t) + \alpha_{2,1} s_2(t - \tau_{2,1}) \otimes s_1^*(-t)$$

$$+ \alpha_{3,1} s_3(t - \tau_{3,1}) \otimes s_1^*(-t)$$

$$(7.9)$$

The first term $\alpha_{1,1} s_1(t - \tau_{1,1}) \otimes s_1^*(-t)$ is the desired matched filtering output, but the remaining terms $\alpha_{2,1} s_2(t - \tau_{2,1}) \otimes s_1^*(-t)$ and $\alpha_{3,1} s_3(t - \tau_{3,1}) \otimes s_1^*(-t)$ are undesired cross-correlation interferences. Since perfect waveform orthogonality cannot be implemented for current radar engineering technology [413], the cross-correlation interferences will significantly degrade the matched filtering performance. To resolve this disadvantage, we introduce the following adaptive matched filtering algorithm.

If we want to extract the return of the first signal $s_1(t)$ from the mixed returns, we firstly matched filtering the mixed returns to the second signal $s_2(t)$

$$z_{1,2}(t) = \left[\sum_{m=1}^{3} \alpha_{m,1} s_m(t - \tau_{m,1})\right] \otimes s_2^*(-t)$$

$$= \alpha_{1,1} s_1(t - \tau_{1,1}) \otimes s_2^*(-t) + \alpha_{2,1} s_2(t - \tau_{2,1}) \otimes s_2^*(-t)$$

$$+ \alpha_{3,1} s_3(t - \tau_{3,1}) \otimes s_2^*(-t)$$

$$(7.10)$$

The second term will be a peak because it is the matched filtering results of the signal $s_2(t)$ and the other two terms are the cross-correlation components. The second term can thus be easily removed by a time-domain filter $h(t)$

$$z_{1,2f}(t) = z_{1,2}(t) \otimes h(t)$$

$$= \alpha_{1,1} s_1(t - \tau_{1,1}) \otimes s_2^*(-t) + \alpha_{3,1} s_3(t - \tau_{3,1}) \otimes s_2^*(-t)$$

$$(7.11)$$

Note that the amplitude changes are ignored. Fourier transforming with the variable t yields

$$Z_{1,2f}(\omega) = \alpha_{1,1} S_1(\omega - \omega_{1,1}) \cdot S_2^*(\omega) + \alpha_{3,1} S_3(\omega - \omega_{3,1}) \otimes S_2^*(\omega) \quad (7.12)$$

where $Z_{1,2f}(\omega)$, $S_1(\omega - \omega_{1,1})$, $S_2(\omega)$ and $S_3(\omega - \omega_{3,1})$ denote respectively the Fourier transform representations of the $z_{1,2f}(t)$, $s_1(t - \tau_{1,1})$, $s_2(t)$ and $s_3(t - \tau_{3,1})$ with $\omega_{1,1}$ and $\omega_{3,1}$ being the frequency shifts associated with the time shift $\tau_{1,1}$ and $\tau_{3,1}$. Inversely filtering the previous equation with $S_2^*(\omega)$ yields

$$Z_{1,2i}(\omega) = \frac{Z_{1,2f}(\omega)}{S_2^*(\omega)}$$

$$\doteq \alpha_{1,1} S_1(\omega - \omega_{1,1}) + \alpha_{3,1} S_3(\omega - \omega_{3,1})$$

$$(7.13)$$

Applying an inverse Fourier transform, we can get

$$z_{1,2i}(t) = \alpha_{1,1}s_1(t - \tau_{1,1}) + \alpha_{3,1}s_3(t - \tau_{3,1}) \tag{7.14}$$

In the same manner, we further matched filtering $z_{1,2i}(t)$ to the third signal $s_3(t)$

$$\begin{aligned} z_{1,2-3}(t) &= z_{1,2i}(t) \otimes s_3^*(-t) \\ &= \alpha_{1,1}s_1(t - \tau_{1,1}) \otimes s_3^*(-t) + \alpha_{3,1}s_3(t - \tau_{3,1}) \otimes s_3^*(-t) \end{aligned} \tag{7.15}$$

Similarly, the second term can be removed by a time-domain filter $h(t)$

$$\begin{aligned} z_{1,2-3f}(t) &= z_{1,2-3}(t) \otimes h(t) \\ &\doteq \alpha_{3,1}s_1(t - \tau_{1,1}) \otimes s_3^*(-t) \end{aligned} \tag{7.16}$$

Fourier transforming with variable t yields

$$Z_{1,2-3f}(\omega) = \alpha_{3,1}S_1(\omega - \omega_{1,1}) \otimes S_3^*(\omega) \tag{7.17}$$

Inversely filtering with $S_3^*(\omega)$ yields

$$Z_{1,2-3i}(\omega) = \frac{Z_{1,2-3f}(\omega)}{S_3^*(\omega)} \doteq \alpha_{1,1}S_1(\omega - \omega_{1,1}) \tag{7.18}$$

The desired return can then be obtained by inverse Fourier transform

$$z_{1,2-3i}(t) = \alpha_{1,1}s_1(t - \tau_{1,1}) \tag{7.19}$$

If over three transmitting beams are employed, the desired return can be extracted in a similar way. Figure 7.2 illustrates the processing steps of the adaptive matched filtering algorithm. Consider a simple example [431]

$$s_1(t) = \text{rect}\left[\frac{t}{T_p}\right] \cdot \exp\left(j2\pi f_c t + j\pi k_r t^2\right) \tag{7.20}$$

$$s_2(t) = \text{rect}\left[\frac{t}{T_p}\right] \cdot \exp\left(j2\pi f_c t - j\pi k_r t^2\right) \tag{7.21}$$

where T_p and k_r denote the pulse duration and chirp rate, respectively. Figure 7.3 shows the comparative processing results between the adaptive matched filtering algorithm and traditional separately matched filtering algorithm. It can be be noticed that the adaptive matched filtering can significantly suppress the sidelobe levels resulting from the cross-interferences due to imperfect waveform orthogonality. The peak-to-sidelobe ratio performance can be improved by a factor of about 10 dB. This is of great interesting for subsequent image formation processing.

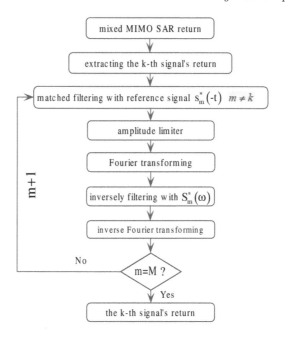

FIGURE 7.2: Illustration of the adaptive matched filtering algorithm.

7.4 Moving Target Indication via Three-Antenna MIMO SAR

Consider the three azimuth-antennas shown in Figure 7.4. For the m-th transmitting antenna, the range pulse compression results at the three receiving antennas are represented, respectively, by

$$s_{1,m}(t,\tau) = \exp\left(-j2\pi f_{c,m}\tau_{1,m}\right) \cdot \text{sinc}[\pi B_r(t - \tau_{1,m})] \qquad (7.22a)$$

$$s_{2,m}(t,\tau) = \exp\left(-j2\pi f_{c,m}\tau_{2,m}\right) \cdot \text{sinc}[\pi B_r(t - \tau_{2,m})] \qquad (7.22b)$$

$$s_{3,m}(t,\tau) = \exp\left(-j2\pi f_{c,m}\tau_{3,m}\right) \cdot \text{sinc}[\pi B_r(t - \tau_{3,m})] \qquad (7.22c)$$

To cancel stationary clutter, we perform the DPCA processing

$$sc_{m,21}(t,\tau) = s_{2,m}(t,\tau) - s_{1,m}(t,\tau + \Delta\tau) \cdot G_0 \qquad (7.23)$$

$$sc_{m,32}(t,\tau) = s_{3,m}(t,\tau - \Delta\tau) \cdot G_0 - s_{2,m}(t,\tau) \qquad (7.24)$$

where $\Delta\tau = d_a/(2v_s)$ is the relative azimuth time delay between two neighboring antennas, and G_0 is used to compensate the corresponding phase shift between two antennas

$$G_0 = \exp\left(-j\frac{2\pi d_a^2}{4R_c\lambda}\right) \qquad (7.25)$$

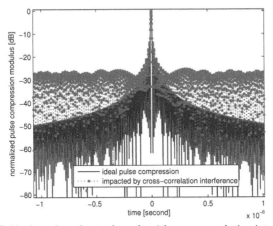

(a) Ideal result and actual result with cross-correlation interference.

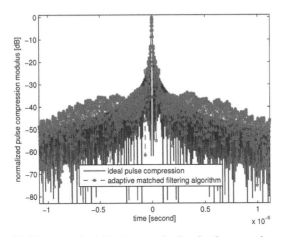

(b) Final matched filtering results for the first waveform.

FIGURE 7.3: Comparative matched filtering results.

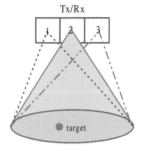

FIGURE 7.4: Scheme of MIMO SAR with three azimuth-antennas for GMTI applications.

Since there is

$$A_0 = \mathrm{sinc}[\pi B_r(t - \tau_{1,m})] \approx \mathrm{sinc}[\pi B_r(t - \tau_{2,m})] \approx \mathrm{sinc}[\pi B_r(t - \tau_{3,m})] \quad (7.26)$$

Eqs. (7.23) and (7.24) can be expanded to

$$
\begin{aligned}
sc_{m,21}(t, \tau) =& A_0 \cdot \exp\left(-j2\pi f_{c,m}\tau_{2,m}\right) \cdot \exp\left(-j\frac{2\pi}{\lambda_m}v_y\Delta\tau\right) \\
& \times \left[1 - \exp\left(-j\frac{2\pi}{\lambda_m}v_y\Delta\tau\right)\right]
\end{aligned}
\quad (7.27)
$$

$$
\begin{aligned}
sc_{m,32}(t, \tau) =& A_0 \cdot \exp\left(-j2\pi f_{c,m}\tau_{2,m}\right) \cdot \exp\left(j\frac{2\pi}{\lambda_m}v_y\Delta\tau\right) \\
& \times \left[1 - \exp\left(-j\frac{2\pi}{\lambda_m}v_y\Delta\tau\right)\right]
\end{aligned}
\quad (7.28)
$$

where λ_m is the corresponding wavelength. It can be noticed that, if $v_y = 0$, there will be $|sc_{m,21}(t, \tau) = 0|$ and $|sc_{m,32}(t, \tau) = 0|$; hence, the clutter has been successfully cancelled by this method. The next problem is to detect the moving targets.

To estimate the moving target's radial velocity v_y, we perform

$$sc_{m,32-21}(t, \tau) = \frac{sc_{m,32}(t, \tau)}{sc_{m,21}(t, \tau)} = \exp\left(j\frac{4\pi}{\lambda_m}v_y\Delta\tau\right) \quad (7.29)$$

We then have

$$sc_{1,32-21}(t, \tau) = \exp\left(j\frac{4\pi}{\lambda_1}v_y\Delta\tau\right) \quad (7.30a)$$

$$sc_{2,32-21}(t, \tau) = \exp\left(j\frac{4\pi}{\lambda_2}v_y\Delta\tau\right) \quad (7.30b)$$

$$sc_{3,32-21}(t, \tau) = \exp\left(j\frac{4\pi}{\lambda_3}v_y\Delta\tau\right) \quad (7.30c)$$

Next, we can get the following interferometry phases

$$\Delta\Phi_1 = \arg\{sc_{2,32-21}(t,\tau) \cdot sc^*_{1,32-21}(t,\tau+\Delta\tau)\}$$
$$= \exp\left[j4\pi v_y\Delta\tau\left(\frac{1}{\lambda_1} - \frac{1}{\lambda_2}\right)\right] \tag{7.31}$$

$$\Delta\Phi_2 = \arg\{sc_{3,32-21}(t,\tau) \cdot sc^*_{2,32-21}(t,\tau+\Delta\tau)\}$$
$$= \exp\left[j4\pi v_y\Delta\tau\left(\frac{1}{\lambda_2} - \frac{1}{\lambda_3}\right)\right] \tag{7.32}$$

$$\Delta\Phi_3 = \arg\{sc_{3,32-21}(t,\tau) \cdot sc^*_{1,32-21}(t,\tau+2\Delta\tau)\}$$
$$= \exp\left[j4\pi v_y\Delta\tau\left(\frac{1}{\lambda_1} - \frac{1}{\lambda_3}\right)\right] \tag{7.33}$$

The radial velocity v_y can then be calculated from Eqs. (7.31) to (7.33). Taking Eq. (7.33) as an example, we can get

$$v_y = \frac{v_s\Delta\Phi_3\lambda_1\lambda_3}{4\pi v_s\Delta\tau(\lambda_1 - \lambda_3)} = \frac{v_s\Delta\Phi_3\lambda_1\lambda_3}{2\pi d(\lambda_1 - \lambda_3)} \tag{7.34}$$

The unambiguous speed estimation range $v_{y_{um}}$ is determined by

$$-\frac{v_s\lambda_3\lambda_1}{2d_a(\lambda_1 - \lambda_3)} \le v_{y_{um}} \le \frac{v_s\lambda_3\lambda_1}{2d_a(\lambda_1 - \lambda_3)}. \tag{7.35}$$

The corresponding estimation variance is

$$\sigma_{v_y} = \frac{\sigma_{\Delta\Phi_3}\lambda_3\lambda_1 v_s}{2\pi d_a(\lambda_1 - \lambda_3)} \tag{7.36}$$

where $\sigma_{\Delta\Phi_3}$ is the phase estimation variance of $\Delta\Phi_3$.

It can be noticed that the unambiguous speed estimation range has been significantly extended when compared to the conventional single-carrier detection method. This is particularly valuable for GMTI applications. However, it is necessary to note that, to avoid possible blind speed problem in target detection, all three interferometry phases should be used to calculate the radial speed. This means that we can get three possible speeds and the true speed is identified with joint processing, so as to avoid loss detection of the moving targets.

The double-interferometry method expressed in Eq. (7.34) has a good unambiguous speed estimate range, which is an advantage for detecting high-speed targets, but its estimate precision may be unsatisfactory for some specific applications. To overcome this disadvantage, we firstly use the speed obtained from the previous double-interferometry method to resolve the ambiguity problem. Next, after range mitigation correction we can perform interferometry processing in range-Doppler domain for the same transmitting antenna [465]

$$\Delta\Phi'_m = \arg\{sc_{m,32}(r,f_d) \cdot sc^*_{m,21}(r,f_d)\} = \frac{2\pi v_{y_e}d_a}{\lambda_m v_s} \tag{7.37}$$

We can then obtain the residual radial speed

$$v_{y_e} = \frac{\lambda_m v_s \Delta \Phi'_m}{2\pi d_a}$$

(7.38)

The corresponding estimation variance is

$$\sigma_{v_{y_e}} = \frac{\sigma_{\Delta\Phi'} \lambda_2 v_s}{2\pi d_a}$$

(7.39)

Therefore, both unambiguous speed estimate range and precision can be improved by this joint processing algorithm.

To estimate the moving targets' Doppler parameters, we transform the range equation expressed in Eq. (7.1) into

$$R_2(\tau) \approx R_c + \frac{x_0^2}{2R_c} + \frac{y_0 v_y + x_0(v_x - v_s)}{R_c}\tau + \frac{v_y^2 + (v_x - v_s)^2}{2R_c}\tau^2$$

(7.40)

and substitute it into Eq. (7.27) which yields

$$s_c(\tau) = A_1 \exp[-j\pi(k_d \tau^2 + 2f_{dc}\tau) + \phi_0]$$

(7.41)

with

$$A_1 = A_0 \exp\left(-j\frac{2\pi}{\lambda_m}v_y\Delta\tau\right) \cdot \left[1 - \exp\left(-j\frac{2\pi}{\lambda_m}v_y\Delta\tau\right)\right]$$

(7.42a)

$$k_d = \frac{2}{\lambda_2 R_c}\left[(v_s - v_x)^2 + v_y^2\right]$$

(7.42b)

$$f_{dc} = \frac{2}{\lambda_2 R_c}\left[y_0 v_y + x_0(v_x - v_s)\right]$$

(7.42c)

$$\phi_0 = -\frac{4\pi}{\lambda}\left(R_c + \frac{x_0^2}{2R_c}\right)$$

(7.42d)

Therefore, once the Doppler parameters k_d and f_{dc} are obtained by estimation algorithms, the parameters v_x, y_0 and R_c can be obtained by Eqs. (7.2) and (7.40) because the h_s in Eq. (7.2) is knowable from the inboard motion measurement or compensation sensors. The close-form solution is

$$\hat{x}_0 = \frac{\left(\lambda_2 \hat{f}_{dc} + 2v_{y_e}\right) R_c}{2v_s}$$

(7.43)

The corresponding estimation variance is

$$\sigma_{x_0} = \frac{\left(\lambda_2 \sigma_{f_{dc}} + 2\sigma_{v_{y_e}}\right) R_c}{2v_s}$$

(7.44)

Now, the parameters (v_x, v_y) and (x_0, y_0) are all determined successfully. Next, the moving targets can then be cooperatively processed with a uniform image formation algorithm, such as range-Doppler (RD) and CS algorithms. Figure 7.5 gives the detailed signal processing steps.

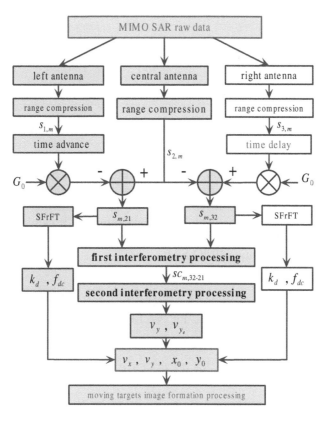

FIGURE 7.5: Azimuth three-antenna MIMO SAR ground moving target imaging with double-interferometry processing algorithm.

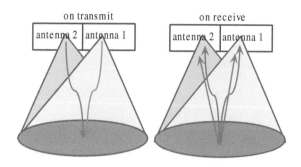

FIGURE 7.6: Illustration of the two-antenna MIMO SAR transmission and reception.

7.5 Moving Target Indication via Two-Antenna MIMO SAR

Moving target indication via two-antenna MIMO SAR is also possible. As shown in Figure 7.6, the two antennas are placed in the azimuth direction. Two orthogonal waveforms are transmitted by the two antennas simultaneously and the returns are also received by the two antennas. It can be easily understood that this configuration has $2 \times 2 = 4$ degrees-of-freedom. GMTI is thus possible for this two-antenna MIMO SAR.

7.5.1 DPCA and ATI Combined GMTI Model

Suppose the subarrays transmit two LFM signals with inverse chirp rate (i.e., $k_{r1} = -k_{r2}$)

$$s_i(t) = \text{rect}\left[\frac{t}{T_p}\right] \exp\left\{j2\pi\left(f_c t + \frac{1}{2}k_{ri}t^2\right)\right\}, \quad \text{i=1, 2} \tag{7.45}$$

Consider the waveform $s_1(t)$ transmitted by the right subarray, the range compressed returns for the two receiving channels can be expressed as

$$\begin{aligned} S_{1,1}(t,\tau) =& \frac{T_p \sin\left[k_{r1}\pi(t\tau - \xi_1)T_p\right]}{k_{r1}\pi(\tau - \xi_1)T_p} \\ &\times \exp\left\{-j2\pi f_c\xi_1 - j\pi k_{r1}\left(\tau - \xi_1\right)^2\right\} \end{aligned} \tag{7.46}$$

$$\begin{aligned} S_{1,2}(t,\tau) =& \frac{T_p \sin\left[k_{r1}\pi(\tau - \xi_2)T_p\right]}{k_{r1}\pi(\tau - \xi_2)T_p} \\ &\times \exp\left\{-j2\pi f_c\xi_2 - j\pi k_{r1}\left(\tau - \xi_2\right)^2\right\} \end{aligned} \tag{7.47}$$

where $\xi_1 = 2R_1(\tau)/c_0$ and $\xi_2 = (R_1(\tau) + R_2(\tau))/c_0$.

Similarly, considering the waveform $s_2(t)$ transmitted by the left subarray, we have

$$S_{2,1}(t, \tau) = \frac{T_p \sin\left[k_{r2}\pi(\tau - \xi_2)T_p\right]}{k_{r2}\pi(\tau - \xi_2)T_p} \times \exp\left\{-j2\pi f_c\xi_2 - j\pi k_{r2}(\tau - \xi_2)^2\right\} \tag{7.48}$$

$$S_{2,2}(t, \tau) = \frac{T_p \sin\left[k_{r2}\pi(\tau - \xi_3)T_p\right]}{k_{r2}\pi(\tau - \xi_3)T_p} \times \exp\left\{-j2\pi f_c\xi_3 - j\pi k_{r2}(\tau - \xi_3)^2\right\} \tag{7.49}$$

with $\xi_3 = 2R_2(\tau)/c_0$.

To cancel the clutter, we perform the DPCA processing on both Eqs. (7.46)-(7.47) and Eqs. (7.48) - (7.49)

$$S_{c1}(t, \tau) = G_1 \cdot S_{1,1}(t, t) - S_{1,2}(t, \tau + \tau_d) \tag{7.50}$$

$$S_{c2}(t, \tau) = S_{2,1}(t, \tau) - G_1 \cdot S_{2,2}(t, \tau + \tau_d) \tag{7.51}$$

where $t_d = d_a/v_a$ is the relative azimuth delay between two apertures, and G_1 is used to compensate the corresponding phase shift

$$G_1 = \exp\left(-j\frac{2\pi d_a^2}{4R_c\lambda}\right) \tag{7.52}$$

Since

$$A_0 = T_p\frac{\sin\left[k_{r1}\pi(\tau - \xi_1)T_p\right]}{k_{r1}\pi(\tau - \xi_1)T_p} \approx T_p\frac{\sin\left[k_{r1}\pi(\tau - \xi_2)T_p\right]}{k_{r1}\pi(\tau - \xi_2)T_p}$$

$$\approx T_p\frac{\sin\left[k_{r2}\pi(\tau - \xi_2)T_p\right]}{k_{r2}\pi(\tau - \xi_2)T_p} \approx T_p\frac{\sin\left[k_{r2}\pi(\tau - \xi_3)T_p\right]}{k_{r2}\pi(\tau - \xi_3)T_p} \tag{7.53}$$

Equation (7.50) can be expanded to

$$S_{c1}(t, \tau) = A_0 \exp\left\{-j\frac{4\pi}{\lambda}\left[R_c + v_yt\right. \right.$$
$$\left. + \frac{x_0^2 + \frac{d_a^2}{4}((v_a - v_x)t)^2 - 2x_0(v_a - v_x)t + (v_yt)^2}{2R_c}\right]\right\}$$
$$\times \left[1 - \exp\left(-j\frac{2\pi}{\lambda}2v_y\tau_d\right)\right] \tag{7.54}$$

It can be noticed that, if $v_y = 0$, there is $|S_{c1}(t, \tau)| = 0$; hence, the clutter has been successfully cancelled at this step. The remaining problem is to detect the moving targets.

Eq. (7.54) can be rewritten as

$$S_{c1}(t, \tau) = A_0' \exp\left[2j\pi\left(f_{dc}\tau + \frac{1}{2}k_d\tau^2\right) + \phi_1\right] \tag{7.55}$$

with

$$A_0' = A_0\left[1 - \exp\left(-j\frac{4\pi}{\lambda}v_y\tau_d\right)\right] \tag{7.56a}$$

$$f_{dc} = -\frac{2}{\lambda}\left[v_y - \frac{(v_a - v_x)x_0}{R_c}\right] \tag{7.56b}$$

$$k_d = -\frac{2}{\lambda R_c}\left[(v_a - v_x)^2 + v_y^2\right] \tag{7.56c}$$

$$\phi_1 = -\frac{4\pi}{\lambda}\left(R_c + \frac{x_0^2 + \frac{d_a^2}{4}}{2R_c}\right) \tag{7.56d}$$

Similarly, from Eq. (7.51) we have

$$S_{c2}(t, \tau) = A_0' \exp\left[2j\pi\left(f_{dc}\tau + \frac{1}{2}k_d\tau^2\right) + \phi_2\right] \tag{7.57}$$

with

$$\phi_2 = -\frac{4\pi}{\lambda}\left(R_c + \frac{x_0^2 - x_0 d_a + \frac{d_a^2}{2}}{2R_c}\right) \tag{7.58}$$

Applying interferometry processing between Eqs. (7.55) and (7.57) yields

$$S_{c1}(t, \tau)S_{c2}^*(t, \tau) = |A_0'|^2 \exp\left(-j\frac{2\pi\left(x_0 d_s - \frac{d_a^2}{4}\right)}{\lambda R_c}\right) \tag{7.59}$$

$$= |A_0'|^2 \exp\left(j\Phi\right)$$

Since $-L_s/2 \leq x_0 \leq L_s/2$ with L_s ($L_s \gg d_a$) being the synthetic aperture length, the interferometric phase Φ can be unambiguously estimated.

$$|\Phi| \ll \pi\left|\frac{\pm L_s d_a - \frac{d_a^2}{2}}{R_c\lambda}\right| \approx \left|\pi\frac{L_s}{R_c\frac{\lambda}{d_a}}\right| \leq \pi \tag{7.60}$$

The azimuth position of the moving target can then be unambiguously determined by

$$x_0 = \frac{\pi d_a^2 - 2R_c\lambda\Phi}{4\pi d_a} \tag{7.61}$$

7.5.2 Estimating the Moving Target's Doppler Parameters

To reconstruct a range compressed moving target signal into a focused target image, the moving target's velocity components v_x and v_y are fundamental parameters of interest. They can be estimated from Eq. (7.55) or Eq. (7.57) by the fractional Fourier transform (FrFT) algorithm.

Applying the FrFT to the Eq. (7.55), we get

$$
\chi_\alpha(\mu) = A'_o \sqrt{\frac{1 - j\cot(\alpha)}{2\pi}} \exp\left[j\frac{\mu^2}{2}\cot(\alpha) + j\phi_1\right]
$$
$$
\times \int_{-T_s/2}^{T_s/2} \exp\left[j2\pi f_{dc}\tau + j\pi k_d\tau^2 + j\frac{\tau^2}{2}\cot(\alpha) - j\mu\tau\csc(\alpha)\right] d\tau
$$

(7.62)

It arrives at its maximum at [406]

$$
\begin{cases} \pi k_d + \frac{\cot(\alpha)}{2} = 0 & \implies k_d = \frac{-\cot(\alpha)}{2\pi}, \\ 2\pi f_{dc} - \mu\csc(\alpha) = 0 & \implies f_{dc} = \frac{\mu\csc(\alpha)}{2\pi}. \end{cases}
$$

(7.63)

Therefore, applying FrFT to the data in which the clutter has been canceled, from the peaks we can get the estimations of $(\hat{\mu}_i, \hat{\alpha}_i)$, where i is the index of the targets. The moving targets' Doppler parameters (k_d, f_{dc}) can then be calculated by Eq. (7.63). Once the Doppler parameters of each target are obtained, substituting them into Eq. (7.56), we can get

$$
v_y = \frac{(v_a - v_x)x_0}{R_c} - \frac{\lambda f_{dc}}{2} \approx \frac{v_a x_0}{R_c} - \frac{\lambda f_{dc}}{2}
$$

(7.64a)

$$
v_x = v_a - \sqrt{-\frac{\lambda k_d R_c}{2} - v_y^2}
$$

(7.64b)

Note that $k_d \leq 0$ is assumed in the above equations.

7.5.3 Focusing the Moving Targets

If the stationary scene reference filter is used to focus the range compressed moving target signal, the phase history will be mismatched. In this case, successfully azimuth correlation processing cannot be obtained. However, a moving target can be focused using a reference filter matched to its motion parameters. Note that if a range compressed signal is focused using a filter matched to the motion of one target, it results in all other signals being mismatched in the processed image. Since the stationary and moving targets have different parameters, they cannot be optimally focused simultaneously.

Using the estimated moving target parameters discussed previously, we can derive a perfectly matched moving target reference filter. Taking the right antenna signal expressed in Eq. (7.48) as an example, the perfectly matched moving reference target filter can be derived as

$$s_{\mathrm{mf}}(t) = \exp\left\{-j\frac{2\pi}{\lambda}\left[\frac{y_0 v_y}{R_c}t + \frac{1}{2R_c}\left[(v_x - v_a)^2 + v_y^2\left(1 - \frac{y_0^2}{R_c^2}\right)\right]t^2\right]\right\}$$

(7.65)

Using a general SAR image formation algorithm and assuming that the signal amplitude is no longer a function of time, the focused moving target image for the right antenna can then be represented by

$$I_1(t, \tau) \doteq \mathrm{sinc}[\pi k_{r1} T_p(\tau - \tau_0)] \cdot \mathrm{sinc}\left[\frac{2\pi}{\lambda}\alpha T_s(t - t_0)\right]$$

(7.66)

with

$$\alpha = \frac{\left[(v_x - v_a)^2 + v_y^2\left(1 - \frac{y_0^2}{R_c^2}\right)\right]}{R_c}$$

(7.67)

where (t_0, τ_0) is the azimuth and range times corresponding to the moving target's position. The moving target has been focused at this step. Figure 7.7 gives a signal processing chart for the two-antenna MIMO SAR GMTI approach. Note that, as the azimuth Doppler bandwidth is determined by $(2\alpha T_s)/\lambda$, variations in the moving target's veocity parameters influences the Doppler characteristics between the target and the radar platform. This may result in a broadening or narrowing of the Doppler bandwidth and consequently impacts the impulse response resolution.

The across-track velocity component can be determined by the ATI phase; however, the estimation performance will be degraded significantly by the clutter. To avoid this disadvantage, we first perform the DPCA clutter suppression processing (i.e., to obtain the $\mathrm{DPCA}_1(t)$ and $\mathrm{DPCA}_2(t)$) and then estimate the v_y and y_0 parameters from them by the FrFT-based algorithm.

7.6   Imaging Simulation Results

Using the parameters listed in Table 7.1 we simulated a near-space vehicle-borne two-antenna stripmap MIMO SAR data reflected from four point targets, two moving targets and two stationary targets. As the response from a moving target is highly dependent upon the target dynamics, the image processed by a stationary scene matched reference filter will be smeared in range and shifted in azimuth due to across-track velocity, and smeared in azimuth due to along-track velocity. Moreover, the presence of clutter will increase the difficulty of detection and subsequent parameter estimation. Figure 7.8 shows the processing results using the chirp scaling-based imaging algorithm. It can be noticed that the imaged two moving targets are overlapped with the two stationary targets and cannot be discernible in the SAR image. As

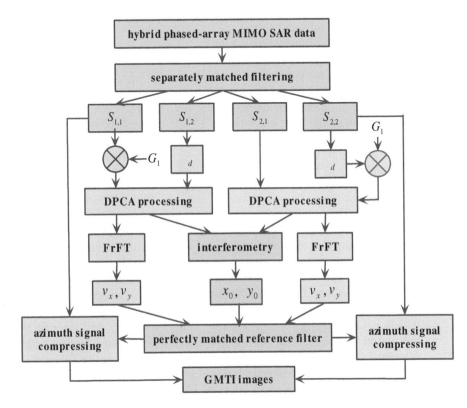

FIGURE 7.7: Signal processing chart for the two-antenna MIMO SAR GMTI algorithm.

TABLE 7.1: Simulation parameters

Parameters	Values and Units
carrier frequency	1.25 GHz
radar platform flying altitude	4000 m
radar platform flying velocity	100 m/s
distance to the central target	4500 m
pulse repeated frequency	150 Hz
pulse duration	5 μ s
signal bandwidth	100 MHz
baseband sampling frequency	250 MHz
azimuth antenna length	1 m
azimuth antenna separation	1 m
position of the moving target 1	($x = 0$ m, $y = 0$ m)
position of the moving target 2	($x = 10$ m, $y = 10$ m)
position of the stationary target 1	($x = 10$ m, $y = 0$ m)
position of the stationary target 2	($x = 0$ m, $y = 10$ m)
velocity of the moving target 1	($v_x = 10$ m/s, $v_y = 5$ m/s)
velocity of the moving target 2	($v_x = 20$ m/s, $v_y = 1$ m/s)

shown in Figure 7.9, after being processed by the proposed clutter suppression algorithm, the two moving targets are discernible, but they are not well focused.

Applying the FrFT algorithm to the clutter suppressed data, from the FrFT peaks shown in Figure 7.10 the Doppler parameters of the moving targets are estimated as ($k_{d1} = 15.0605, f_{dc1} = 0$) and ($k_{d2} = 11.8534, f_{dc2} = 0.4524$). The moving targets' velocity components, v_x and v_y, can then be calculated by the relations developed in Section 7.5 and a perfectly matched reference filter can be constructed. Since the moving targets cannot be optimally focused simultaneously because if a moving target is focused using a reference filter matched to its motion, it results in all other moving targets that have different motions being mismatched. The CLEAN technique [371], which can be regarded as an iterative processing technique, is employed in the simulation. Figure 7.11 shows the focused moving targets' image. Note that perfect waveform orthogonality between the two subarrays is assumed in the simulation. Another note is that the moving targets are located in different resolution cells. This is a rather practical hypothesis.

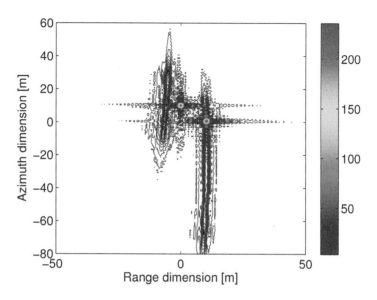

FIGURE 7.8: The focusing results by a stationary scene matched filter.

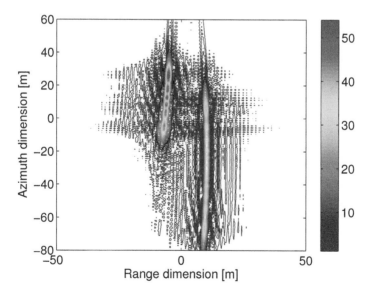

FIGURE 7.9: The focusing results after processed by the clutter suppression algorithm.

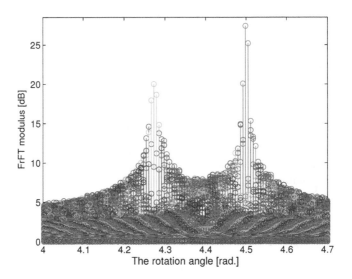

FIGURE 7.10: The peaks in the FrFT domain from which the targets' Doppler parameters can be estimated.

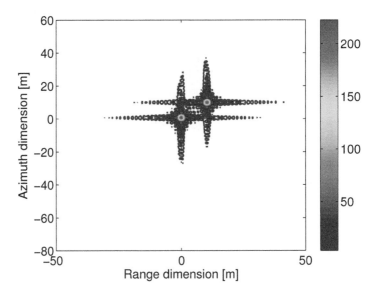

FIGURE 7.11: The final focused moving targets SAR image via the CLEAN technique.

7.7 Conclusion

Moving target indication, especially ground moving target imaging, is another application of MIMO SAR. In this chapter, we introduced the azimuth three-antenna and azimuth two-antenna SAR for ground moving target imaging. The signal models, Doppler parameter estimation and moving target focusing are discussed. To efficiently separate the mixed multichannel MIMO SAR data, we presented an adaptive matched filtering algorithm which can extract the desired returns for subsequent image formation processing. It is necessary to note that, although this chapter considers only azimuth multi-antenna MIMO SAR, elevation or two-dimensional multi-antenna MIMO SAR is also possible for ground moving target imaging.

8

Distributed Multi-Antenna SAR Time and Phase

Synchronization

Distributed multi-antenna SAR, in which the transmitter and receiver are mounted on separate platforms [279], will play a great role in future radar applications [218]. Compared to monostatic systems, such a spatial separation can offer many benefits, such as jamming and antiradiation weapons [416], as well as increased slow-moving target detection and identification capability via "clutter tuning" (in which the receiver maneuvers so that its motion compensates for the transmitter motion, creating a zero Doppler shift for the area being searched) [446]. This could be worthwhile, e.g., for topographic features, surficial deposits, and drainage, to show the relationships that occur between forest, vegetation, and soils. Even for objects that show a low radar cross section (RCS) in monostatic SAR imagery, one can find distinct bistatic angles that increase their RCS and make these objects visible in the final SAR imagery. The strong reflections happen mainly in urban areas due to the dihedral and polyhedral effects can also be reduced by using separate transmitter and receiver. Furthermore, the distributed functionality opens up many new possibilities. For instance, future distributed SARs could use small "receive-only" sensors mounted on airplanes, unmanned aerial vehicles (UAVs), or mountains in conjunction with a transmitting satellite, or both the transmitter and receiver could be mounted on airplanes. Various spaceborne and airborne distributed SAR missions have been suggested, and some are now under development or in planning.

However, distributed SAR is subject to the problems and special requirements that are either not encountered or encountered in less serious forms in monostatic SAR. In a monostatic SAR, the phase only decorrelates over short periods of time, because the colocated transmitter and receiver use the same oscillator. In contrast, in a distributed SAR the receiver uses an oscillator that is spatially displaced from that of the transmitter; hence, the phase noise of the two independent oscillators does not cancel out. This superposed phase noise corrupts the received radar signal over the whole coherent integration time, and may significantly degrade subsequent imaging performance. In the case of indirect frequency synchronization using identical local oscillators (LOs) in the transmitter and receiver, frequency stability is required over the coherent integration time. Even when low-frequency or quadratic phase errors as large as $45°$ in one coherent processing interval can be tolerated, the requirement of frequency stability is only achievable with an ultra high-quality oscillator [434]. Moreover, aggravating circumstances are accompanied for airborne platforms due to different platform motions, further degrading the frequency stability.

Thus, the biggest technical challenge for distributed SAR lies in synchronizing the independent transmitter and receiver. For time synchronization, the receiver must precisely know when the transmitter fires; for phase synchronization, the receiver and transmitter must be coherent over extremely long periods of time. Additionally, spatial synchronization of antenna footprints is also a technical challenge (will be discussed in next chapter). This chapter concentrates on time and phase synchronization for general distributed multi-antenna SAR (especially bistatic SAR), aimed at developing a practical solution without too much alteration to existing radars.

This chapter is organized as follows. Section 8.1 introduces several basic concepts associated with frequency stability which is of great importance for distributed SAR time and phase synchronization. Section 8.2 outlines the motivation of time and phase synchronization for distributed SAR systems. Section 8.3 analyzes the impacts of oscillator frequency instability on distributed SAR imaging. Next, direct-signal-based time and phase synchronization is discussed in Sections 8.4, followed by the global positioning system (GPS)-based time and phase synchronization in Section 8.5. Finally, synchronization link-based and transponder-based solutions are discussed in Sections 8.6 and 8.7, respectively.

8.1   Frequency Stability in Frequency Sources

High-precision frequency sources have undergone tremendous advances during the decades since the advent of the first laboratory cesium beam clock in 1955. Thousands of atomic clocks, such as the cesium beam and the optically pumped rubidium sources manufactured by industry, are routinely used today. The ultrastable hydrogen master is also used on a large scale for very demanding applications. Quality quartz-crystal-controlled oscillators have also shown such progress in stability that they can sometimes compete with rubidium clocks [335].

Ideal sinusoidal oscillators of frequency f_0 have a mathematically strict spectrum in the form of delta functions, centered at $-f_0$ and f_0. However, practical oscillators seldom exhibit this kind of clean spectrum. They tend to have spreading of spectral energy around the carrier frequency, as shown in Figure 8.1. The spreading or spillage of energy to neighboring points around

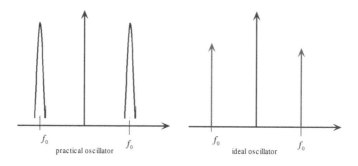

FIGURE 8.1: Ideal sinusoidal oscillator and practical oscillators.

f_0 can cause unwanted behavior at both the transmitter and receiver mixers. The practical problem is how to characterize the properties of the output signal from a real oscillator. Phase noise and jitter are two related quantities associated with a noisy oscillator. Phase noise is a frequency-domain view of the noise spectrum around the oscillator signal, while jitter is a time-domain measure of the timing accuracy of the oscillator period.

8.1.1 Oscillator Output Signal Model

A relatively simple model that was introduced in the early 1960s and has found wide acceptance is

$$V(t) = [V_0 + \varepsilon(t)] \sin [2\pi\nu_0 + \varphi(t)] \tag{8.1}$$

where $\varphi(t)$ is a random process denoting phase noise [19, 334], V_0 and ν_0 are the nominal amplitude and frequency, respectively; and amplitude noise characterized by $\varepsilon(t)$ that can usually be neglected in high-performance frequency sources. In this treatment we assume that frequency shift, if any, has been removed.

$$V(t) = V_0 \sin [2\pi\nu_0 + \varphi(t)] \tag{8.2}$$

Such a quasi-sinusoidal signal has an instantaneous frequency defined as

$$\nu(t) = \frac{1}{2\pi} \cdot \frac{d}{dt} (2\pi\nu_0 t + \varphi(t)) = \nu_0 + \frac{1}{2\pi} \cdot \frac{d\varphi(t)}{dt} \tag{8.3}$$

Frequency noise is the random process defined by

$$\Delta\nu(t) \doteq \nu(t) - \nu_0 = \frac{1}{2\pi} \cdot \frac{d\varphi(t)}{dt} \tag{8.4}$$

which exists simultaneously with and has properties similar to phase noise. Very often it is useful to introduce the normalized dimensionless frequency fluctuations

$$y(t) = \frac{\Delta\nu(t)}{\nu_0} \tag{8.5}$$

This quantity remains unchanged under frequency multiplication or division and can be used as a basis for comparisons of oscillators at different nominal frequencies.

If we model phase and frequency fluctuations by random processes, various statistical tools, such as correlation function, spectral density, average, standard deviation and variance etc., can be employed.

8.1.2   Frequency-Domain Representation

The frequency content of a specific deterministic signal can be found through direct application of the Fourier transform. This frequency-domain analysis is considerably more involved if the signal contains random features. The spectral nature of random signals is normally described in terms of their power spectral density (PSD).

The PSD of a stochastic signal $x(t)$ is specifically defined as

$$S_x\left(f\right) = \lim_{T_m \to \infty} \left\{ \frac{1}{T_m} \mathbf{E}\left[\left|X_{T_m}\left(f\right)\right|^2\right] \right\} \tag{8.6}$$

where the finite-time Fourier transform of $x(t)$ is given by

$$X_{T_m}\left(f\right) = \int_{-T_m/2}^{+T_m/2} x(t)e^{-j2\pi ft}\mathrm{d}t \tag{8.7}$$

Note that ν is the time-dependent instantaneous frequency of the oscillator, and f is the time-independent Fourier frequency that appears in any spectral density.

Although the definition of PSD in Eq. (8.6) involves statistical expectation, the basic operation of a classical spectrum analyzer provides a clear indication that the PSD can be obtained (for most signals encountered in everyday practice) by exclusively using time-averages rather than ensemble averages [140, 274]. This viewpoint is advantageous since the theory of random process based on time-averages is considerably more developed than the theory based on statistical ensemble-averages. Time averages represent the physical behavior of most wireless systems more closely as well. Time-average-based concepts have historically been known as generalized harmonic analysis.

A random process is said to be ergodic if all of the statistics from the ensemble of possible signals can be determined using only a single member of the ensemble, in which case ensemble averaging can be replaced with time-averaging. Ergodicity also requires that the time-domain signal exhibit stationary behavior up through its fourth-order moments.

When $x(t)$ is a wide-sense stationary random process having an autocorrelation function $R_x(\tau)$, the spectrum $S_x\left(f\right)$ can be given by the Wiener-Khintchine theorem as

$$S_x\left(f\right) = \int_{-\infty}^{+\infty} R_x(\tau)e^{-j2\pi f\tau}\mathrm{d}\tau \tag{8.8}$$

where $R_x(\tau)$ and $S_x(f)$ constitute a Fourier transform pair.

According to Eq. (8.7), the Fourier transform of Eq. (8.2) over a time interval T_m can be written as

$$V_{T_m}(f) = \frac{V_0}{j2} \int_{-T_m/2}^{+T_m/2} \{\exp[j2\pi\nu_0 t + j\varphi(t)] - \exp[-j2\pi\nu_0 t - j\varphi(t)]\} e^{-j2\pi ft} dt$$

(8.9)

Performing a transformation of variables, applying the statistical expectation, and finally taking the limit over T_m as directed by Eq. (8.6), the resulting PSD is given by

$$P_V(f) = \frac{1}{2}U(f - \nu_0) + \frac{1}{2}U(f + \nu_0)$$

(8.10)

where

$$U(f) = \int_{-\infty}^{+\infty} u(\tau)e^{-j2\pi f\tau} d\tau$$

(8.11a)

$$u(\tau) = \mathbf{E}\{\exp[j\varphi(t + \tau) - j\varphi(t)]\}$$

(8.11b)

This is simply another restatement of the Wiener-Khintchine relationship, where $u(\tau)$ is the autocorrelation function of the phase noise process in exponential form.

Expanding the complex exponential in Eq. (8.11b) into the first few terms of its Taylor series expansion yields

$$u(\tau) \approx \mathbf{E}\left\{1 + j[\varphi(t + \tau) - \varphi(t)] - \frac{[\varphi(t + \tau) - \varphi(t)]^2}{2}\right\}$$

$$\approx 1 - R_\varphi(0) + R_\varphi(\tau)$$

(8.12)

where $R_\varphi(\tau)$ is the autocorrelation function of the phase noise process $\varphi(t)$. Note that the imaginary portion has been dropped out because the process is mean-zero. Substituting this result back into Eq. (8.11a), we can get

$$P_V(f) = \frac{[1 - R_\varphi(0)]}{2}\delta(f - \nu_0) + \frac{1}{2}P_\varphi(f - \nu_0) + [\text{negative frequency terms}]$$

(8.13)

where $R_\varphi(\tau)$ and $P_\varphi(f)$ constitute a Fourier transform pair. The first term is a carrier frequency term that has a slightly reduced amplitude due to the phase noise presence. The second term represents the continuous phase noise spectrum. If the small-angle assumption $\frac{\varphi(t)}{\nu_0} \ll 1$ is sufficiently accurate, the PSD of $V(t)$ is a scaled-equivalent of the PSD of the underlying phase noise process $\varphi(t)$.

Therefore, phase and frequency fluctuations can be characterized by the

TABLE 8.1: $S_y(f)$ region shape and associated frequency dependence.

Type of noise	α	$S_y(f)$
white phase modulation noise	2	$\propto f^2$
flicker phase modulation noise	1	$\propto f$
white frequency modulation noise	0	$\propto f^0$
flicker frequency modulation noise	-1	$\propto f^{-1}$
random-walk frequency modulation noise	-2	$\propto f^{-2}$

respective one-sided spectral densities, $S_\varphi(f)$ and $S_{\Delta\nu}(f)$, which are related by

$$S_{\Delta\nu}(f) = f^2 S_\varphi(f) \tag{8.14}$$

which corresponds to the time derivative between $\varphi(t)$ and $\Delta\nu(t)$. The spectral density $S_y(f)$ is also widely used and is related to $S_\varphi(f)$ by

$$S_y(f) = \frac{1}{\nu_0^2} S_{\Delta\nu}(f) = \frac{f^2}{\nu_0^2} S_\varphi(f) \tag{8.15}$$

It has been shown from both theoretical considerations and experimental measurements, that the spectral densities due to random noise of all high stability frequency sources can be represented as a straight-line template in log-frequency, dB/Hz format. More specifically, $S_y(f)$ can be written as

$$S_y(f) = \sum_{\alpha=-2}^{+2} h_\alpha f^\alpha \tag{8.16}$$

for $0 \le f \le f_h$ with f_h being an upper cutoff frequency. For a given type of oscillator two or three terms of the sum are usually dominant. Each term is related to a given noise source in the oscillator. The most common power-law regions that are encountered in practical sources are given in Table 8.1.

Spectral densities of phase or frequency can be measured by a spectrum analyzer following some kind of demodulation of $\varphi(t)$ or $\Delta\nu(t)$. Numerous experimental tests have been developed for that purpose [226, 389, 387]. Figure 8.2 gives an example phase noise density measured by a spectrum analyzer. Specific techniques for measuring $S\varphi(f)$ (or equivalently $S_y(f)$) are described in [20, 390, 421]. Particular attention is focused on describing the errors in such measurements and in determining $S\varphi(f)$ or $S_y(f)$ for a single oscillator.

The use of phase noise densities is a frequent point of confusion in the literature. $S\varphi(f)$ is the one-sided spectral density of phase fluctuations. The range of frequencies f span from 0 to ∞, and the dimensions are rad^2/Hz. It is measured by passing the signal through a phase detector and measuring the PSD at the detector output. In contrast, $L(f)$ is the single-sideband phase noise level which has units of dBc/Hz. It is the ratio of the power spectral

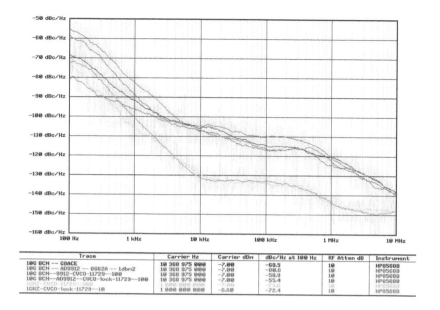

Trace	Carrier Hz	Carrier dBm	dBc/Hz at 100 Hz	RF Atten dB	Instrument
10G BCN -- GBACE	10 368 975 000	-7.00	-68.5	10	HP8568B
10G BCN --- AD9912 -- 8662A -- 1dbn2	10 368 975 000	-7.00	-88.6	10	HP8568B
10G BCN--9912-CVCO-11729--100	10 368 975 000	-7.00	-59.9	10	HP8568B
10G BCN--AD9912--CVCO-lock-11729--100	10 368 975 000	-7.00	-55.4	10	HP8568B
1GHZ-CVCO-11729--100	1 000 000 000			10	HP8568B
1GHZ-CVCO-lock-11729--10	1 000 000 000	-6.60	-72.4	10	HP8568B

FIGURE 8.2: Example phase noise density.

density in one phase modulation sideband, referred to the carrier frequency on a spectral density basis, to the total signal power, at a frequency offset f (corresponds to a frequency offset from the carrier in Hertz). $L(f)$ is the phase noise spectrum that would be measured and displayed on a spectrum analyzer. Having recognized the carrier and continuous spectrum portions within Eq.(8.13), it is possible to equate

$$L(f) \approx P_\varphi(f) \qquad (8.17)$$

$L(f)$ is therefore a two-sided spectral density and is also called single-sideband phase noise, being defined for positive as well as negative frequencies. Normally, the approximation

$$L(f) \approx \frac{1}{2}S\varphi(f), \quad \mathrm{rad}^2/\mathrm{Hz} \qquad (8.18)$$

is made in industry, but it is only valid when the lower frequency bound f_1 satisfy

$$\int_{f_1}^{\infty} S\varphi(f) \ll 1, \quad \mathrm{rad}^2 \qquad (8.19)$$

One troublesome issue is the use of one-sided versus two-sided measures. Two-sided spectral densities are defined such that the frequency range of integration is from $-\infty$ to $+\infty$. Normally, the one-sided PSDs are taken to be simply twice as large as the corresponding two-sided spectral densities. Two-sided spectral densities are primarily useful in mathematical analyses involving Fourier transforms. It should be emphasized that the terminology for single-sideband signals versus double-sideband is totally distinct from one-sided and two-sided spectral density terminology [171].

In previous discussions, the random process $\varphi(t)$ is assumed to be stationary to at least second-order and as much as fourth-order. In the rigourous sense, the theoretical analysis of internal white noise in oscillators leads to a stationary phase diffusion process analogous to Brownian noise. In this case, it is impossible to use the autocorrelation/PSD theory.

8.1.3   Time-Domain Representation

Time-domain characterization of frequency stability is widely used since it answers the obvious question: what is the stability over a time interval for a given application? In order to assess frequency stability over a time interval τ, it is necessary to make a series of measurements, y_i with $i = 1, 2, \ldots, N$. Several kinds of variances have been introduced.

8.1.3.1   True Variance

The true variance is a theoretical parameter denoted as $I^2(\tau)$ and defined as $I^2(\tau) = \langle \bar{y}_k^2 \rangle$. When $y(t)$ has a zero mean, the bracket $\langle \rangle$ denotes an infinite time average made over one sample of $y(t)$. Despite its mathematical simplicity, the true variance is not really useful for characterizing frequency instability since it approaches infinity for all real oscillators. Practical estimators of the time-domain stability relying on the sample variance concept were introduced in [5] to avoid the divergence of the true variance observed in most sources.

8.1.3.2   Sample Variance

The sample variance is a more practical estimate of time-domain stability based upon a finite number of N samples $\bar{y}_k$ rather than the true variance. Each sample has a duration τ, the k-th sample begins at t_k; the $(k + 1)$-th sample begins at $t_{k+1} = t_k + T$; the dead time between two successive samples is then $T - \tau$. The quantity T is the repetition interval for individual measurements of duration τ. The sample variance is defined as

$$\sigma_y^2(N, T, \tau) = \frac{1}{N-1} \sum_{i=1}^{N} \left(\bar{y}_i - \frac{1}{N} \sum_{j=1}^{N} \bar{y}_j \right)^2 \qquad (8.20)$$

where N is the sample size.

8.1.3.3 Allan Variance

The Allan variance is based on the average of the variance with $N = 2$ and adjacent samples (i.e, $T = \tau$). The Allan variance is defined as [5]

$$\sigma_y^2(\tau) = \frac{1}{2} \left\langle (\bar{y}_2 - \bar{y}_1)^2 \right\rangle \tag{8.21}$$

It is a theoretical measure since infinite duration is implied in the average; however, it has a much greater practical utility than the true variance. When m values of $\bar{y}_k$ are used to estimate $\sigma_y^2(\tau)$, a widely used estimator is

$$\sigma_y^2(\tau, m) = \frac{1}{2(m-1)} \sum_{i=1}^{m-1} (\bar{y}_{i+1} - \bar{y}_i)^2 \tag{8.22}$$

m should be stated with the results to avoid ambiguity and to allow for meaningful comparisons. The Allan variance is a very efficient estimator for noise types $\alpha = 0, -1, -2$ but diverges for $\alpha \leq -3$.

The Allan variance can be calculated from the power density $S_y(f)$ [80]

$$\sigma_y^2(\tau) = 2 \int_0^{+\infty} S_y(f) \frac{\sin^4 (\pi f \tau)}{(\pi f \tau)^2} \, df \tag{8.23}$$

It is necessary to specify the value and shape of the low-pass filter for noise types with $\alpha > 0$. The most common shapes are the infinitely sharp low-pass filter with cut-off frequency f_h, and a single pole filter of equivalent noise bandwidth f_h.

8.1.3.4 Modified Allan Variance

Since the Allan variance is not useful for distinguishing white and flicker phase noise, the modified Allan variance was introduced in [6].

$$\sigma_y^2(n\tau_0) = \frac{1}{2n^2\tau_0^2} \left\langle \left[\frac{1}{n} \sum_{i=1}^{n} (x_{i+2n} - 2x_{i+n} + x_i) \right]^2 \right\rangle \tag{8.24}$$

where x_i are the time variations measured at intervals τ_0 and t_i. The modified Allan variance behaves very similarly to the Allan variance for $\alpha = 0, -1, -3$, and can easily separates the noise types $\alpha = 1$ and $\alpha = 2$.

The modified Allan variance can be calculated from the power density $S_y(f)$ [233, 388]

$$\sigma_y^2(n\tau_0) = 2 \int_0^{+\infty} S_y(f) \frac{\sin^6 (\pi f n \tau_0)}{(\pi f n^2 \tau_0)^2 \sin^2 (\pi f \tau_0)} \, df \tag{8.25}$$

It is also necessary to specify the value and shape of the low-pass filter for noise types $\alpha > 0$.

8.2 Time and Phase Synchronization Problem in Distributed SAR Systems

Since the reference for both the local oscillator and pulse repetition frequency (PRF) timing is provided by the master oscillator, any frequency instability on the master oscillator output is magnified by the frequency multiplication. In a monostatic SAR, the same oscillator is used for both the transmitter and the receiver. Thus, the master oscillator phase noise, which degrades subsequent SAR imaging performance, is the result of a first-difference operation on the local oscillator with a time offset equal to the round-trip time delay. This provides significant cancellation of the low-frequency components of phase noise in the master oscillator. However, for a distributed SAR system, the separation of the transmitter and receiver results in additional phase noise between the two systems instead of cancellation. Consequently, the requirement of frequency stability on the local oscillators for distributed SAR systems is much more stringent than in the monostatic case. For these reasons, oscillator frequency instability deserves special attention in distributed SAR systems, so it is also the subject of a rapidly growing number of publications [106, 220, 423, 474].

For simplicity and without loss of generality, only distributed bistatic SAR systems are considered in the following discussions. To complete coherent accumulation on azimuth, the radar echoes of the same range but different azimuth should have the same phase after range compression and range migration correction. Since only oscillator phase noise is of interest, the modulation waveform used for range resolution can be ignored and the bistatic SAR signal model can be simplified into an "azimuth only" system [14]. Figure 8.3 shows the bistatic SAR signal model, where f_T/f_R denotes the transmitter/receiver carrier frequency with φ_T/φ_R being the corresponding phase noise.

Suppose the transmitted signal is a sinusoid with phase argument

$$\Phi_T(t) = 2\pi f_T t + \varphi_{T_0} \tag{8.26}$$

The first term is the carrier frequency, and the second term is an initial phase. The f_T can be represented by

$$f_T = f_0 \left[1 + \delta_T(t)\right] \tag{8.27}$$

where f_0 is the error-free carrier frequency. After reflection from a scatterer, the phase of the received signal is that of the transmitted signal delayed by the round-trip time t_d.

Similarly, for the receiver we have

$$\Phi_R(t) = 2\pi f_R t + \varphi_{R_0} \tag{8.28}$$

$$f_R = f_0 \left[1 + \delta_R(t)\right] \tag{8.29}$$

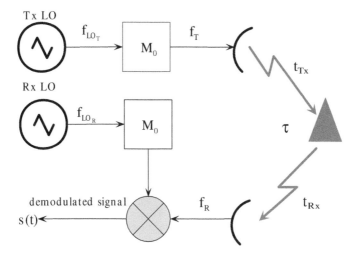

FIGURE 8.3: Distributed bistatic SAR signal model.

Suppose the transmitting time is t_0, we can obtain the results by demodulating the received signal with the receiver oscillator

$$\Psi_{t_0} = \int_0^{t_0} 2\pi(f_R - f_T)dt + \int_{t_0}^{t_0+t_d} 2\pi f_R dt + \varphi_{R_0} - \varphi_{T_0} \tag{8.30}$$

$$=\varphi_{t_0} + \varphi_{d_0} + \varphi_0$$

where

$$\varphi_{t_0} = \int_0^{t_0} 2\pi(f_R - f_T)dt \tag{8.31a}$$

$$\varphi_{d_0} = \int_{t_0}^{t_0+t_d} 2\pi f_R dt \tag{8.31b}$$

$$\varphi_0 = \varphi_{R_0} - \varphi_{T_0} \tag{8.31c}$$

Analogously, assuming the transmitting time is t_1, we then have

$$\Psi_{t_1} = \int_0^{t_1} 2\pi(f_R - f_T)dt + \int_{t_1}^{t_1+t_d} 2\pi f_R dt + \varphi_{R_0} - \varphi_{T_0} \tag{8.32}$$

$$=\varphi_{t_1} + \varphi_{d_1} + \varphi_0$$

where

$$\varphi_{t_1} = \int_0^{t_1} 2\pi(f_R - f_T)\mathrm{d}t \tag{8.33a}$$

$$\varphi_{d_1} = \int_{t_1}^{t_1+t_d} 2\pi f_R \mathrm{d}t \tag{8.33b}$$

From Eqs.(8.30) and (8.32), we can express the phase synchronization errors as

$$\varphi_e = \int_{t_0}^{t_1} 2\pi(f_R - f_T)\mathrm{d}t + \varphi_{d_1} - \varphi_{d_0} \tag{8.34}$$

Since

$$|\varphi_{d_1} - \varphi_{d_0}| = \left| \int_{t_1}^{t_1+t_d} 2\pi f_R \mathrm{d}t - \int_{t_0}^{t_0+t_d} 2\pi f_R \mathrm{d}t \right|$$

$$\leq 4\pi \cdot \Delta f_{R_{\max}} \cdot t_d \leq \frac{\pi}{4} \tag{8.35}$$

with $\Delta f_{R_{\max}}$ being the maximum frequency synchronization error, Eq. (8.34) can be approximated as

$$\varphi_e = \int_0^{t_1-t_0} 2\pi(f_R - f_T)\mathrm{d}t \tag{8.36}$$

SAR imaging requires a frequency coherence for at least one aperture time T_s, i.e., $t_1 - t_0 \geq T_s$.

8.3   Impacts of Oscillator Frequency Instability

The impacts of limited oscillator stability in distributed SAR systems are discussed in [212], which employed a second-order stationary stochastic process and pointed out that uncompensated phase noise may cause a time variant shift, spurious sidelobes and a deterioration of the impulse response, as well as a low frequency phase modulation of the focused SAR imagery. In [173], a white phase noise model is discussed, but it cannot describe the statistical process of phase noise. In [130], a Wiener phase noise model is discussed, but it cannot describe the low-frequency phase noise components, since this part of the phase is an unstationary process. As different phase noise will bring

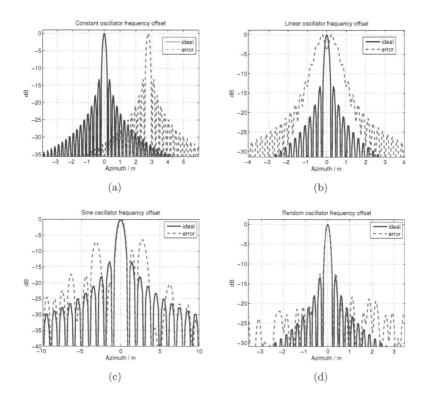

FIGURE 8.4: Impacts of various oscillator frequency offsets: (a) constant off-set, (b) linear offset, (c) sine offset and (d) random offset.

different effects on bistatic SAR (see Figure 8.4), the practical problem is how to develop an useful and comprehensive model of frequency instability that can be understood and applied in bistatic SAR processing.

8.3.1 Analytical Model of Phase Noise

Suppose the oscillator frequency centroid is f_0, the phase noise spectrum can be analytically represented by [333]

$$L_\varphi(f) = f_0^2 h_{-2} f^{-4} + f_0^2 h_{-1} f^{-3} + f_0^2 h_0 f^{-2} + f_0^2 h_1 f^{-1} + f_0^2 h_2 \qquad (8.37)$$

It can be simplified to

$$L_\varphi(f) = a f^{-4} + b f^{-3} + c f^{-2} + d f^{-1} + e \qquad (8.38)$$

A pity is that it is a frequency domain expression and is not convenient in analyzing its impacts on bistatic SAR imaging. To overcome this disadvantage, we proposed a statistical model to generate time-domain phase noise.

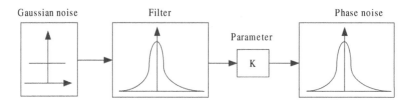

FIGURE 8.5: Statistical time-domain phase noise generator.

As shown in Figure 8.5, the statistical model uses Gaussian noise as the input of a hypothetical low-pass filter. This model may represent the output signal of a hypothetical filter with impulse response $h(t)$ receiving an input signal $x(t)$. Since the transforms in Figure 8.5 are all linear transforms, the obtained phase noise is a kind of Gaussian stationary stochastic process.

It is well known that the output signal PSD is given by the product $L_x(f)|H(f)|^2$ with $L_x(f)$ being the PSD of Gaussian white noise and also the input of filter, where the filter transfer function $H(f)$ is the Fourier transform of $h(t)$. Note that, here, $|H(f)|^2$, must be satisfactory with

$$|H(f)|^2 = \begin{cases} L_\varphi(f), & f_l \leq |f| \leq f_h \\ L_\varphi(f_l), & |f| \leq f_l \\ 0, & else \end{cases} \qquad (8.39)$$

where a sharp up cutoff frequency f_h and a sharp down cutoff frequency f_l are introduced. Note that time domain stability measures sometimes depend on f_h and f_l which must be given with any numerical result, although no recommendation has been made for this value. Here, $f_h = 3kHz$ and $f_l = 0.01Hz$ are adopted. Thereby, the PSD of phase noise can be analytically expressed as

$$L_\varphi(f) = KL_x(f)|H(f)|^2 \qquad (8.40)$$

An inverse Fourier transform yields

$$\varphi(t) = \sqrt{K}x(t) \otimes h(t) \qquad (8.41)$$

where $\varphi(t)$ is just the phase noise in time domain.

Taking one typical 10 MHz local oscillator (see Table 8.2) as an example, from Eq. (8.38) we can obtain that $a = -115$ dB, $b = -135$ dB, $c = -194$ dB, $d = -214$ dB and $e = -375$ dB. Figure 8.6 shows the corresponding phase noise PSD. This oscillator can be regarded as a representative example of ultra stable oscillator used in current airborne SAR systems. The simulated time-domain phase noises in the local oscillator are shown in Figure 8.7 for a time interval of 50 seconds. Note that, for better illustrations for different transmitter and receiver oscillator frequencies, a linear phase ramp has been subtracted in Figure 8.8. From the reconstructed phase noise PSD shown in

TABLE 8.2: Phase noise parameters of a typical local oscillator.

Frequency (Hz)	1	10	100	1k	10k
$L_\varphi(f)(dBc/Hz)$	-125	-135	-140	-150	-160

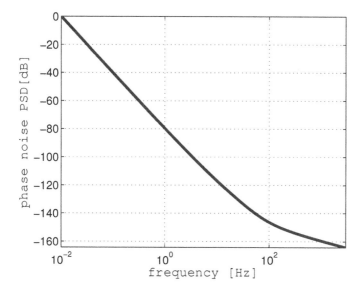

FIGURE 8.6: Phase noise PSD of one typical 10 MHz local oscillator.

Figure 8.9, we can make a conclusion that the phase noise model is a reasonable statistical model. In a typical airborne SAR system, one synthetic aperture time can be designed as long as 15 seconds. The residual phase noise of one typical 10 MHz oscillator in this time interval with ten realizations of the stochastic process are shown in Figure 8.10.

8.3.2   Impact of Phase Synchronization Errors

Suppose the transmitter oscillator frequency is f_{LO_T} with phase noise $\varphi_T(t)$, Eq. (8.26) can be rewritten as

$$\Phi_T(t) = 2\pi f_{LO_T} t + M\varphi_T(t) \qquad (8.42)$$

with M being the ratio of the radar carrier frequency to the local oscillator frequency. Similarly, for the receiver oscillator, we have

$$\Phi_R(t) = 2\pi f_{LO_R} t + M\varphi_R(t) \qquad (8.43)$$

The receiver output signal phase $\Delta\Phi(t)$ is

$$\Delta\Phi(t) = 2\pi \left(f_{LO_R} - f_{LO_T} \right) + 2\pi f_{LO_T} \tau + M \left[\varphi_R(t) - \varphi_T(t - \tau) \right] \qquad (8.44)$$

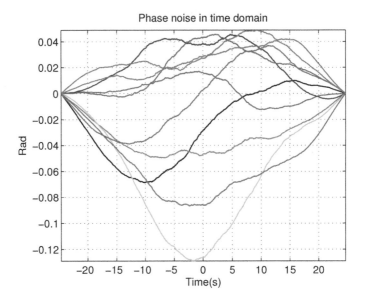

FIGURE 8.7: Phase noise in 50 seconds.

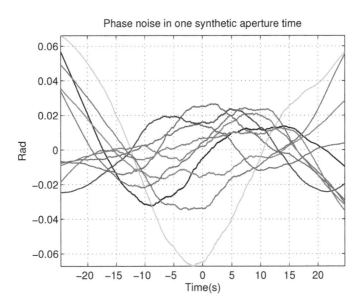

FIGURE 8.8: Phase noise in 50 seconds after subtraction of a linear phase ramp.

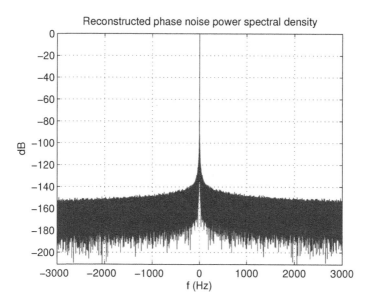

FIGURE 8.9: Reconstructed phase noise PSD.

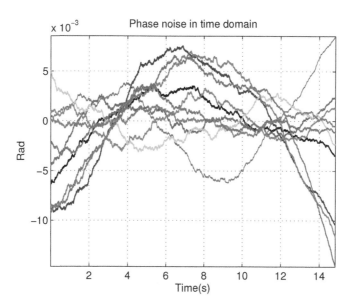

FIGURE 8.10: Residual phase noise in 15 seconds.

The first term is associated with the frequency offset arising from non-identical local oscillator frequencies, which will result in a drift on the SAR imagery. Since this drift can easily be corrected using a ground calibrator, it is ignored in subsequent discussions. The second term forms the usual Doppler term due to the varying round-trip time to the target, it should be preserved. The last term represents the impact of local frequency instability which is of interest for bistatic SAR synchronization processing.

Generally, a typical local oscillator used in current SAR systems has a frequency accuracy (δf) about 10^{-9} Hz/s or better [434]. Assuming a typical X-band bistatic SAR system with the following parameters: carrier frequency is 1×10^{10} Hz and the round-trip from radar to target is 12000 m, then the phase synchronization error in fast-time is found to be

$$M[\Phi_T(t) - \Phi_T(t)(t - \tau)] = 2\pi \cdot \delta f \cdot \tau$$
$$\ll 2\pi \times 10^{10} \times 10^{-9} \times \frac{6000}{3^8} = \frac{4\pi}{3} \times 10^{-4}(rad.)$$

(8.45)

which has negligible effects on the synchronization phase. Hence, we have an approximative expression

$$\Phi_T(t) \doteq \Phi_T(t - \tau)$$

(8.46)

That is to say, the phase noise of oscillator in fast-time is negligible, we can consider only the phase noise in slow-time.

Accordingly, the phase synchronization error in a bistatic SAR can be modeled as

$$\Phi_B(t) = M[\varphi_T(t) - \varphi_R(t)]$$

(8.47)

It is assumed that $\varphi_T(t)$ and $\varphi_R(t)$ are independent random variables having identical PSD $L_\varphi(f)$. Then the PSD of phase noise in the bistatic SAR is

$$L_{\varphi_B}(f) = 2M^2 L_\varphi(f)$$

(8.48)

where the factor 2 arises from the addition of two uncorrelated but identical PSD. This is true in those cases where the effective division ratio in a frequency synthesizer is equal to the small integer fractions exactly. In other instances, an experimental formula is [224]

$$L_{\varphi_B}(f) = 2\left[M^2 L_\varphi(f) + \frac{10^{-8}}{f} + 10^{-14}\right]$$

(8.49)

Take a typical 10 MHz oscillator as an example which phase noise parameters are listed in Table 8.3. Simulation examples of the predicted phase errors are shown in Figure 8.11 for a time interval of 10 seconds. The impacts of phase synchronization errors on bistatic SAR compared with the ideal compression results in azimuth can be found in Figure 8.12(a). We can draw a conclusion that oscillator phase instabilities in a bistatic SAR manifest themselves as a deterioration of the impulse response function. It is also evident

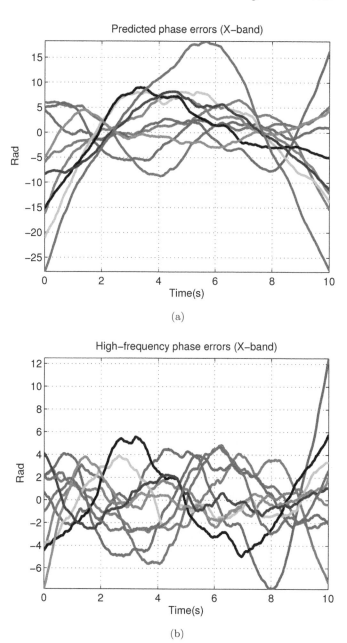

FIGURE 8.11: Simulation results of oscillator phase instabilities (noises) with ten realisations: (a) predicted phase noises in 10 seconds in X-band (linear phase ramp corresponding to a frequency offset has been removed), (b) predicted high-frequency including
cubic and more phase errors.

TABLE 8.3: Phase noise parameters of a typical oscillator.

Frequency (Hz)	1	10	100	1k	10k
$S_\varphi(f)(dBc/Hz)$	-80	-100	-130	-145	-160

that oscillator phase noise may not only defocus the SAR image, but also introduce significant positioning errors along the scene extension.

Furthermore, it is known that high frequency phase synchronization errors will cause spurious sidelobes in the impulse function. This deterioration can be characterized by the integrated sidelobe ratio (ISLR) which measures the transfer of signal energy from the mainlobe to the sidelobes. For an azimuth integration time T_s, the ISLR contribution due to phase errors can be computed as [14]

$$\text{ISLR} = 10 \log \left[\int_{1/T_s}^{\infty} 2M^2 L_\varphi(f) df \right] \tag{8.50}$$

A typical requirement for the maximum tolerable ISLR is -20 dB, which enables a maximum coherent integration time T_s of 2 seconds in this example as shown in Figure 8.12(b). This result is coincident with that of [212].

Generally, for $f \leq 10$ Hz, the region of interest for SAR operation, $L(f)$ can be modeled as [446]

$$L(f) = L_1 10^{-3 \log f} \tag{8.51}$$

Note that L_1 is the value of $L(f)$ at $f = 1$ Hz for a specific oscillator. As the slope of Eq. (8.51) is so large, we have

$$\log \left(\int_{1/T_s}^{\infty} L(f) df \right) \simeq L \left(\frac{1}{T_s} \right) \tag{8.52}$$

Hence the deterioration of ISLR may be approximated as

$$\text{ISLR} \simeq 10 \log(4M^2 L_1) + 30 \log T_s \tag{8.53}$$

It was concluded in [446] that, the error in this approximation is less than 1 dB for $T_s > 0.6$ s.

8.3.3   Impact of Time Synchronization Errors

Clock time jitter is the term most widely used to describe time synchronization errors, which are a measurement of the variations in time domain, and essentially describe how far the signal period wanders from its ideal value [278]. Due to its practical importance in communication and radar systems,

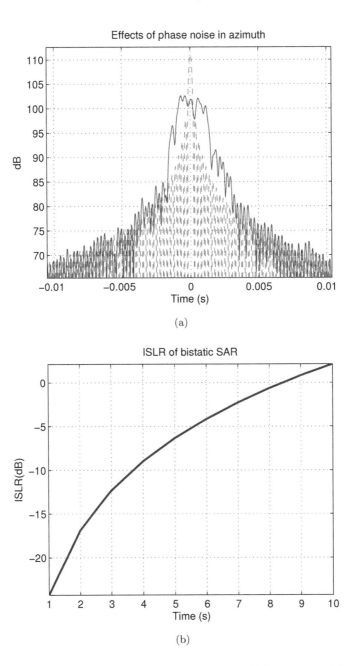

FIGURE 8.12: Impacts of phase noise on bistatic SAR systems: (a) impact of predicted phase noise in azimuth, (b) impact of integrated sidelobe ratio in X-band.

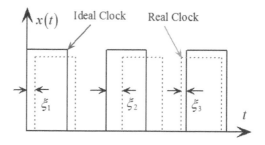

FIGURE 8.13: Illustration of clock signal with jitter.

clock time jitter in oscillators is the object of extensive research [108]. A special attention should be given to study the effects of clock time jitter in order to predict the possible degradations on the behavior of bistatic SAR systems. The impacts of linear and random time synchronization errors were discussed in [473]. Here, we reconsider it in a more general way.

Generally, we can model jitter in a signal by starting with a noise-free signal ν and displacing time with a stochastic process ζ. The noisy signal becomes [296]

$$\nu_n(t) = \nu(t + \zeta(t)) \tag{8.54}$$

The ζ, has units of seconds and can be interpreted as a noise in time, is assumed to be a zero-mean process, and ν is assumed to be a periodic function. Figure 8.13 shows a square wave with jitter compared to an ideal signal at the same long-term frequency. The instabilities can eventually cause slips or missed signals that result in loss of radar echoes.

Without loss of generality, the relative receiver-transmitter position estimation errors are ignored in the following discussions, because they can be compensated with motion compensation sensors. The typical requirement of time synchronization is about the tenth of the transmitted pulse width [433]

$$\delta t = \frac{T_p}{20 T_{\sqcap}} = \frac{1}{20 B_r T_{\sqcap}} \tag{8.55}$$

where δt is the time difference, T_p is the pulse width, B_r is the frequency bandwidth and $\tau_{\sqcap}$ is the update rate of the clocks. For example, when a bistatic SAR uses a frequency bandwidth of $B_r = 1.2$ GHz and the update rate of the clock is $T_{\sqcap} = 15$ seconds, the time synchronization precision has to be better than $2.8 \cdot 10^{-12}$ s. Certainly, this is still a technical challenge for current radar engineering.

Because bistatic SAR is a coherent system, to complete the coherent accumulation in azimuth, the signals of same range but different azimuths should have the same phase after range compression and range migration correction. If time synchronizes strictly, intervals between the echo window and the PRF

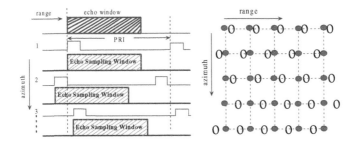

FIGURE 8.14: Illustration of impact of clock jitter on radar data.

of the receiver system would be a fixed value to preserve a stable phase relationship. But once there is clock timing jitter, the start time of the echo sampling window changes with certain time difference between the echo sampling window (or PRI) and the real echo signal, as shown in Figure 8.14. Notice that, the symbol • represents the jitter-free echo, while ∘ denotes the echo with jitter. Consequently, the phase relation of the sampled data would be destroyed.

To find an analytical expression for the impact of time synchronization errors on bistatic SAR imaging, we suppose the transmitted radar signal is

$$S_T(t) = \text{rect}\left[\frac{t}{T_p}\right] \exp\left(j2\pi f_c t + j\pi k_r t^2\right) \tag{8.56}$$

Let $\zeta(t)$ denote the clock timing jitter, the radar echo from a given scatterer is given by

$$
\begin{aligned}
S_R(t) = {} & \text{rect}\left[\frac{t - R_{\text{ref}}/c_0 - \zeta(t)}{T_W}\right] \text{rect}\left[\frac{t - \tau_r}{T_p}\right] \\
& \times \exp\left[j\pi f_c(t - \tau_r) + j\pi k_r(t - \tau_r)^2\right]
\end{aligned}
\tag{8.57}
$$

where the first term is the range sampling window centered at R_{ref}, having a length of T_W, and τ_r is the delay corresponding to the time it takes the signal to travel the distance transmitter-target-receiver distance, R_B.

Considering only time synchronization error, i.e., phase synchronization error is ignored here, we can obtain the demodulated signal as

$$S'_r(t) = \text{rect}\left[\frac{t - R_{\text{ref}}/c_0 - \zeta(t)}{T_W}\right] \text{rect}\left[\frac{t - \tau_r}{T_p}\right] \exp\left[j2\pi f_c\tau_r + j\pi k_r(t - \tau_r)^2\right] \tag{8.58}$$

Suppose the range reference signal is

$$S_{\text{ref}}(t) = \exp\left(j\pi k_r t^2\right) \tag{8.59}$$

The signal, after range compression, can be expressed as

$$S_R(r) = \text{rect}\left[\frac{r - R_{\text{ref}}}{c_0 T_W}\right] \text{sinc}\left[\frac{B(r - R_B + \Delta R)}{c_0}\right] \exp\left[-j\frac{2R_B\pi f_0}{c_0}\right] \tag{8.60}$$

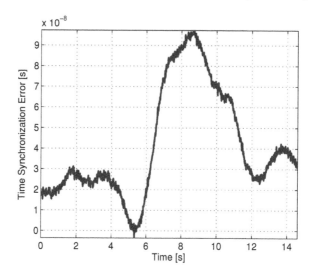

FIGURE 8.15: Predicted time synchronization error in a typical oscillator.

It can noticed from Eq. (8.60) that, if the two clocks deviate a lot, the radar echoes will be lost due to the drift of echo sampling window. Fortunately, such a case hardly occurs for current radars. Hence we considered only the case that each echo can be successfully received but be drifted because of clock timing jitter. In other words, the collected data with the same range but different azimuths are not on the same range any more, which means different phases and results in mismatch with the reference signal in azimuth.

To derive quantitative estimation, we have predicted the time synchronization accuracy by using the model described previously, as shown in Figure 8.15. As a typical example, assuming a S-band airborne SAR with the following parameters: range resolution is 1 m, azimuth resolution is 1 m, pulse duration is 10 μs, and the transmitter and receiver are moving in parallel tracks with constant identical velocity 90 m/s and altitude 8 km. From Figure 8.16 we can conclude that time synchronization errors will result in unfocused images, drift of radar echoes and displacement of targets. Therefore some time synchronization compensation techniques must be applied to focus the raw data.

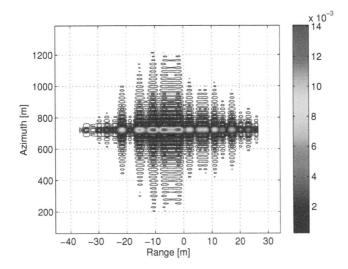

FIGURE 8.16: Impact of time synchronization error on SAR imaging.

8.4 Direct-Path Signal-Based Time and Phase Synchronization

Figure 8.17 shows a time and phase synchronization approach via direct-path signal. The direct-path signal of transmitter is received with one appropriative antenna and divided into two channels, one is passed through an envelope detector and used to synchronize the sampling clock, and the other is down-converted and used to compensate the phase synchronization errors. The residual time synchronization errors are compensated with range alignment, and the residual phase synchronization errors are compensated with GPS/INS(inertial navigation system)/inertial measurement units(IMU) information.

8.4.1 Time Synchronization

To complete coherent accumulation in azimuth, the signals of equal range but different azimuths should have the same phase after range compression and range migration correction. If time synchronizes strictly, intervals between the echo window and the PRF of the receiver would be a fixed value to preserve a stable phase relation. But once there are time synchronization errors, the start time of the echo sampling window will change with certain time difference between the echo sampling window and the real echo signal. As a consequence, the phase relation of the sampled data will be destroyed.

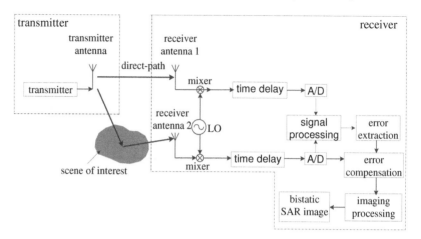

FIGURE 8.17: Illustration of the direct-path signal-based time and phase synchronization.

It is well known that, for monostatic SAR, the azimuth processing operates upon the echoes which come from the targets with equal range. Since time synchronization errors (without considering phase synchronization which are compensated separately in subsequent phase synchronization processing) have no impact on the initial phase of each echo, time synchronization errors can be compensated separately with range alignment. Here the spatial domain realignment [56] is used. That is, let $f_{t_1}(r)$ and $f_{t_2}(r)$ denote the recorded complex echo from adjacent pulses where $t_2 - t_1 = \Delta t$ is the PRI and r is the range assumed within one PRI. If we consider only the magnitude of the echoes, then

$$m_{t_1}(r + \Delta r) \approx m_{t_2}(r) \tag{8.61}$$

where $m_{t_1}(r) \doteq |f_{t_1}(r)|$, and Δr is the amount of misalignment that should be estimated. Define a correlation function between the two waveforms $m_{t_1}(r)$ and $m_{t_2}(r)$ as [56]

$$R(s) \triangleq \frac{\int_{-\infty}^{\infty} m_{t_1}(r) m_{t_2}(r - s) dr}{\sqrt{\int_{-\infty}^{\infty} m_{t_1}^2(r) dr \int_{-\infty}^{\infty} m_{t_2}^2(r) dr}} \tag{8.62}$$

According to the Schwartz inequality, $R(s)$ will be maximal at $s = \Delta r$ and the amount of misalignment can thus be determined. Note that some other range alignment methods may also be adopted, such as the frequency domain realignment [56], recursive alignment [87], and minimum entropy alignment [394]. As an example, for one point target, this process is shown in Figure 8.18. Another note is that sensor motion error will also result in the drift of echo envelope, which can be corrected with motion compensation algorithms. When the transmitter and receiver are moving in non-parallel trajectories, the phase errors of the normal channel and synchronization channel must be compensated separately. This compensation can be achieved with motion sensors and effective image formation algorithms.

8.4.2   Phase Synchronization

After time synchronization compensation, the primary causes of phase errors include uncompensated target or sensor motion and residual phase synchronization errors. Practically, the receiver of direct-path can be regarded as a strong scatterer in the process of phase compensation. To the degree that motion sensor is able to measure the relative motion between the targets and SAR sensor, the image formation processor can eliminate undesired motion effects from the collected signal history with GPS/INS/IMU and autofocus algorithms. Thereafter, the focusing of bistatic SAR image can be achieved with autofocus image formation algorithms, e.g., [385].

Suppose the n-th transmitted pulse is

$$x_n(t) = s(t)\exp\left(j2\pi f_{Tn}t\right)\exp\left(j\varphi_{d(n)}\right) \tag{8.63}$$

where $\varphi_{d(n)}$ is the original phase, f_{Tn} is the carrier frequency and $s(t)$ is the radar signal in baseband

$$s(t) = \text{rect}\left[\frac{t}{T_p}\right]\exp\left(j\pi k_r t^2\right) \tag{8.64}$$

Let t_{dn} denote the delay time of direct-path signal, the received direct-path signal is

$$s'_{dn}(t) = s(t - t_{dn})\exp\left[j2\pi\left(f_{Tn} + f_{dn}\right)(t - t_{dn})\right]\exp\left(j\varphi_{d(n)}\right) \tag{8.65}$$

where f_{dn} is the Doppler frequency for the n-th transmitted pulse.

Suppose the demodulating signal in receiver is

$$s_f(t) = \exp\left[-j\left(2\pi f_{Rn}t\right)\right] \tag{8.66}$$

The received signal in baseband can be expressed as

$$\begin{aligned} s_{dn}(t) =\ &s(t - t_{dn})\exp\left[-j2\pi\left(f_{Tn} + f_{dn}\right)t_{dn}\right] \\ &\times\exp\left(j2\pi\Delta f_n t\right)\exp\left(j\varphi_{d(n)}\right) \end{aligned} \tag{8.67}$$

where

$$\Delta f_n = f_{Tn} - f_{Rn} \tag{8.68}$$

with $\varphi_{d(n)}$ being the term to be extracted to compensate the phase synchronization errors in the reflected signal.

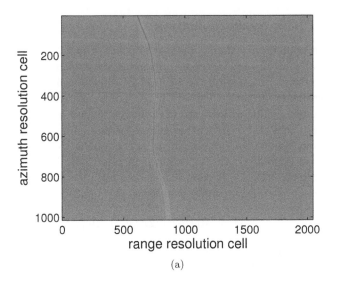

(a)

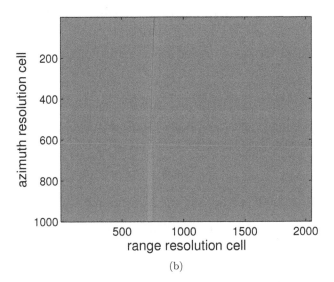

(b)

FIGURE 8.18: Range compression results for one single target: (a) results of range compression with time synchronization errors, (b) results of range compression with range alignment.

Fourier transforming Eq. (8.67) yields

$$
\begin{aligned}
S_{dn}(f) =&\mathrm{rect}\left[\frac{f - \Delta f_n}{k_r T_p}\right] \exp\left[-j2\pi(f - \Delta f_n)t_{dn}\right] \\
&\times \exp\left[\frac{-j\pi\,(f - \Delta f_n)^2}{k_r}\right] \exp\left[-j2\pi(f_{Tn} + f_{dn})t_{dn} + j\varphi_{d(n)}\right]
\end{aligned}
\tag{8.69}
$$

Suppose the range reference function is

$$
s_{\mathrm{ref}}(t) = \mathrm{rect}\left[\frac{t}{T_p}\right] \exp\left(-j\pi k_r t^2\right)
\tag{8.70}
$$

Range compressing Eq. (8.69) with Eq. (8.70) yields

$$
\begin{aligned}
y_{dn}(t) =&(k_r T_p - \Delta f_n)\mathrm{sinc}\left[(k_r T_p - \Delta f_n)\left(t - t_{dn} + \frac{\Delta f_n}{k_r}\right)\right] \\
&\times \exp\left[j\pi\Delta f_n\left(t - t_{dn} + \frac{\Delta f_n}{k_r}\right)\right] \\
&\times \exp\left\{-j\left[2\pi(f_{dn} + f_{Rn})t_{dn} - \frac{\pi\Delta f_n^2}{k_r} + \varphi_{d(n)}\right]\right\}
\end{aligned}
\tag{8.71}
$$

When $t = t_{dn} - \Delta f_n/k_r$, it arrives the maxima

$$
\exp\left[j\pi\Delta f_n\left(t - t_{dn} + \frac{\Delta f_n}{k_r}\right)\right]\bigg|_{t=t_{dn}-\Delta f_n/k_r} = 1
\tag{8.72}
$$

Hence, the residual phase term in Eq. (8.71) is

$$
\psi(n) = -2\pi(f_{dn} + f_{Rn})t_{dn} - \frac{\pi\Delta f_n^2}{k_r} + \varphi_{d(n)}
\tag{8.73}
$$

As Δf_n and k_r are typically on the orders of 1 kHz and 1×10^{13} Hz/s respectively, $\frac{\pi\Delta f_n^2}{k_r}$ has negligible impacts. Eq. (8.73) can be simplified as

$$
\psi(n) = -2\pi(f_{dn} + f_{Rn})t_{dn} + \varphi_{d(n)}
\tag{8.74}
$$

In a like manner, we have

$$
\psi(n + 1) = -2\pi(f_{d(n+1)} + f_{R(n+1)})t_{d(n+1)} + \varphi_{d(n+1)}
\tag{8.75}
$$

Let

$$
f_{d(n)} = f_{d0} + \delta f_{dn}
\tag{8.76}
$$

$$
f_{R(n)} = f_{R0} + \delta f_{Rn}
\tag{8.77}
$$

where f_{d0} and f_{R0} are the original Doppler frequency and error-free demodulating frequency in the receiver, respectively. Accordingly, δf_{dn} and δf_{Rn} are the frequency errors for the n-th pulse. We then have

$$
\begin{aligned}
\varphi_{d(n+1)} - \varphi_{d(n)} = [\psi(n+1) - \psi(n)] - 2\pi \left(f_{R0} + f_{d0} \right) \left(t_{d(n+1)} - t_{dn} \right) \\
- 2\pi \left(\delta f_{dn} + \delta f_{Rn} \right) \left(t_{d(n+1)} - t_{dn} \right)
\end{aligned}
\tag{8.78}
$$

Generally, $\delta f_{dn} + \delta f_{Rn}$ and $t_{d(n+1)} - t_{dn}$ are typically on the orders of 10 Hz and $10^{-9}s$ [218, 434], respectively, then $2\pi(\delta f_{dn} + \delta f_{Rn})(t_{d(n+1)} - t_{dn})$ is found to be smaller than $2\pi \times 10^{-8}$ radian, which has negligible effects. Furthermore, since $t_{d(n+1)}$ and t_{dn} can be known from GPS/INS/IMU, Eq. (8.78) is simplified into

$$
\varphi_{d(n+1)} - \varphi_{d(n)} = \psi_e(n)
\tag{8.79}
$$

where

$$
\psi_e(n) = [\psi(n+1) - \psi(n)] - 2\pi(f_{R0} + f_{d0})(t_{d(n+1)} - t_{dn})
\tag{8.80}
$$

We then have

$$
\begin{cases}
\varphi_{d(2)} - \varphi_{d(1)} = \psi_e(1) \\
\varphi_{d(3)} - \varphi_{d(2)} = \psi_e(2) \\
\cdots\cdots\cdots\cdots \\
\varphi_{d(n+1)} - \varphi_{d(n)} = \psi_e(n)
\end{cases}
\tag{8.81}
$$

From Eq. (8.81) we can get $\varphi_{d(n)}$, then the phase synchronization compensation for the reflected channel can be achieved with this method. Notice that the remaining motion compensation errors are usually low frequency phase errors, which can be compensated with autofocus image formation algorithms.

In summary, the time and phase synchronization compensation process may include the following steps:

1) Extract one pulse from the direct-path channel as the range reference function.

2) Direct-path channel range compression.

3) Estimate time synchronization errors with range alignment.

4) Direct-path channel motion compensation.

5) Estimate phase synchronization errors from direct-path channel.

6) Reflected channel time synchronization compensation.

7) Reflected channel phase synchronization compensation.

8) Reflected channel motion compensation.

9) Bistatic SAR image formation.

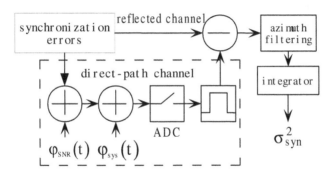

FIGURE 8.19: Illustration of the direct-path signal-based time and phase synchronization.

8.4.3 Prediction of Synchronization Performance

We consider only the phase synchronization performance since the time synchronization requirement (0.1 range resolution cell) can be more easily achieved with the technique described above. In a manner like [468], we can find several factors (see Figure 8.19) such as receiver noise and hardware devices as well as contributions known from sampling theory may influence the phase synchronization performance.

8.4.3.1 Receiver Noise

The receiver noise, consisting of thermal noise and the noise collected by antenna, will introduce both amplitude and phase variations to the radar signal. Here, only the phase variations described by $L_{\varphi_{\mathrm{SNR}}}(f)$ are of interest. For band-limited white Gaussian noise, it is related to the SNR by [468, 358]

$$L_{\varphi_{\mathrm{SNR}}}(f) = \frac{1}{2B_w \cdot \mathrm{SNR}} \tag{8.82}$$

where B_w denotes the receiver noise bandwidth. As an example, assuming a bistatic SAR system with the following typical parameters: $B_w = 100$ MHz, PRF = 1000 Hz, $\tau = 16~\mu s$, $T_s = 5$ s, then $L_{\varphi_{\mathrm{SNR}}}(f)$ is found to be smaller than -120 dBc/Hz. Additionally, the SNR will be further improved by image formation processing about 50 dB. More importantly, uncorrelated noise and equal SNR values can be assumed for both direct-path channel and reflected channel.

8.4.3.2 Amplifiers

For RF amplifier noise, where we consider only a narrow bandwidth around the carrier, one-half of the thermal white noise contributes to the amplitude noise modulation. The theoretical limit of the phase noise at the RF amplifier output is [172]

$$L_{\varphi AMP}(f) = 4\frac{K_0 T_0 R}{V_{rms}^2} \tag{8.83}$$

where K_0 and T_0 are the Boltzmann's constant and temperature in Kelvin, respectively. As an example, suppose $R = 50$ Ω and $V_{rms} = 1$ V, we found for $L_{\varphi AMP}(f)$

$$L_{\varphi AMP}(f) \geq -184dBc/Hz \tag{8.84}$$

which has negligible effects. Note that, generally, a noise factor of about 10-20 dB may be added in practical systems.

8.4.3.3 Analog-Digital-Converter (ADC)

ADC quantization errors may result in what appears to be a white noise floor, but is actually a "sea" of very finely spaced discrete spurs. The amplitude quantization errors e_A can be assumed to be totally uncorrelated and uniformly distributed within each quantization step

$$-\frac{\Delta_A}{2} \leq e_A \leq \frac{\Delta_A}{2} \tag{8.85}$$

For a D-bit ADC, the quantization step size is

$$\Delta_A = \frac{1}{2^{D-1}} \tag{8.86}$$

Then the amplitude error power is [25]

$$E\left(e_A^2\right) = \frac{1}{\Delta_A} \int_{-\frac{\Delta_A}{2}}^{\frac{\Delta_A}{2}} e_A^2 de_A = \frac{\Delta_A^2}{12} \tag{8.87}$$

Correspondingly, the noise variance is

$$\sigma = \sqrt{\frac{\Delta_A^2}{12}} = \frac{\Delta_A}{2\sqrt{3}} \tag{8.88}$$

Since the signal power is

$$S = \left(2^D \Delta_A\right)^2 \tag{8.89}$$

The noise-signal-ratio can then be expressed as

$$\frac{N}{S} = 10\log\frac{\sigma^2}{S} = 20\log\frac{\Delta_A/2\sqrt{3}}{2^D \Delta_A} \tag{8.90}$$

Suppose the sampling noise bandwidth is B_n, we then have

$$L_{\varphi AD}(f) = \frac{N}{S} \cdot \frac{1}{B_n} = 20\log\frac{\Delta_A/2\sqrt{3}}{2^D \Delta_A} - 20\log(B_n) \tag{8.91}$$

TABLE 8.4: Possible phase errors caused by ADC.

D	4	6	8	10	12	14	16
e_A	0.03125	0.0078	0.0020	4. 9e-3	1.2e-3	3.1e-5	7.6e-6
$\delta\varphi$	3.8°	0.9°	0.22°	0.056°	0.014°	0.0035°	0.00087°

As an example, assuming the quantization bits is are 12-bit and the sampling rate is 300MHz, $L_{\varphi_{AD}}(f)$ can then be found to be -167.77 dBc/Hz. Experimentally, the phase errors caused by ADC can be modeled as [2]

$$\delta\varphi = \tan^{-1}\left(\frac{1}{2^D - 1}\right) \tag{8.92}$$

It can be concluded from Table 8.4 that ADC has negligible effect on phase synchronization errors.

8.4.4 Other Possible Errors

The transmitted signal (from the waveform generator through frequency conversions, amplification, and transmission by the transmit antenna) has also some unwanted phase characteristics. Sources of unwanted phase noise include nonlinearity in the FM characteristic of the generated waveform, antenna, link path, frequency dependent phase effects in filters, and waveguide dispersion. The impact of hardware system, dominated by active and passive radar RF components, will change within the duration of data collection. Another problem is the potential degradation due to aircraft motion and vibrations. It is well demonstrated that an oscillator's nominal frequency reacts sensitively to accelerations in proportion to its gravity sensitivity [119]. Time-dependent accelerations or vibrations will create additional frequency modulations or phase noise in an oscillator. This will generate discrete sidebands in the case of sinusoidal vibration or an increase in the noise floor with random vibration [156]. Fortunately, these effects are generally small given today's hardware technology [51]. More importantly, if we down-convert the both channels using a common oscillator, the contributions from components common to the direct-path channel and reflected channel will cancel out.

The remaining phase error contributions generated by hardware devices can be expressed as

$$\Delta\varphi_{\text{sys}}(t) = \varphi_{\text{syss}}(t_k, \tau_{\text{sys}} + \tau_d + \tau_a) - \varphi_{\text{sysd}}(t_k, \tau_{\text{sys}} + \tau_d) \tag{8.93}$$

where $\varphi_{\text{syss}}(t_k, \tau_{\text{sys}} + \tau_d + \tau_a)$ and $\varphi_{\text{sysd}}(t_k, \tau_{\text{sys}} + \tau_d)$ are the phase errors of reflected channel and direct-path channel for a given transmitted pulse, respectively. Here τ_{sys}, τ_d, and τ_a denote the delay time of hardware system, direct-path, and difference between two channels, respectively. For a typical

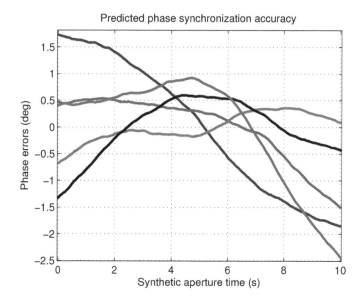

FIGURE 8.20: Predicted phase synchronization accuracy via direct-path signal with five realizations.

bistatic SAR, τ_a is about 0.1 μs, hence $\Delta\varphi_{\mathrm{sys}}(t)$ can be ignored within this periods for current high performance frequency synthesizer [415]. Additionally, there may be some other possible errors such as aliasing and interpolation errors, which have been detailed in [468]. We denote these possible errors as $L_{\varphi_{oth}}(f)$.

In a conclusion, the phase synchronization performance may be evaluated by

$$L_{\varphi_{\mathrm{total}}}(f) = L_{\varphi_{\mathrm{SNR}}}(f) + L_{\varphi_{\mathrm{AMP}}}(f) + L_{\varphi_{\mathrm{ADC}}}(f) + L_{\varphi_{\mathrm{oth}}}(f) \qquad (8.94)$$

With the statistical model of phase noise described previously, we predicted the possible contributions to the phase synchronization performance, as shown in Figure 8.20. We predicted that the total phase synchronization errors can be smaller than $10°$. Notice that, this is a statistical result, hence there may be some variation in actual systems. Moreover, some other factors such as atmospheric or relativistic effects, may cause the performance of synchronization to be worse than predicated.

8.5   GPS-Based Time and Phase Synchronization

For the direct-path signal-based synchronization approach, the receiver must fly with a sufficient altitude and position to maintain a line-of-sight contact with the transmitter. To get around this disadvantage, a GPS signal disciplined synchronization was proposed in [407].

The wide availability of GPS signal enables worldwide dissemination of high-accuracy time and frequency [85]. Due to their excellent long-term accuracy and stability compared with crystal oscillators, GPS-disciplined oscillators are widely used as standards of time and frequency [58]. Standards laboratories and research facilities often rely on GPS-disciplined oscillators as their primary reference for time and frequency calibration [177], and some cases as the frequency reference for the Josephson voltage standard [369] and for the mode-locked lasers used in length metrology [131]. Using GPS-disciplined oscillator as a frequency reference for code division multiple access (CDMA) communication system was investigated in [314]. The use of GPS-disciplined references for multistatic radar time and frequency synchronization was investigated in [189]. Electric power grids use time information from GPS-disciplined oscillators to rapidly locate faults was proposed in [230]. These applications generally treat GPS-disciplined oscillators as self-calibrating standards, where it is assumed that the devices perform according to the manufacturer's specification, and that they do not need to be periodically calibrated or compared to another standard.

GPS-disciplined oscillators are very interesting thanks to their continuous observability, nevertheless they raise another problem. Since GPS-disciplined oscillators are continuously adjusted to agree with signals broadcasted by the GPS satellite constellation, it is true that they are self-calibrating standards. Even so, when GPS signals are lost due to either deliberate interference or malfunctioning GPS equipment, the disciplined oscillator will loose its ability to update and continually adjust the output of the internal oscillator. Specifications provided by the manufacturer might provide only a rough indication of their actual performance [261]. In this case, the local oscillator is necessary to provide a holdover of time or frequency in the absence of GPS signal and to suppress short-term jitter in the GPS signal recovered from the receiver, so that the oscillator is held at the predicted control value and free-runs until the return of GPS signal allows a new correction to be calculated.

8.5.1   System Architecture

Because of their excellent long-term frequency accuracy, GPS-disciplined rubidium oscillators are widely used as standards of time and frequency [252]. Here, selection of a crystal oscillator instead of rubidium is based on the superior short-term accuracy of the crystal. As such, we use high-quality space-qualified 10 MHz quartz crystal oscillators, which have a typical short-term stability of $\sigma_{\text{Allan}}(\Delta t = 1s) = 10^{-12}$ and an accuracy of $\sigma_{\text{rms}}(\Delta t = 1s) = 10^{-11}$. In addition to good time-keeping ability, these oscillators have good phase noise performance.

The transmitter/receiver contains the high-performance quartz crystal oscillator, direct digital synthesizer (DDS) and GPS receiver. The antenna collects the GPS L1 (1575.42 MHz) signals and, if dual frequency capable, L2 (1227.60 MHz) signals. The radio frequency (RF) signals are filtered though a preamplifier, then down-converted to intermediate frequency (IF). The IF section provides additional filtering and amplification of the signal to levels more amenable to signal processing. The GPS signal processing component features most of the core functions of the receiver, including signal acquisition, code and carrier tracking, demodulation and extraction of the pseudo-range and carrier phase measurements. The details can be found in many textbooks on GPS [312].

The ultra-stable oscillator (USO) is disciplined by the output pulse-per-second (PPS), and frequency trimmed by varactor-diode tuning, which allows a small amount of frequency control on either side of the nominal value. Next, a narrow-band high-resolution DDS is applied, which allows the generation of various frequencies with extremely small step size and high spectral purity. This technique combines the advantages of the good short-term stability of high-quality USO with the advantages of GPS signals over the long term. When GPS signals are lost because of deliberate interference or malfunctioning GPS equipment, the oscillator is held at the best control value and free-runs until the return of GPS allows new corrections to be calculated.

8.5.2 Frequency Synthesis

Figure 8.21 shows the functional block diagram of GPS-disciplined frequency synthesis. To avoid phase noise and spurs being amplified by frequency multiplication factor if multiplier is used directly, the DDS output is not used as a reference, but is instead mixed to the local oscillator. The frequency of the local oscillator output sinewave signal is 10 MHz plus a drift Δf, which feeds into a double-balanced mixer. The advantage of double-balanced mixer over other forms of mixer is that it can suppress unwanted components in the output signal, so pass only the sum and difference signals. The other input of the mixer receives the filtered DDS output sinewave signal. The mixer outputs an upper and a lower sideband carrier (10 MHz $\pm\Delta f$). The wanted lower sideband is selected by a 10 MHz filter, and any other sidebands are rejected.

The DDS output frequency is determined by its clock frequency f_{dclk} and an D-bit tuning word 2^d ($d \in [1, D]$) written to its register, wherein D is the register length. The value of 2^d is added to an accumulator at each clock update, and the resulting digital wave is converted into an analog wave by digital-to-analog converter (DAC). The DDS output frequency is then deter-

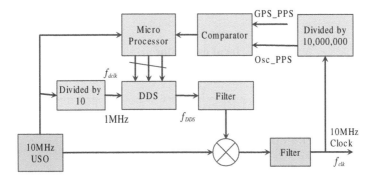

FIGURE 8.21: Block diagram of GPS-disciplined frequency source.

mined by [384]

$$f = \frac{f_{\text{dclk}} \cdot 2^d}{2^D}, d \in [1, 2, 3, \ldots, D-1]. \tag{8.95}$$

Clearly, a small frequency step requires a low clock frequency, but the lower the clock frequency the harder it becomes to filter out the unwanted clock components in the DDS output. As a good compromise, we use a clock at about 1 MHz, obtained by dividing the nominal 10 MHz local oscillator by 10. Then, the approximate resolution of the DDS output frequency is

$$\delta f = \frac{1MHz}{2^{48}} = 3.55 \cdot 10^{-9} Hz. \tag{8.96}$$

Here $M = 48$ is assumed. This frequency is subtracted from the oscillator output frequency. The corresponding minimum frequency step of the frequency corrector is $3.55 \cdot 10^{-16}$ Hz.

As shown in Figure 8.22, the 1-PPS epoch is used to trigger the local timing generator which thus precisely synchronized with the GPS timing. We can find an exact value of the 48-bit DDS to correct the oscillator frequency drift by measuring the LO_{PPS}, divided from the $10MHz$ oscillator output, against the GPS_{PPS} coming from the tracked GPS satellites. According to the PLL principle, the output signal of the GPS-disciplined oscillator will be adaptively synchronized with the GPS clock which has an ultra-high frequency stability. This approach is of interest because the quartz oscillator can be automatically steered to obtain a high frequency or phase accuracy and stability in the short term, as well as in the long term.

Thereafter, the DDS is able to exert control over a much wider frequency range with high-resolution and remove the oscillator calibration errors successfully. In this way, we can find an exact value of the 48-bit DDS to correct the exact drift to zero by measuring the Osc_PPS, divided from the 10 MHz oscillator output, against the GPS_PPS coming from the tracked GPS satellites. This technique is interest because that the local oscillator is free-running,

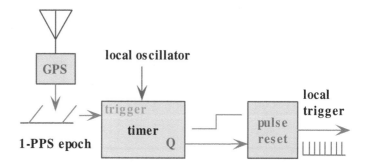

FIGURE 8.22: Illustration of the GPS-disciplined scheme using the PPS epoch.

and we know the frequency of the correcting offset for a given code into the DDS, it is easy to deduce the exact behavior of an oscillator that is under closed loop control.

Finally, the time and phase synchronization can be realized by generating all needed frequencies by dividing, multiplying or phase-locking to the GPS disciplined USO at the transmitter and receiver. To further reduce the phase noise and spurious level due to DDS, the triple tuned frequency synthesizer [415] can be adopted. The DDS output frequency (f_d) is converted to the PLL reference frequency (f_r) by a mixer and a variable frequency divider (VFD) 1. We then have

$$f_0 = \frac{N}{R} \cdot \left[f_{\text{clk}} + \frac{k \cdot f_{\text{clk}}}{2^L} \right] \tag{8.97}$$

where R and N are the frequency division ratios of the VFD1 and VFD2, respectively; f_{clk} is the clock frequency, k is the frequency setting data, L is the word length of DDS. We have demonstrated in [399] that, this configuration can set fine frequency tuning of f_0 by altering R and N. The values of k, N and R are determined to avoid high level spurious components from the PLL's loop bandwidth. By continuously increasing the phase increments themselves a linearly frequency modulated signal with excellent linearity can be obtained. Generally, the limited bandwidth of the DDS output signal has to be extended for high resolution SAR, which can be achieved with a frequency multiplier.

8.5.3 Measuring Synchronization Errors between Osc_PPS and GPS_PPS Signals

For continuous frequency correction the synchronization errors between GPS_PPS and Osc_PPS signals (i.e., their time intervals) must be measured. To reach this aim, we apply a high-precision time-to-phase conversion measurement technique, as shown in Figure 8.23. This technique uses the two

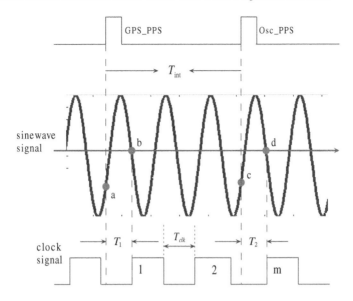

FIGURE 8.23: Measurement of synchronization errors between *GPS_PPS* and *Osc_PPS* signals.

GPS_PPS and *Osc_PPS* signals to trigger an DAC to sample a sinewave signal generated from the local oscillator directly.

To describe this approach, we starting from a sinewave signal represented by

$$s(t) = a_0 \cos(2\pi f_{lo} t + \phi(t) + \phi_0) \tag{8.98}$$

where a_0 is the amplitude, f_{lo} is the center frequency, $\phi(t)$ is the phase fluctuation and ϕ_0 is the starting phase. We then have

$$s_a = a_0 \cos(2\pi f_{lo} t_a + \phi(t_a) + \phi_0) \tag{8.99}$$

$$s_b = a_0 \cos(2\pi f_{lo}(t_a + T_1) + \phi(t_a + T_1) + \phi_0) \tag{8.100}$$

$$s_c = a_0 \cos(2\pi f_{lo} t_c + \phi(t_c) + \phi_0) \tag{8.101}$$

$$s_d = a_0 \cos(2\pi f_{lo}(t_c + T_2) + \phi(t_c + T_2) + \phi_0) \tag{8.102}$$

where t_a and t_c are the time at a and c, respectively, T_1 is the time interval between a and b, and T_2 is the time interval between c and d.

We can further obtain that

$$\Phi_a = 2\pi f_{lo} t_a + \phi(t_a) + \phi_0 = \arccos(\frac{s_a}{a_0}) + 2k_1\pi \tag{8.103}$$

$$\Phi_b = 2\pi f_{lo}(t_a + T_1) + \phi(t_a + T_1) + \phi_0 = \arccos(\frac{s_b}{a_0}) + 2k_1\pi \tag{8.104}$$

$$\Phi_c = 2\pi f_{lo} t_c + \phi(t_c) + \phi_0 = \arccos(\frac{s_c}{a_0}) + 2k_2\pi \tag{8.105}$$

$$\Phi_d = 2\pi f_{lo}(t_c + T_2) + \phi(t_c + T_2) + \phi_0 = \arccos(\frac{s_d}{a_0}) + 2k_2\pi \qquad (8.106)$$

After removing the 2π-phase ambiguities in Φ_a, Φ_b, Φ_c, and Φ_d, we can get

$$T_1 = \frac{(\Phi_b - \Phi_a) - [\phi(t_a + T_1) - \phi(t_a)]}{2\pi f_{lo}} \qquad (8.107)$$

$$T_2 = \frac{(\Phi_d - \Phi_c) - [\phi(t_c + T_2) - \phi(t_c)]}{2\pi f_{lo}} \qquad (8.108)$$

Hence, the time interval between GPS_PPS and Osc_PPS signals is

$$\begin{aligned}
T_{int} =& mT_{clk} - T_1 + T_2 \\
=& mT_{cl} - \frac{(\Phi_b - \Phi_a) - [\phi(t_a + T_1) - \phi(t_a)]}{2\pi f_{lo}} \\
& + \frac{(\Phi_d - \Phi_c) - [\phi(t_c + T_2) - \phi(t_c)]}{2\pi f_{lo}}
\end{aligned} \qquad (8.109)$$

As the parameters T_{clk}, Φ_a, Φ_b, Φ_c, Φ_d, m and f_{lo} are all measurable or calculable, and the parameters $\phi(t_a + T_1) - \phi(t_a)$ and $\phi(t_c + T_2) - \phi(t_c)$ are neglectable, the synchronization errors between GPS_PPS and Osc_PPS signals can be calculated from Eq. (8.109). The corresponding measure accuracy is determined by the sampling quantization errors. Our simulation results show that a measure accuracy of several ps can be achieved for an 14-bit ADC. Thus, this technique can achieve a satisfied accuracy for distributed SAR applications.

8.5.4   *GPS_PPS* Prediction in the Presence of GPS Signal

Because GPS-disciplined oscillators are adjusted to agree with the GPS signals, it is true that they are self-calibrating standards. Even so, differences in GPS_PPS signal fluctuations will be observed because of uncertainties in the satellite signals and the measurement process in the GPS receivers [169]. Measurements with today commercial GPS receivers which use the $L1$-signal at 1575.42 MHz, a standard deviation of 15 ns may be observed [433]. Using differential GPS (DGPS) or GPS common-view, one can expect a standard deviation of less than 10 ns [261]. But, in some GPS receivers the worst-case short-term period fluctuations of the GPS_PPS signal is in the order of some tens of ns. As a typical example, Figure 8.24 shows the stability of GPS clock bias observed on May 22, 2008 and May 23, 2008. Note that the data is obtained from the IGS website [1]. In essence GPS_PPS signal constitutes a perfect frequency signal with some residual jitter. The amount of jitter can be reduced by averaging the time values over a certain time interval [238]. A jitter of ± 100 ns is suitable for most consumer and industrial applications [144]. Nonetheless, a much lower jitter is required for distributed SAR imaging applications.

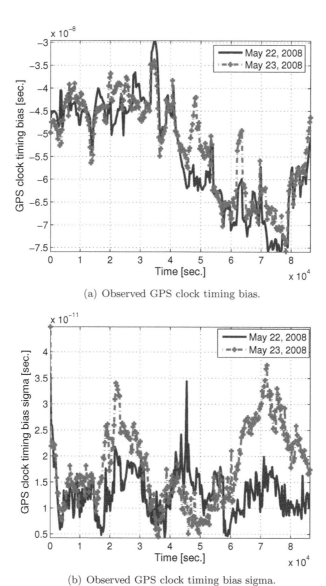

(a) Observed GPS clock timing bias.

(b) Observed GPS clock timing bias sigma.

FIGURE 8.24: Example observation of GPS clock signal stability over two days.

More unfortunately, when GPS signals are lost, the disciplined reference loose its ability to update and continually adjust the output of the internal oscillator. The control parameters will be fixed and the reference enters into a so-called holdover or free-flying mode. In this case, the oscillator will not maintain its initial frequency value but aging effects and temperature changes cause the frequency to drift. Drift in frequency hence results in an accumulating time error. To provide a successful holdover of frequency in the absence of GPS signal, an *GPS_PPS* signal prediction algorithm can be employed.

We start with the derivation of general linear least-square (LS) filters that are implemented using the linear combiner structure

$$y(n) = a_1 y(n-1) + a_2 y(n-2) + \ldots + a_N y(n-N) + \varepsilon(n) \qquad (8.110)$$

where $y(n-1)$, $y(n-2)$, ..., $y(n-N)$ are the past N *GPS_PPS* values, a_1, a_2, ..., a_N are the coefficients, and $\varepsilon(n)$ is the noise with mean value of 0 and squared variance of σ_ε^2. Eq. (8.110) can be rewritten more compactly as

$$y(n) = \boldsymbol{\varphi}^T(n)\boldsymbol{\theta} + \varepsilon(n) \qquad (8.111)$$

where $\boldsymbol{\theta} = [a_1, a_2, \ldots, a_N]^T$, $\boldsymbol{\varphi}^T(n) = [y(n-1), y(n-2), \ldots, y(n-N)]$ with T a transpose operator. We define the estimation error as

$$\varepsilon(n) = y(n) - \boldsymbol{\varphi}^T(n)\boldsymbol{\theta} \qquad (8.112)$$

and the coefficients of the combiner are determined by minimizing the sum of the squared errors

$$E = \sum_{n=0}^{N-1} |\varepsilon(n)|^2 \qquad (8.113)$$

From multiple observations we then have

$$\boldsymbol{Y}(n) = \boldsymbol{\Phi}(n)\boldsymbol{\theta} + \boldsymbol{\Sigma}(n) \qquad (8.114)$$

where $\boldsymbol{\Phi}(n) = [\boldsymbol{\varphi}(1), \boldsymbol{\varphi}(2), \ldots, \boldsymbol{\varphi}(n)]$, $\boldsymbol{Y}(n) = [y(1), y(2), \ldots, y(n)]^T$, and $\boldsymbol{\Sigma}(n) = [\varepsilon(1), \varepsilon(2), \ldots, \varepsilon(n)]$. To obtain an optimum MMSE (minimum mean square error) estimation, we define

$$\boldsymbol{J} = [\boldsymbol{Y}(n) - \boldsymbol{\Phi}(n)\boldsymbol{\theta}]^T[\boldsymbol{Y}(n) - \boldsymbol{\Phi}(n)\boldsymbol{\theta}]. \qquad (8.115)$$

Let $\partial \boldsymbol{J}/\partial\boldsymbol{\theta} = 0$, we have

$$\boldsymbol{\Phi}^T(n)\boldsymbol{\Phi}(n)\boldsymbol{\theta} = \boldsymbol{\Phi}^T(n)\boldsymbol{Y}(n). \qquad (8.116)$$

The LS estimation is

$$\hat{\boldsymbol{\theta}}(n) = [\boldsymbol{\Phi}^T(n)\boldsymbol{\Phi}(n)]^{-1}\boldsymbol{\Phi}^T(n)\boldsymbol{Y}(n). \qquad (8.117)$$

Denoting

$$\boldsymbol{P}(n) = [\boldsymbol{\Phi}^T(n)\boldsymbol{\Phi}(n)]^{-1}, \qquad (8.118)$$

we have

$$\hat{\boldsymbol{\theta}}(n+1) = \hat{\boldsymbol{\theta}}(n) + \frac{\boldsymbol{P}(n)\boldsymbol{\varphi}(n+1)}{1 + \boldsymbol{\varphi}^T(n+1)\boldsymbol{P}(n)\boldsymbol{\varphi}(n+1)},$$
$$\times [y(n+1) - \boldsymbol{\varphi}^T(n+1)\hat{\boldsymbol{\theta}}(n)] \qquad (8.119)$$

$$\boldsymbol{P}(n+1) = \boldsymbol{P}(n) + \frac{\boldsymbol{P}(n)\boldsymbol{\varphi}(n+1)\boldsymbol{\varphi}^T(n+1)\boldsymbol{P}(n)}{1 + \boldsymbol{\varphi}^T(n+1)\boldsymbol{P}(n)\boldsymbol{\varphi}(n+1)}, \qquad (8.120)$$

$$\hat{\boldsymbol{\theta}}(0) = \boldsymbol{\theta}(0), \boldsymbol{P}(0) = \boldsymbol{P}_0, y(i) = 0, i \le 0. \qquad (8.121)$$

The corresponding estimation variance is

$$\hat{\sigma}_\varepsilon^2(n) = \frac{1}{n}\sum_{i=1}^{n}\hat{\varepsilon}^2(i) = \frac{1}{n}\sum_{i=1}^{n}[y(n) - \boldsymbol{\varphi}^T(n)\hat{\boldsymbol{\theta}}(n-1)]^2. \qquad (8.122)$$

It is a common practice to apply this algorithm iteratively. On each iteration, the algorithm forms an estimate and applies this estimate to the input data. Iteration of the procedure greatly improves the final estimation accuracy, because iteration enhances the algorithm's ability to identify and discard the drifts that, for one reason or another, provide anomalous estimate of the current iteration. The iterative processing process is shown in Figure 8.25.

To evaluate the prediction performance, supposing $N = 2$ and using the actual GPS data obtained from the IGS website [1] many processing examples have been performed. Figure 8.26 shows the algorithm convergence for estimating the parameter or coefficient vector of the estimator. From the positive results we can conclude that the parameter or coefficient vector can be successfully estimated with this prediction algorithm. But a relative slow velocity of estimation convergence is found. This disadvantage can be overcome in two ways: one is to use more advanced but more complex prediction algorithms [271]; the other is to use some historical data to train the prediction algorithm. In this way, Figure 8.27 shows the predicted GPS clock bias while compared with the actual GPS clock bias observed on May 18, 2008. Note that the data is also obtained from the IGS website [1]. The corresponding prediction variance (sigma) as a function of the statistical samples is shown in Figure 8.28. It can be thus concluded that *GPS_PPS* signal fluctuation can be successfully predicted with this prediction algorithm. In this way, possible problems when the GPS signals are lost can be overcome.

8.5.5   Compensation for Residual Time Synchronization

Errors

Because time synchronization errors (without considering phase synchronization errors which are compensated separately) have no effect on the initial phase of each echo, we have concluded that time synchronization errors can be

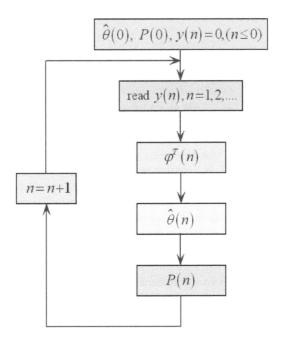

FIGURE 8.25: Iterative processing procedure for predicting the *GPS_PPS* signals.

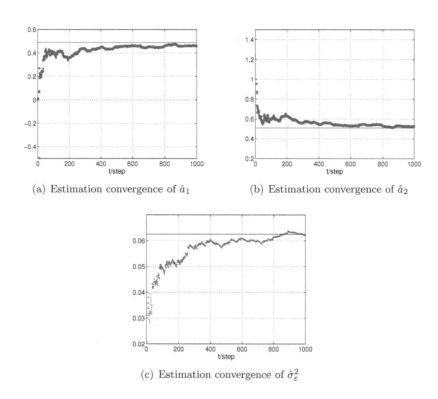

(a) Estimation convergence of $\hat{a}_1$ (b) Estimation convergence of $\hat{a}_2$

(c) Estimation convergence of $\hat{\sigma}_\varepsilon^2$

FIGURE 8.26: Estimation convergence of the estimator.

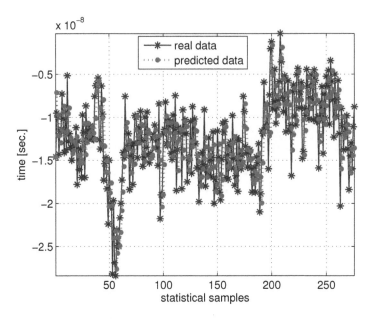

FIGURE 8.27: Predicted GPS clock bias while compared with the real data observed on May 18, 2008.

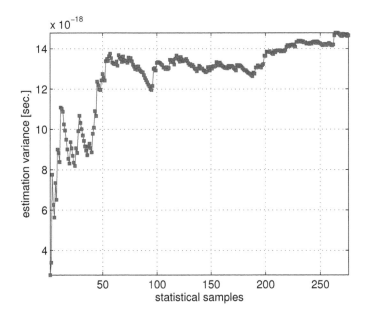

FIGURE 8.28: Estimation variance (sigma) as a function of statistical samples.

compensated with range alignment. Range alignment is a basic processing step in inverse synthetic aperture radar (ISAR) image formation [56]. Normally range alignment is performed with respect to the amplitude of the profiles, followed by the phase compensation. Some range alignment algorithms such as recursive alignment[87], cross-correlation[56] and minimum entropy[394], have been proposed for ISAR.

Let $f_{t_1}(r)$ and $f_{t_2}(r)$ denote the recorded complex echo from adjacent pulses where $t_2 - t_1 = \Delta t$ is the PRI and r is the range assumed within one PRI. If we consider only the magnitude of the echoes, then $m_{t_1}(r + \Delta r) \simeq m_{t_2}(r)$, where $m_{t_1}(r) \triangleq |f_{t_1}(r)|$. The Δr is the amount of misalignment which we would like to estimate. Define a correlation function between the two waveforms $m_{t_1}(r)$ and $m_{t_2}(r)$ as[56]

$$R(s) \triangleq \frac{\int_{-\infty}^{\infty} m_{t_1}(r) m_{t_2}(r - s) dr}{[\int_{-\infty}^{\infty} m_{t_1}^2(r) dr \int_{-\infty}^{\infty} m_{t_2}^2(r) dr]^{1/2}} \qquad (8.123)$$

From Schwartz inequality we have that $R(s)$ will be maximal at $s = \Delta r$, and the amount of misalignment can thus be determined. However the misalignment errors may accumulate across the coherent integration integration interval. In range bins without high energy relative to clutter energy, the performance will be further deteriorated. Additionally, for bistatic SAR, unideal antenna pointing synchronization will result in unsymmetrical radar echo amplitude. In this case, conventional cross-correlation methods are not usable any more. As a consequence, conventional range alignment algorithms are not effective for bistatic SAR time synchronization compensation due to their limited accuracy (about one range resolution cell).

Figure 8.29 shows the range alignment algorithm. The input signal $S_i(t)$ is modeled as the radar echoes. The "early-gate" is a gated integrator summing up the echoes, that is

$$y_{i1}(\tau, \Delta) = \frac{1}{T} \int_0^T S_i(\tau - t) S_0 \left(t + \frac{\Delta}{2} \right) dt = R_{i1} \left(\tau + \frac{\Delta}{2} \right) \qquad (8.124)$$

where $R(\cdot)$ and $S_0(t)$ denote the matched filter output and the first echo, respectively. Similarly, the "late gate" is another integrator, we then have

$$y_{i2}(\tau, \Delta) = \frac{1}{T} \int_0^T S_i(\tau - t) S_0 \left(t - \frac{\Delta}{2} \right) dt = R_{i2} \left(\tau - \frac{\Delta}{2} \right) \qquad (8.125)$$

When echo timing is correct, the outputs of both integrators are identical, the output signals $y_{i1}(\tau, \Delta)$ and $y_{i2}(\tau, \Delta)$ are equal then, and the error signal $e_i(\tau, \Delta)$ is zero. In contrast, while there is an offset in echo timing, the outputs of the integrators becomes unequal, and the error signal $e(\tau, \Delta)$ becomes positive or negative. Its polarity depends on the sign of the timing error. With the approximation

$$R_{i1}(\tau) = R_{i2}(\tau) = R_i(\tau) \qquad (8.126)$$

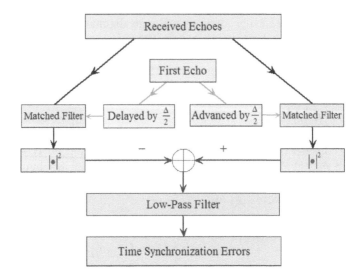

FIGURE 8.29: Block diagram of the proposed range alignment.

we have

$$e_i(\tau, \Delta) = R_i^2\left(\tau - \frac{1}{2}\Delta\right) - R_i^2\left(\tau + \frac{1}{2}\Delta\right),$$
$$\text{subject to} \quad e(\tau + \varsigma, \Delta) \cdot e(\tau, \Delta) \le 0,$$
$$\text{and} \quad e(\tau, \Delta) \cdot e(\tau - \varsigma, \Delta) \ge 0. \tag{8.127}$$

where ς denotes an infinitesimal value.

The above analysis suggests the following way to estimate the range mis-alignment τ. After matched filtering by using the first echo as the reference function, and find the time difference of adjacent echoes from Eq. (8.127) which gives an estimate of range bin shift required to realign the range bins, thereby we can estimate the time synchronization errors. More importantly, the impact of noise can be averaged out, and the shortcoming of accumulate errors unavoidable for conventional range alignment algorithms are avoided in this method.

8.5.6 Compensation for Residual Phase Synchronization

Errors

The differences in PPS fluctuations will result in linear phase synchronization errors, $\varphi_0 + 2\pi\Delta f \cdot t = a_0 + a_1 t$, in one synchronization period, i.e., in one second. Even the USO has a very good short-term time-keeping ability, frequency drift may be observed in one second. We can model them as quadratic phase errors. The residual phase errors in the i-th second can then be modeled as

$$\varphi_i(t) = a_{i0} + a_{i1}t + a_{i2}t^2, 0 \le t \le 1. \tag{8.128}$$

Motion compensation is ignored here because it can be addressed with motion sensors. Thus, after time synchronization compensation, the next step is residual phase error compensation, i.e., autofocus processing.

We use the Mapdrift autofocus algorithm described in [270]. The mapdrift technique divides the i-th second data into two nonoverlapping subapertures of duration of 0.5 second. This concept uses the fact that a quadratic phase error across one second (in one synchronization period) has a different functional form across two half-length subapertures [51]. The phase error across each subapertures consists of a quadratic component, a linear component, and an inconsequential constant component of $\Omega/4$ radians. The quadratic phase components of the two subapertures are identical with a center-to-edge magnitude of $\Omega/4$ radians. The linear phase components of the two subapertures have identical magnitudes but slopes of opposite sign.

Divide the i-th second azimuthal data into two nonoverlapping subapertures. There is an approximately linear phase throughout the subaperture

$$\varphi_{ei}(t + t_j) \approx b_{0j} + b_{1j}t, \quad |t| < \frac{T}{4},$$
$$\text{where} \quad t_j \equiv \frac{1}{2}\left(\frac{2j-1}{2} - 1\right), j \in \{1, 2\} \tag{8.129}$$

The model for the first subaperture $g_1(t)$ is the product of error-free signal history $s_1(t)$ and a complex exponential with linear phase

$$g_1(t) = s_1(t) \exp\left[j\left(b_{01} + b_{11}t\right)\right] \tag{8.130}$$

Similarly, for the second subaperture we have

$$g_2(t) = s_2(t) \exp\left[j(b_{02} + b_{12}t)\right] \tag{8.131}$$

Let

$$g_{12}(t) = g_1(t)g_2^*(t) = s_1(t)s_2^*(t) \exp\left[j(b_{01} - b_{02}) + j(b_{11} - b_{12})t\right] \tag{8.132}$$

Applying the Fourier transform yields

$$G_{12}(\omega) = \int_{-1/4}^{1/4} s_1(t)s_2^*(t) \exp\left]j(b_{01} - b_{02}) + j(b_{11} - b_{12})t\right] e^{-j\omega t} dt$$
$$= e^{j(b_{01}-b_{02})} S_{12}(\omega - b_{11} + b_{12}) \tag{8.133}$$

where $S_{12}(\omega)$ denotes the error-free cross-correlation spectrum. The relative shift between the two apertures is $\Delta_\omega = b_{11} - b_{12}$, which is directly proportional to the coefficient a_{i2} in Eq. (8.128) as

$$a_{i2} = \Delta_\omega = b_{11} - b_{12} \tag{8.134}$$

Various methods are available to estimate this shift. The most common method is to measure the peak location of the cross-correlation of the two subaperture images.

After compensating the quadratic phase errors (a_{i2}) in each second, Eq. (8.128) can be changed into

$$\varphi_i^c(t) = a_{i0} + a_{i1}t, 0 \leq t \leq 1 \tag{8.135}$$

Apply again the Mapdrift described above to the i-th and (i+1)-th second data, the coefficients in Eq. (8.135) can be derived. That is, define a mean value operator $< \varphi >_2$ as

$$< \varphi >_2 \equiv \int_{-1/2}^{1/2} \varphi dt \tag{8.136}$$

We can get [46]

$$a_{1i} = \frac{< (t - \bar{t})(\varphi_{ei} - \bar{\varphi}_{ei}) >_2}{< (t - \bar{t})^2 >_2} \tag{8.137}$$

$$a_{0i} = \bar{\varphi}_{ei} - b_{1i} < t >_2 \tag{8.138}$$

where $\bar{\varphi}_{ei} \equiv < \varphi_{ei} >_2$. In this step, the coefficients in Eq. (8.128) are derived, i.e., the residual phase errors are successfully compensated.

Notice that a typical implementation applies the algorithm to only a small subset of available range bins based on peak energy. An average of the individual estimates of the error coefficient from each of these range bins provides a final estimate. This procedure naturally reduces the computational burden of the algorithm. The range bins with the most energy are likely to contain strong, dominant scatterers with high signal energy relative to clutter energy. The signatures from such scatterers typically show high correlation between the two subaperture images while clutter is poorly correlated between the two images.

It is a common practice to apply this algorithm iteratively. On each iteration, the algorithm forms an estimate and applies this estimate to the input signal data. Typically, two to six iterations are sufficient to yield an accurate error estimate which does not change significantly on subsequent iterations. Iteration of the procedure greatly improves the accuracy of the final error estimate for two reasons [51]. First, iteration enhances the algorithm's ability to identify and discard those range bins that, for one reason or another, provide anomalous estimate for the current iteration. Second, the improved focus of the image data after each iteration results in a narrower cross-correlation peak, which leads to a more accurate determination of its location. Notice that the Mapdrift algorithm can be extended to estimate high-order phase error by dividing the azimuthal signal history in one second into more than two subapertures. Generally speaking, N subapertures are adequate to estimate the coefficients of an Nth-order polynomial error. However, this decrease in subaperture length degrades both the resolution and the SNR of the targets in the images, which results in degrade estimation performance.

8.5.7 Synchronization Performance Analysis

One importance performance measure for time synchronization compensation is the mean-square delay error σ_ε^2. Under the assumption that the early-late loop behaves like a linear filter and the error is in the linear tracking area, the output noise variance σ_n^2 is[139]

$$\sigma_n^2 = 2A^2 R_s^2(\Delta/2)N_0 B_n(\Delta/2T_c) + 2N_0^2 \omega_L B_n(\Delta/2T_c)(1 - \Delta/2T_c) \quad (8.139)$$

where ω_L, $A^2/2$, $N_0/2$ and Δ are the loop bandwidth, signal power, noise spectral density and early-late delay spacing, respectively. The $\Delta/2T_c$ and $(1 - \Delta/2T_c)$ terms appear because of the effective time-gating of the noise by the reference waveform. The second term is the noise $\times$ noise term. To translate this noise power to σ_δ^2, we can divide the noise power by the squared slope in the effective discriminator characteristic; namely, $D_s'(0)$:

$$\left(\frac{\sigma_\delta^2}{T_c}\right)^2 = \frac{\sigma_n^2}{\left[\frac{P_0 D_s'(0) T_c}{2}\right]^2} \quad (8.140)$$

where P_0 is the signal power. It can be proved that, the value of $D_s'(0)$ attains its maximum magnitude at $\Delta = \delta T_c$, i.e., the spacing equals to the rise-time. Hence, we have

$$\sigma_\delta = T_c \sqrt{\frac{\Delta \cdot \omega_L}{42 \cdot \text{SNR}}} \quad (8.141)$$

We suppose there are time synchronization errors caused by fluctuations of the GPS PPS, as shown in Figure 8.30. The corresponding estimates using our time synchronization algorithm are also illustrated in this figure. Notice that no interpolation was applied in this example. With the model of residual phase synchronization errors derived in Eq. (8.128), Figure 8.31 shows the corresponding phase compensation errors.

8.6 Phase Synchronization Link

If a dedicated phase synchronization link is established to exchange the oscillator signals, then by adequate processing, a correction signal can be derived to compensate the oscillator phase noise in the distributed SAR signal. The use of continuous duplex synchronization link for oscillator drift compensation was proposed in [106, 434]. The concept is somehow similar to microwave ranging used to determine the separation between platforms [200, 268]; however, phase synchronization requires a different processing approach. Depending on the SAR hardware and the affordable synchronization system complexity, various hardware configurations may be used to establish the synchronization link. In all cases, the aim is to exchange signals containing information on the oscillator noise between radar instruments.

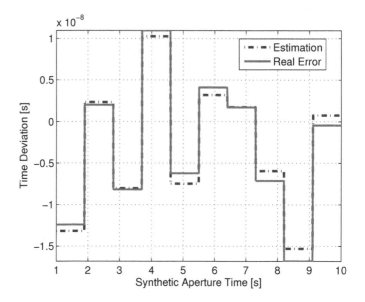

FIGURE 8.30: Prediction of timing jitter in the GPS PPS.

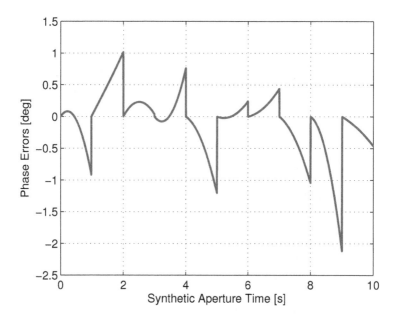

FIGURE 8.31: Typical phase synchronization compensation errors (are the difference in the two curves in Figure 8.30).

8.6.1   Two-Way Synchronization Link

The use of a dedicated two-way synchronization link to quantify and compensate oscillator phase noise is investigated in [468]. The two-way synchronization link can be further classified into continuous synchronization and pulsed synchronization. In the case of continuous duplex synchronization, both transmitter and receiver, continuously transmit and receive their local oscillator signals. Alternatively, in the pulsed synchronization, each radar platform repeatedly transmits its synchronization signal. For the pulsed duplex synchronization, both radar platforms transmit synchronization pulses at the same time instances. In the case of pulsed alternate synchronization, there is a time delay between the transmit instances of the two radar platforms.

Suppose the i-th radar platform transmits a synchronization signal, which is received by the j-th radar platform. The frequency of the i-th oscillator at the start time of data take t_0 is $f_i = f_0 + \Delta f_i$, with the nominal frequency f_0 and a constant frequency offset Δf_i. The phase $\varphi_i(t)$ at time t is the integration over frequency

$$\varphi_i(t) = 2\pi \int_{t_0}^{t} f_i(t)\mathrm{d}t + \varphi_{\mathrm{ini},i} + n_{\varphi_i}(t) \tag{8.142}$$

where $\varphi_{\mathrm{ini},i}$ is the time-independent initial phase and $n_{\varphi_i}(t)$ is the corresponding oscillator phase noise.

The j-th radar platform receives the signal after a delay τ_{ij} corresponding to the time it takes the signal to travel the distance r between the two platforms. At the receive instance $t + \tau_{ij}$, the phase $\varphi_j(t + \tau_{ij})$ of the j-th oscillator is

$$\varphi_j(t + \tau_{ij}) = 2\pi \int_{t_0}^{t+\tau_{ij}} f_j(t)\mathrm{d}t + \varphi_{\mathrm{ini},j} + n_{\varphi_j}(t + \tau_{ij}) \tag{8.143}$$

The demodulated phase $\varphi_{ji}(t)$ available at the j-th radar platform for a signal transmitted by the i-th radar platform is

$$\varphi_{ji}(t) = \varphi_j(t + \tau_{ij}) - \varphi_i(t) \tag{8.144}$$

The data take start time t_0 can be set to zero without restricting generality. The compensation phase $\varphi_c(t)$ is obtained by computing the difference

$$\varphi_c(t) = \frac{1}{2}\left[\varphi_{21}(t) - \varphi_{12}(t)\right] \tag{8.145}$$

where $\varphi_{21}(t)$ and $\varphi_{12}(t)$ are the demodulated phases of the synchronization signals recorded by the second and first radar platforms, respectively. The advantage is that the antenna, link path, and all common Tx/Rx system phase variations will cancel out as long as their contributions do not change within the time a pair of synchronization signals are exchanged. The compensation phase can then be used to correct the time-varying oscillator phase noise errors and the frequency offset of the SAR signal.

8.6.2 Synchronization Performance

A general synchronization and SAR signal processing flow block diagram for *priori* phase correction was investigated in [468]. Several factors will influence the phase of the synchronization link (see Figure 8.19). The transfer function $H_{LP}(f)$ describes the effect of alternate pulse synchronization. The impact of the receiver noise is described by the receiver phase noise spectral density function, $L_{\varphi_{SNR}}(f)$. Further, for the pulsed synchronization, the phase is sampled, which requires a later interpolation of the compensation phase. The compensation phase can be filtered with an arbitrary transfer function $H_{\text{syn}}(f)$. Finally, the compensated SAR phase is filtered through the transfer function $H_{az}(f)$ (it is applied on the signal's phase instead of the signal itself) and is dependent on azimuth processing [220]. The performance of the synchronization link is determined by the quality of the phase error reconstruction. The figure-of-merit is the phase standard deviation σ_{link}^2 after subtracting the oscillator phase noise and performing azimuth compression.

8.6.2.1 Continuous Duplex Synchronization

Both radar platforms continuously transmit the synchronization signal during the data take. The system hardware must be capable of simultaneous transmission and reception, and the signals must be sufficiently decoupled, which may cause problems when using the same carrier frequency. The single pulse length is equal to the data take time $T = T_{\text{data}}$, the system delay vanishes at $\tau_{\text{sys}} = 0$, and the signal travel times are equal $\tau_{21} = \tau_{12} = \tau = l_0/c_0$, where l_0 is the distance between the two platforms. The compensation phase is [468]

$$
\begin{aligned}
\varphi_c(t) =&\, 2\pi \left(\Delta f_2 - \Delta f_1\right) t + n_{\varphi_2}(t + \tau) - n_{\varphi_1}(t + \tau) \\
&+ \pi \left(\Delta f_2 - \Delta f_1\right) \tau + \varphi_{\text{ini},1} - \varphi_{\text{ini},2} \\
&+ \frac{1}{2} \left[\varphi_{\text{SNR},1}(t + \tau) - \varphi_{\text{SNR},2}(t + \tau)\right]
\end{aligned}
\tag{8.146}
$$

The first line contains the time-varying terms used to compensate the SAR phase noise and frequency offset. The second line is constant for constant platform separation and thus irrelevant for link performance. The main error contribution, dictating the performance of the synchronization link, is the phase variation due to receiver noise (see Eq. (8.82)).

Since the SNR is improved through azimuth compression, the link phase error variance is [468]

$$
\sigma_{\text{link}}^2 = \frac{1}{4B_w \text{SNR}} \int\limits_{-B_w/2}^{+B_w/2} |H_{\text{syn}}(f)H_{az}(f)|^2 \, df
\tag{8.147}
$$

where uncorrelated noise and equal SNR values are assumed for both receivers.

8.6.2.2 Pulsed Duplex Synchronization

Suppose synchronization pulses are simultaneously transmitted by both radar platforms every $1/f_{\text{syn}}$ seconds. Since the operation is duplex, the system time delay vanishes at $\tau_{\text{sys}} = 0$ and the signal travel times are equal $\tau_{21} = \tau_{12} = \tau = l_0/c_0$. The compensation phase is

$$\varphi_c(t) = \frac{1}{2} \left[\varphi_{\text{SNR},1}(t + \tau) - \varphi_{\text{SNR},2}(t + \tau) \right]$$
$$\pi \left(\Delta f_2 - \Delta f_1 \right) \left(\tau + 2t_k \right) + n_{\varphi_2}(t_k + \tau) - n_{\varphi_1}(t_k + \tau) \tag{8.148}$$

where the discrete sample instances t_k are given by

$$t_k = \frac{k}{f_{\text{syn}}}, \qquad k = 0, 1, \ldots, \lfloor T_{\text{data}} f_{\text{syn}} \rfloor \tag{8.149}$$

with $\lfloor T_{\text{data}} f_{\text{syn}} \rfloor$ being the total number of synchronization pulses during data take.

As the time-continuous compensation phase should be recovered from the discrete samples, the compensation phase will contain interpolation and aliasing errors. The link phase error variance is the sum of the interpolating, aliasing and receiver noise variances. They are expressed, respectively, by [468]

$$\sigma_{\text{inter}}^2 = 2\gamma^2 \int_{f_{\text{syn}}/2}^{+\infty} L_\varphi(f) \left| H_{az}(f) \right|^2 \mathrm{d}f \tag{8.150}$$

$$\sigma_{\text{alias}}^2 = 2\gamma^2 \sum_{i=1}^{\infty} \int_{-f_{\text{syn}}/2}^{+f_{\text{syn}}/2} L_\varphi(f + i \cdot f_{\text{syn}}) \left| H_{\text{syn}}(f) H_{az}(f) \right|^2 \mathrm{d}f \tag{8.151}$$

$$\sigma_{\text{SNR}}^2 = \frac{1}{4 f_{\text{syn}} \text{SNR}} \int_{-f_{\text{syn}}/2}^{+f_{\text{syn}}/2} \left| H_{\text{syn}}(f) H_{az}(f) \right|^2 \mathrm{d}f \tag{8.152}$$

8.6.2.3 Pulsed Alternate Synchronization

The transmit instance of the second radar platform is delayed by τ_{sys} with respect to the first radar platform. The pulsed alternate scheme has been suggested within the context of navigation satellite state determination in [107]. The time-discrete compensation phase is

$$\varphi_c(t) = \frac{1}{2} \left[\varphi_{\text{SNR},1}(t + \tau + \tau_{\text{sys}}) - \varphi_{\text{SNR},2}(t + \tau) \right]$$
$$\pi \left(\Delta f_2 - \Delta f_1 \right) \left(\tau + \tau_{\text{sys}} + 2t_k \right) - \pi f_d \tau_{\text{sys}}$$
$$+ n_{\varphi_2}(t_k + \tau) + n_{\varphi_2}(t_k + \tau_{\text{sys}})$$
$$- n_{\varphi_1}(t_k + \tau + \tau_{\text{sys}}) - n_{\varphi_1}(t_k) \tag{8.153}$$

where f_d is the Doppler frequency due to the relative velocity between the two platforms because of the unequal travel times $\tau_{12} \neq \tau_{21}$ due to the changing platform separation between the transmit instances t and $t + \tau_{\text{sys}}$. The Doppler phase contribution is constant for constant relative velocity. Only a relative acceleration will cause a phase error. For severe intersatellite acceleration, a Doppler phase compensation that requires the satellite separation to be known is necessary.

The total phase error variance is

$$\sigma_{\text{link}}^2 = \sigma_{\text{inter}}^2 + \sigma_{\text{alias}}^2 + \sigma_{\text{filter}}^2 + \frac{1}{2}\sigma_{\text{SNR}}^2 \qquad (8.154)$$

with interpolation and receiver noise variances given by Eqs. (8.150) and (8.152), respectively. The aliasing variance is

$$\sigma_{\text{alias}}^2 = 2\gamma^2 \sum_{i=1}^{\infty} \int_{-f_{\text{syn}}/2}^{+f_{\text{syn}}/2} L_\varphi(f + i \cdot f_{\text{syn}}) \left| H_{LP}(f + i \cdot f_{\text{syn}}) H_{\text{syn}}(f) H_{az}(f) \right|^2 \mathrm{d}f$$

$$(8.155)$$

where $H_{LP}(f)$ is a low-pass comb filter because each oscillator's phase noise is sampled at different time instances (this is equivalent to a low-pass comb filter). Additionally, a filter mismatch error σ_{filter} caused by the distortions of the synchronization link transfer function $H_{\text{syn}}(f)$ may appear. This is given by [468]

$$\sigma_{\text{filter}}^2 = 2\gamma^2 \int_0^{+f_{\text{syn}}/2} L_\varphi(f) \left| H_{az}(f) \right|^2 \left| H_{LP}(f) H_{\text{syn}}(f) - 1 \right|^2 \mathrm{d}f \qquad (8.156)$$

The $H_{\text{syn}}(f)$ is chosen to eliminate the filter mismatch error; hence, $H_{LP}(f)H_{\text{syn}}(f) = 1$ for $|f| \leq (1/2) f_{\text{syn}}$. The main difference between alternate and duplex synchronizations is the aliasing error. This error is reduced for f_{syn} values near $(1/2)\tau_{\text{sys}}$.

8.6.3   One-Way Synchronization Link

The two-way (duplex) synchronization link will destroy the passive characteristics of the receiver and make real-time imaging impossible, which allows the enemy to detect its presence while used in wartime. Moreover, the two-way operation is too complex to be applied to multistatic SAR systems. To get around this disadvantage, an one-way synchronization link (Figure 8.32) was proposed in [405] for distributed SAR real-time imaging.

8.6.3.1   Synchronization Scheme

If high-gain narrow-beam scanning antennas are used by both the transmitter and receiver, inefficient use is made of the radar energy because only the

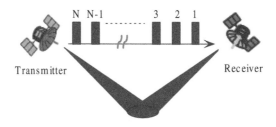

FIGURE 8.32: Illustration of the one-way synchronization link.

volume common to both beams can be observed by the receiver at any given time. Targets illuminated by the transmitting beam outside of this volume are lost to the receiver. To produce high-quality bistatic SAR imagery, accurate spatial synchronization or precise position and alignment estimation of both platforms must be ensured. The required accuracy should be an eighth of the wavelength, measuring about 3.8 mm at X-band [110]. To reach this aim, the high precision knowledge of the attitude and flight trajectories of transmitter, derived from a network of GPS receivers, gyroscopes, and accelerometers, as well as other redundant position and attitude information (e.g., from base-line measurement, other GNSS and appropriate services), is converted into a binary-valued number and transmitted in real-time to the passive receiver during data acquisition time.

Without direct-path measurement, a unique determination of transmitter position and velocity requires, at least, six measurements (three using for position and three using for velocity) [126]. They are converted into a binary-valued number and transmitted from the transmitter to the receiver. The location and velocity of the illuminator can then be determined in the receiver, which provides a workable basis for a SAR image and enables a determination of the location of a target relative to the receiver.

Additionally, some all-one data used as an initial timing signal, which helps in achieving successful demodulation, must also be added. Thus, the unipolar baseband data signal in the transmitter is given by

$$m(t) = \{C_1(t_1), A(t_2), B(t_3), C_2(t_4), [1, 1, \ldots, 1]\} \qquad (8.157)$$

where $A(t_2)$ and $B(t_3)$ are the spatial synchronization information and motion compensation information, and $C_1(t_1)$ and $C_2(t_2)$ are the initial timing signals, respectively.

As shown in Figure 8.33, the synchronization signal is modulated with a modified amplitude modulation (AM), which consists of a linearly frequency modulated carrier (not a sinusoid signal) with two amplitudes corresponding to a unipolar binary signal. This modified AM signal is represented by

$$s(t) = A_c [1 + m(t)] \exp \left(2\pi f_c t + k_r t^2\right) \qquad (8.158)$$

where $m(t)$ denotes the unipolar baseband data signal [spatial synchronization

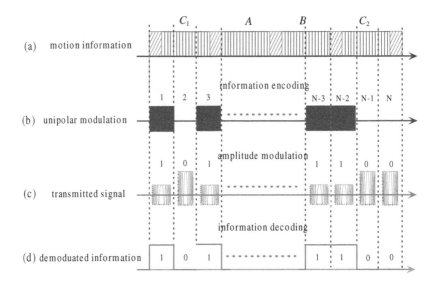

FIGURE 8.33: Time diagram for the synchronization pulses with spatial synchronization information including motion compensation information.

information including motion compensation information, as shown in Figure 8.33(a)].

The complex envelope is s

$$g(t) = A_c \left[1 + m(t)\right] \tag{8.159}$$

It may be detected using either an envelope detector or a product detector because it is a form of amplitude modulation signaling. Here, a superheterodyne receiver is used, where one detector is placed after the intermediate-frequency output stage.

After range realignment and motion compensation, the next step is phase synchronization compensation. Because the synchronization signal must be sufficiently decoupled from the normal SAR signal, we suppose two different carrier frequencies, $f_{\mathrm{T,syn}}$ and $f_{\mathrm{T,sar}}$, are used. Correspondingly, the phase arguments of the synchronization signal and normal SAR signal at the start of data take are given, respectively, by

$$s_{\mathrm{T,syn}}(t) = 2\pi \int_{t_0}^{t} f_{\mathrm{T,syn}} \mathrm{d}t + \phi_{\mathrm{T,ini,syn}} + \varphi_{\mathrm{T,syn}}(t) \tag{8.160}$$

$$s_{\mathrm{T,sar}}(t) = 2\pi \int_{t_0}^{t} f_{\mathrm{T,sar}} \mathrm{d}t + \phi_{\mathrm{T,ini,sar}} + \varphi_{\mathrm{T,sar}}(t) \tag{8.161}$$

with the initial–time-independent–phase $\phi_{T,\mathrm{ini,syn}}$ and $\phi_{T,\mathrm{ini,sar}}$, and the oscillator phase noise $\varphi_{T,\mathrm{syn}}(t)$ and $\varphi_{T,\mathrm{sar}}(t)$.

Receiver receives the synchronization signal after a delay τ_{syn} corresponding to the time it takes the signal to travel the distance between the transmitter and receiver. At the receive instance $t + \tau_{\mathrm{syn}}$, the phase argument of the synchronization signal in receiver is

$$s_{R,\mathrm{syn}}(t + \tau_{\mathrm{syn}}) = 2\pi \int_{t_0}^{t+\tau_{\mathrm{syn}}} f_{R,\mathrm{syn}}(t)dt \tag{8.162}$$

$$+ \phi_{R,\mathrm{ini,syn}} + \varphi_{R,\mathrm{syn}}(t + \tau_{\mathrm{syn}})$$

The demodulated phase $\psi_{syn}(t)$ is the difference between Eqs. (8.160) and (8.162) after including the system and path contributions $\varphi_{\mathrm{syn}}(t)$. The data take start time t_0 can be set to zero without restricting generality. Hence, we have

$$\psi_{\mathrm{syn}}(t) = \varphi_{R,\mathrm{syn}}(t + \tau_{\mathrm{syn}}) - \varphi_{T,\mathrm{syn}}(t) + \varphi_{\mathrm{syn}}(t) \tag{8.163}$$

Note that the initial phase is dropped here since it has no effect on the following processing results.

In a similar manner, for the normal SAR signal, we have

$$\psi_{\mathrm{sar}}(t) = \varphi_{R,\mathrm{sar}}(t + \tau_{\mathrm{sar}}) - \varphi_{T,\mathrm{sar}}(t) + \varphi_{\mathrm{sar}}(t) \tag{8.164}$$

where τ_{sar} denotes the delay time. The phase compensation term is

$$\psi_{\mathrm{comp}}(t) = \psi_{\mathrm{syn}}(t) \cdot \frac{f_{T,\mathrm{sar}}}{f_{T,\mathrm{syn}}} \tag{8.165}$$

This phase can then be used to correct the time-varying oscillator phase noise errors and the frequency offsets of the SAR signal. The corresponding synchronization accuracy is decided by

$$\psi_{\mathrm{err}}(t) = \psi_{\mathrm{sar}}(t) - \psi_{\mathrm{syn}}(t) \cdot \frac{f_{T,\mathrm{sar}}}{f_{T,\mathrm{syn}}} \tag{8.166}$$

Autofocus algorithms can be further employed in subsequent image formation processing. Here, autofocus algorithms improve imaging performance by removing residual phase errors. Autofocus algorithms for monostatic SAR have been fairly well developed, and a variety of examples can be found in literature. These algorithms possess varying degrees of sophistication, maturity, and practicality. Essentially these algorithms either utilize strong point-like scatterers or the ensemble of weak scatterers. The most popular technique for monostatic SAR autofocus is the phase gradient autofocus algorithm [385], which exhibits an excellent capability to remove higher-order phase errors over a variety of scenes.

In summary, the adaptive synchronization compensation process may include the following steps:

1) Transmit the synchronization information of the transmitter though amplitude modulation.

2) Extract the first pulse as the range reference signal.

3) Receiver spatial synchronization with demodulated information.

4) Estimate the residual time synchronization errors with range re-alignment.

5) Synchronize channel motion compensation.

6) Estimate the phase synchronization errors in the synchronization channel.

7) Scattered channel time synchronization compensation.

8) Scattered channel motion compensation.

9) Scattered channel phase synchronization compensation.

10) Bistatic SAR image formation with autofocus algorithm.

8.6.3.2 Synchronization Performance

Suppose the radar and synchronization signals are generated from the same oscillator signal though PLL. Figure 8.34 shows a fairly general PLL arrangement with a phase detector (PD), a low-pass loop filter $H_L(s)$, a voltage controlled oscillator (VCO) in the forward path and a mixer, an IF filter $H_M(s)$, and a divider ($\div N$). Additionally, a divider ($\div Q$) and a multiplier ($\times N$) are also placed. Since all the noise generated or added in individual PLL blocks are small compared with the useful signals, the small signal theory makes it possible to use the Laplace transform approach to find the output noise of the considered PLL system or, more exactly, the respectively power spectral densities.

According to the signal and system theory, we can get [222]

$$
\begin{aligned}
N_{\mathrm{PLL}}(s) = \Bigg[& N_{in}(s)\left(M + \frac{N}{Q}\frac{1}{F_M(s)} \right) + (N_{DQ}(s) - N_{DN}(s) \\
& + \frac{V_{\mathrm{PDn}}(s) + V_{Fn}(s)}{K_d} \right)\frac{N}{F_M(s)} + N_{mu}(s) - N_{mi}(s) \Bigg] H(s) \\
& + N_{\mathrm{osc}}(s)\left[1 - H(s)\right]
\end{aligned}
$$

(8.167)

where $H(s)$ denotes the effective loop transfer function. Since most of the noise components are random and uncorrelated, the power spectral density of

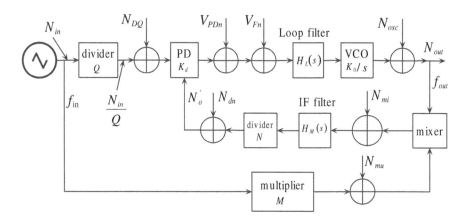

FIGURE 8.34: Model of a general PLL with additive noise.

the PLL output phase noise is

$$
\begin{aligned}
S_{\varphi,\mathrm{PLL}}(f) = \Bigg\{ & S_{\varphi,\mathrm{in}}(f) \left(M + \frac{N}{Q} \right)^2 + S_{\varphi,\mathrm{mu}}(f) + S_{\varphi,\mathrm{mi}}(f) \\
& + \left[S_{\varphi,\mathrm{DQ}}(f) + S_{\varphi,\mathrm{dn}}(f) + \frac{S_{\varphi,\mathrm{PDn}}(f) + S_{\varphi,\mathrm{Fn}}(f)}{K_d^2} \right] N^2 \Bigg\} \\
& \times |H(jf)|^2 + S_{\varphi,\mathrm{osc}}(f) \left| 1 - H(jf) \right|^2
\end{aligned}
$$

$$(8.168)$$

The first term in the brace is inevitable since it is merely a multiplied reference oscillator noise. The second and third terms are the multiplier and mixer noises. The remaining terms in the brace include the divider noise, phase detector noise and loop filter noise, all multiplied by the division ratio N.

There are both theoretical and experimental evidence that the additive noise due to the mixers is quite small and of the order of the loading circuit noise. Experimental results show that the best phase detector is double-balanced mixer [421]. Measurements reveal that the phase noise in a double-balanced mixer is approximated by [280]

$$
S_{\varphi,\mathrm{PDn}}(f) \approx \frac{10^{-14\pm1}}{f} + 10^{-17} \qquad (8.169)
$$

and for a phase detector based on CMOS logic family is

$$
S_{\varphi,\mathrm{PDn}}(f) \approx \frac{10^{-12.7}}{f} + 10^{-16.2} \qquad (8.170)
$$

Theoretically, the division process reduces the input PSD in proportion to the square of the division factor N^2. However, investigation of the divider output phase noise performed by Kroupa [225] reveals that the output phase noise is

$$S_{\varphi,\text{dn}}(f) \approx \frac{S_{\varphi,\text{dn,in}}(f)}{N^2} + \frac{10^{-10\pm1} + 10^{-27\pm1} f_0^2}{f} + 10^{-16\pm1} + 10^{-22\pm1} f_0 \quad (8.171)$$

The phase noise PSD at the output of a frequency multiplier is equal to its input multiplied by the square of multiplication factor plus an additive term

$$S_{\varphi,\text{mu}}(f) \approx S_{\varphi,\text{mu,in}}(f) \cdot N^2 + \frac{10^{-13\pm2}}{f} + 10^{-16\pm1} \quad (8.172)$$

The VCO phase noise improves as one goes to farther offsets from the carrier. Although there could be more regions with different slopes to the phase noise, a reasonable model for this is to divide this noise into three regions. A fairly general VCO phase noise equation is[223]

$$S_{\varphi,\text{vco}}(f) \approx \frac{f_0^2 \cdot 10^{-11.6}}{f^3 \cdot Q_L^2} + \frac{f_0^2 \cdot 10^{-15.6}}{f^2 \cdot Q_L^2} + \frac{10^{-11}}{f \cdot Q_L^2} + 10^{-15} \quad (8.173)$$

where Q_L is the loaded quality factor.

Suppose the carrier frequencies of radar and synchronization channels are 10 GHz and 2.5 GHz, respectively. Using the statistical models of phase noise, we can get the comparative results of output phase noise between radar channel and synchronization channel, as shown in Figure 8.35. The observation could be made that, within the loop bandwidth, the PLL phase detector is typically the dominant noise source, and outside the loop bandwidth, the VCO noise is often the dominant noise source.

8.7   Transponder-Based Phase Synchronization

A novel transponder used for calibrating high-resolution microwave remote sensing radar was proposed in [435], which retransmits the original radar signal with two artificial Doppler frequency shifted signals. Thereafter, the transponder signal is amplified to an appropriate level and retransmitted towards the radar transmitter. If the artificial Doppler shift is chosen to be larger than the Doppler bandwidth of the raw data, the transponder signal can be separated out during subsequent radar signal processing. The details can be found in [435]. In the following, we present a transponder-based phase synchronization for distributed SAR systems. Figure 8.36 gives the system scheme and geometry.

Without loss of generality, suppose the transmitted radar signal is a LFM pulse signal

$$s_0(t) = \exp\left[j\omega_T t + j\pi k_r t^2\right] \quad (8.174)$$

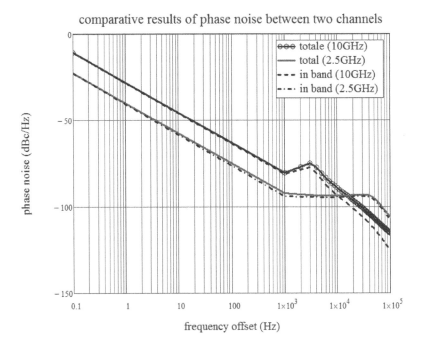

FIGURE 8.35: Comparative results of output phase noise between synchronization channel and SAR channel.

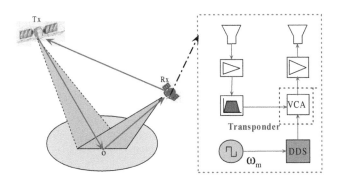

FIGURE 8.36: The transponder-based phase synchronization, where VCA and DDS denote the voltage controlled attenuator and direct digital synthesizer, respectively.

where $\omega_T = 2\pi f_T$. The signal arriving at the transponder can be represented by

$$s_r(t) = \sum_i s_0(t - \tau_i) \tag{8.175}$$

where $i = 1, 2, \ldots$ denotes the scatters in the radar observing scene, and τ_i is the corresponding time delay due to the transmitter-scatter-receiver distance.

As the transponder can be seen as an amplitude modulator, for the m-th pulsed signal transmitted by the master element, the signal modulated by the transponder can be represented by

$$s_{\mathrm{sl}}(t) = [\alpha + \beta \cos(\omega_m + \varphi_m)]\, s_r(t) \tag{8.176}$$

where α and β are constants determined by the transponder, ω_m is the transponder modulation frequency, and $\varphi_e(m)$ is the starting phase to be estimated for compensating the phase synchronization errors. The ω_m and φ_m have direct relation to the receiver oscillator frequency, from which we can extract the phase synchronization errors. The signal coming to the master element is

$$s_{\mathrm{ma}}(t) = s_{\mathrm{un}}(t - \tau_{rt}) + [\alpha + \beta \cos(\omega_m + \varphi_m)]\, s_r(t - \tau_{rt}) \tag{8.177}$$

where $s_{\mathrm{un}}(t)$ denotes the normal radar returns unmodulated by the transponder and τ_{rt} is the time-delay required for the signal transmitting from the Rx to the Tx.

Applying Fourier transform to Eq. (8.177), we get

$$\begin{aligned} S_{\mathrm{ma}}(\omega) =\, & S_{\mathrm{un}}(\omega) + \alpha S_{rt}(\omega) \\ & + \frac{\beta}{2} e^{-j\varphi_m} S_{rt}(\omega + \omega_m) + \frac{\beta}{2} e^{j\varphi_m} S_{rt}(\omega - \omega_m) \end{aligned} \tag{8.178}$$

where $S_{rt}(\omega)$ is the Fourier transform of $s_r(t - \tau_{rt})$. For simplicity and without loss of generality, we assume

$$S_{\mathrm{un}}(\omega) = S_{rt}(\omega) \tag{8.179}$$

The upper and lower side bands of the spectra can then be acquired by the following filters [437]

$$H_{\mathrm{up}}(\omega) = \mathrm{rect}\left(\frac{\omega - \omega_m}{B_r}\right) \tag{8.180a}$$

$$H_{\mathrm{dn}}(\omega) = \mathrm{rect}\left(\frac{\omega + \omega_m}{B_r}\right) \tag{8.180b}$$

where $\mathrm{rect}(\cdot)$ is a gate function, and B_r is the filter bandwidth which has to be chosen according to the signal bandwidth of the $s_r(t)$ and the modulation frequency ω_m.

Then, the upper and lower side bands are represented in frequency domain, respectively, by

$$S_{\text{up}}(\omega) = \frac{\beta}{2} e^{j\varphi_m} S_{\text{rt}}(\omega - \omega_m) \tag{8.181a}$$

$$S_{\text{dn}}(\omega) = \frac{\beta}{2} e^{-j\varphi_m} S_{\text{rt}}(\omega + \omega_m) \tag{8.181b}$$

The corresponding time-domain representations can be obtained by inverse Fourier transform

$$s_{\text{up}}(t) = \frac{\beta}{2} e^{j\varphi_m} s_{\text{rt}}(t) e^{j\omega_m t} \tag{8.182a}$$

$$s_{\text{dn}}(t) = \frac{\beta}{2} e^{-j\varphi_m} s_{\text{rt}}(t) e^{-j\omega_m t} \tag{8.182b}$$

We then have

$$y_m(t) = \frac{s_{\text{up}}(t)}{s_{\text{dn}}(t)} = e^{j2[\omega_m t + \varphi_m]} \tag{8.183}$$

The estimates of φ_m and ω_m denoted as $\hat{\varphi}_m$ and $\hat{\omega}_m$ can then be used for the phase synchronization process.

In summary, the phase synchronization approach is summarized as follows.

(1) The Tx broadcasts the radar signal $s_0(t)$ to the Rx.

(2) Upon receiving the transmitted signal, the Rx uses the transponder to amplitude modulate the received radar signal.

(3) The Rx retransmits the amplitude modulated radar signal back to the transmitter.

(4) Upon receiving the returned signal from Rx, Tx estimates the φ_m and ω_m according to Eq. (8.183).

(5) Phase unwrapping occurs at the Tx and adjusts its carrier frequency/phase correspondingly.

It is evident from the above procedures that this approach is a closed-loop phase synchronization scheme. Since the channel between the transmitter and receiver may vary from time to time, though at a slow rate, the values of $\varphi_e(m)$ and ω_m have to be estimated and updated in each pulse period. Generally the estimates can be obtained by PLL; however, digital processing estimation is more preferred when better accuracy is desired [461]. We can apply the maximum likelihood estimator instead to obtain estimates of these unknown but deterministic parameters.

8.8   Conclusion

Distributed multi-antenna SAR systems is subject to time and phase synchronization problem that is not encountered in colocated multi-antennas SAR systems. In this chapter, we analyzed the impacts of time and phase synchronization errors on distributed multi-antenna SAR imaging by using the developed statistical phase noise model. This model can predict time-domain phase errors in practical oscillators. Since effective time and phase synchronization compensation must be performed for current radar hardware systems, we introduced four types of time and phase synchronization methods. They are direct-path signal-based time and phase synchronization, GPS-based time and phase synchronization, synchronization link-based phase synchronization, and transponder-based phase synchronization.

9

Distributed Multi-Antenna SAR Antenna Synchronization

Besides time and phase synchronization, antenna synchronization (spatial synchronization, i.e., the receiving and transmitting antennas must simultaneously illuminate the same region on the ground) is also required for distributed multi-antenna SAR systems. There are two parameters on the transmitter side, which have a high influence on the time during which the area of interest can be illuminated. The first parameter is the velocity of the transmitter platform, which is very high relative to the receiver platform's velocity. The second parameter is the very small antenna steering range in azimuth. The requirement of antenna directing synchronization for interferometric SAR was analyzed in [277].

Several altitude and antenna directing strategies for satellite formation configuration were proposed in [93]. If high-gain narrow-beam scanning antennas are used at both the transmitter and receiver in distributed SAR systems, inefficient use is made of the radar energy because only the volume common to both beams can be observed by the receiver at any given time. Targets illuminated by the transmitting beam outside of this volume are lost to the receiver, even the targets outside of this volume but still in the receiving beam. Figure 9.1 shows the distributed SAR beam coverage geometry. Four remedies are considered in [446] to mitigate the beam scan-on-scan problem: step scan, floodlight beams, multiple simultaneous beams-receivers-signal processors, and time-multiplexed multiple-receivers-signal processors, which in the limit is called pulse chasing. In this approach, the single beam chases one pulse at a time. This imposes a limit on the maximum allowable PRF.

The geometrical setup of an airborne/spaceborne SAR experiment is described in [149]. Due to the high difference between the velocities of the platforms, both systems must perform antenna steering to achieve an appropriate scene length in azimuth [112, 113]. Sliding spotlight geometry was thus investigated in [149]. The comparison of altitude and antenna pointing design strategies of noncooperative spaceborne bistatic SAR were investigated in [178].

This chapter is organized as follows. Section 9.1 analyzes the impacts of antenna directing synchronization errors on distributed multi-antenna SAR systems. Section 9.2 introduces the beam scan-on-scan technique, followed by the pulse chasing in Section 9.3. Next, several sliding spotlight geometries are

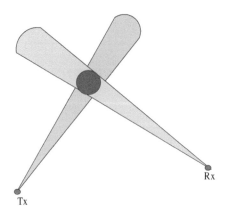

FIGURE 9.1: Distributed SAR beam coverage problem.

presented in Section 9.4, followed by the multibeam forming on receiver in Section 9.5. Finally, the determination of baseline and orientation of distributed SAR platforms are introduced in Section 9.6.

9.1 Impacts of Antenna Directing Errors

Since SAR is a two-dimensional (range and azimuth) imaging technique, we analyze the impacts of antenna directing errors from range dimension and azimuth dimension separately.

9.1.1 Impacts of Range Antenna Directing Errors

Without loss of generality, we consider a rather general bistatic SAR geometry, as shown in Figure 9.2. Suppose the normalized transmitting antenna gain is

$$G_T(\theta_T, \phi_T) = \exp\left\{-2\alpha\left[\left(\frac{\theta_T}{\theta_{T_{3dB}}}\right)^2 + \left(\frac{\phi_T}{\phi_{T_{3dB}}}\right)^2\right]\right\} \qquad (9.1)$$

where α is the Gauss parameter which is often assumed to be 1.3836, θ_T and ϕ_T are the antenna beamwidth in range and azimuth, respectively. Correspondingly, $\theta_{T_{3dB}}$ and $\phi_{T_{3dB}}$ are their 3 dB beamwidth. If there are range antenna directing errors $\Delta\theta_T$, we then have

$$\frac{\Delta G_T(\theta_T, \phi_T)}{G_T(\theta_T, \phi_T)} = -\frac{4\alpha\theta_T\Delta\theta_T}{\theta_{T_{3dB}}^2} \qquad (9.2)$$

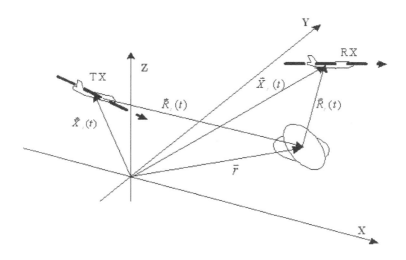

FIGURE 9.2: A general bistatic SAR geometry.

Similarly, for the receiving antenna we have

$$\frac{\Delta G_R(\theta_R, \phi_R)}{G_R(\theta_R, \phi_R)} = -\frac{4\alpha\theta_R\Delta\theta_R}{\theta_{R3dB}^2} \tag{9.3}$$

Then there is [370]

$$\frac{\Delta A}{A} = \frac{1}{2}\left[\frac{\Delta G_T(\theta_T, \phi_T)}{G_T(\theta_T, \phi_T)} + \frac{\Delta G_R(\theta_R, \phi_R)}{G_R(\theta_R, \phi_R)}\right]$$
$$= -2\alpha\left[\frac{\theta_T}{\theta_{T3dB}^2}\Delta\theta_T + \frac{\theta_R}{\theta_{R3dB}^2}\Delta\theta_R\right] \tag{9.4}$$

To evaluate the impact of antenna directing synchronization errors on SAR imaging performance, linear and quadratic errors are usually assumed in literature. In fact, antenna directing error usually is oscillatory for practical systems. Thus, we use an oscillatory model

$$\Delta\theta_T(t) = \Delta\theta_R(t) = A_r\cos(\omega_r t + \varphi_0) \tag{9.5}$$

where A_r, ω_r and φ_0 are the amplitude, frequency and initialization angle in range, respectively. Take an azimuth-invariant X-band bistatic SAR, in which the transmitter and receiver are moving in parallel tracks with constant identical velocities $v = 150$ m/s using the following parameters: $\Delta\theta_T(t) = \theta_T(t) = 8°$, $\Delta\theta_R(t) = \theta_R(t) = 5°$, $A_r = 1$, $\omega_r = 1.98$ and $\varphi_0 = 0$, as an example, the corresponding impacts are illustrated in Figure 9.3, respectively. It is seen that antenna directing synchronization errors manifest themselves as a deterioration of the impulse response function. They may defocus bistatic SAR images and introduce significant increase of the sidelobes.

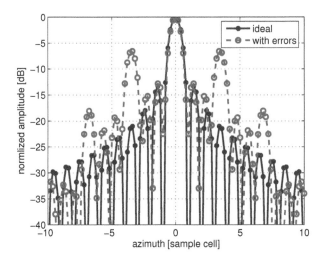

FIGURE 9.3: Impact of antenna directing synchronization errors in range.

9.1.2   Impacts of Azimuth Antenna Directing Errors

Similarly, suppose the azimuth antenna directing synchronization errors are represented by

$$\Delta\beta_T = \Delta\beta_R = A_a \sin(\omega_a t + \beta_0) \tag{9.6}$$

where A_a, ω_a and β_0 are the amplitude, frequency and initialization angle in azimuth, respectively. The corresponding transmit/receive azimuth antenna figure can be represented by

$$w(t) = w_a[t - A_a \sin(\omega_a t + \beta_0)] \tag{9.7}$$

with

$$w_a(t) = \text{sinc}^2\left[\frac{\pi L_a v_s}{\lambda R_c}(t - t_c)\right] \tag{9.8}$$

where L_a, v_s, λ, R_c and t_c are the antenna length, platform velocity, wave length, nearest range from the platform to the scene and its corresponding time, respectively.

Suppose the transmitted signal is

$$S_t(t) = w(t) \exp\left[j2\pi\left(f_c t + \frac{1}{2}k_r t^2\right)\right] \tag{9.9}$$

where f_c and k_r are the carrier frequency centra and chirp rate, respectively. From SAR processing procedure, we know the azimuth signal after range and azimuth compressions is

$$S_{out}(t) = \exp\left(j2\pi f_c t - j\pi k_r t^2\right) \cdot \int_{-T_s/2}^{T_s/2} w(\tau) \exp\left(2\pi k_r t\tau\right) d\tau \tag{9.10}$$

with T_s being the synthetic aperture time. Since

$$t \gg A_a \sin\left(\omega_a t + \beta_0\right) \tag{9.11}$$

Eq. (9.7) can be rewritten as

$$w(t) = w_a(t) - A_a \sin(\omega_a t + \beta_0) \cdot w'_a(t) \tag{9.12}$$

Substituting Eq. (9.12) to Eq. (9.10) yields

$$
\begin{aligned}
S_{out}(t) =& \exp\left(j2\pi f_c t - j\pi k_r t^2\right) \cdot \int_{-T_s/2}^{T_s/2} w_a(\tau) \exp\left(2\pi k_r t\tau\right) d\tau \\
& - A_a \exp\left(j2\pi f_c t - j\pi k_r t^2\right) \\
& \times \int_{-T_s/2}^{T_s/2} \sin\left(\omega_a \tau + \beta_0\right) \exp\left(2\pi k_r t\tau\right) d\tau \\
=& S_{\text{ideal}}(t) - S_{\text{err}}(t)
\end{aligned}
\tag{9.13}
$$

where $S_{\text{ideal}}(t)$ and $S_{\text{err}}(t)$ are the processed results without and with antenna directing synchronization errors. We can see that, in a like manner as the range antenna directing synchronization errors, azimuth antenna directing synchronization errors also introduce paired echoes. The amplitude ratio between the paired echoes and the ideal central one can be evaluated by

$$K_{am} = \frac{\pi L_a A_a}{4\lambda} \tag{9.14}$$

To evaluate the quantitative impact, we suppose an antenna length $L_a = 1$ (the other parameters are defined as the same as previously). Figure 9.4 shows the ratios between the paired echoes and the ideal central one as a result of azimuth antenna directing synchronization errors as a function of the oscillatory amplitude. We can conclude that, to obtain satisfactory imaging performance, a high-precision antenna directing synchronization should be ensured for bistatic SAR high-resolution imaging.

9.1.3 Impacts of Antenna Directing Errors on Distributed InSAR Imaging

Distributed multi-antenna SAR provides a potential to multi-baseline InSAR imaging [7, 153, 213, 219, 281], but the measurement accuracy and azimuth resolution will be impacted by antenna directing errors.

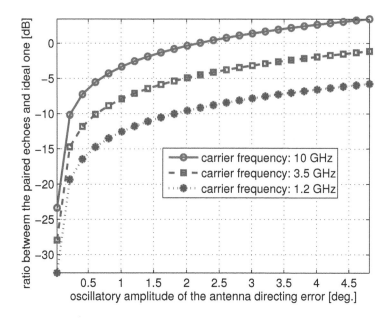

FIGURE 9.4: Impact of antenna directing synchronization errors in azimuth.

InSAR channel signals may be decorrelated by antenna directing errors [470]. The impacts of channel correlation parameter c_γ on distributed InSAR phase estimation accuracy is expressed as [328, 469]

$$m_\phi = \pm \frac{\sqrt{1 - c_\gamma^2}}{c_\gamma \sqrt{2N_0}} \qquad (9.15)$$

where N_0 is the multilook number. The corresponding InSAR measurement error is

$$m_h = \pm \frac{\lambda R_c \left(\sin\theta + \cos\theta \tan\xi \right)}{2\pi B_l \cos(\theta - \beta)} \qquad (9.16)$$

where R_c is the nearest slant range, θ is the incidence angle for the transmitter, ξ is the Earth grade, B_l is the baseline length and β is the air base inclination angle.

Suppose the azimuth antenna directing errors are $\Delta\alpha$, the corresponding Doppler frequency drift is

$$\Delta f_{\mathrm{dc}} \approx \frac{2v_s}{\lambda} \Delta\alpha \qquad (9.17)$$

The InSAR azimuth resolution

$$\rho_a = \frac{v_s}{B_a - \Delta f_{\mathrm{dc}}} \qquad (9.18)$$

can then be rewritten as

$$\rho_a = \frac{v_s \lambda}{\lambda B_a - 2 v_s \Delta \alpha} \tag{9.19}$$

where B_a is the azimuth Doppler bandwidth.

9.2 Beam Scan-On-Scan Technique

The step scan approach consists of fixing the transmitting beam position and scanning the receiving beam across the transmitting beam. The transmitting beam is then stepped one beamwidth and the receiving beam scan is repeated, and so forth until the transmitting beam has stepped across the surveillance sector. This approach increases the surveillance frame time by approximately the number of required transmitting beam steps, depending on the geometry, and is usually not acceptable for area surveillance. It can be considered for limited surveillance regions, for example, in an over-the-shoulder geometry, where surveillance is only required near the extended baseline.

The floodlight and multiple simultaneous beam (multibeam) approaches can be evaluated by using a variant of the monostatic surveillance radar range equation. The details can be found in [349, 446]. When floodlight and multibeam approaches are considered, their range performance will equal, or be less than, this base case performance depending on the required beam configuration. Five transmitting-receiving beam combinations are of interest [446].

9.2.1 One Transmitting Beam and Multiple Receiving Beams

For this case, the single transmitting beam gain G_T is assumed to satisfy

$$G_T = \frac{4\pi}{\Omega} \left(\frac{T_a}{T_0} \right) \tag{9.20}$$

where Ω, T_a and T_0 denote the required angular region to be searched, surveillance frame time and required time on target, respectively. When T_a/T_0 multiple, identical, receiving beams are designed to cover the search region, each receiving beam will have the gain of the single transmitting beam. Consequently, the bistatic radar maximum range equation can now be written as

$$(R_T \cdot R_R)_{\max} = \kappa \tag{9.21}$$

where κ is the bistatic maximum range product, and the term in parentheses is the reciprocal of the beam scan-on-scan loss.

Since each receiving need process only those range cells illuminated by the transmitting beam at any time, the processing load can be reduced by identifying these cells, given the geometry and transmitting beam position, and time multiplexing the signal processor to operate in these range cells. Cost can be further reduced by time multiplexing a cluster of receiving beams to cover only the volume illuminated by the transmitting beam at any time.

9.2.2   One Transmitting Beam and Flood Receiving Beam

The single transmitting beam gain is also assumed to satisfy Eq. (9.15). When the flood receiving beam covers the search region Ω, we have

$$G_R = \frac{4\pi}{\Omega} \tag{9.22}$$

Consequently, we can get

$$(R_T \cdot R_R)_{\max} = \kappa \sqrt{G_T G_R} \left(\frac{\Omega}{4\pi} \right) \left(\frac{T_0}{T_s} \right) = \kappa \sqrt{\frac{T_0}{T_a}} \tag{9.23}$$

Thus, when compared with the last approach, the bistatic maximum range product is reduced.

Moreover, mainbeam clutter levels are significantly increased, and angle measurement accuracy and target resolution are decreased. Consequently, this approach is not usually suitable for surveillance.

9.2.3   Flood Transmitting Beam and One Receiving Beam

The flood transmitting beam is assumed to cover the search region, Ω, and the single receiving beam is assumed to satisfy Eq. (9.20). Thus,

$$(R_T \cdot R_R)_{\max} = \kappa \sqrt{\frac{T_0}{T_a}} \tag{9.24}$$

In contrast to the one transmitting beam and flood receiving beam case, mainbeam clutter levels are reduced and angle measurement accuracy and target resolution are restored. However, sidelobe clutter levels remain high [125].

In special bistatic geometries, the flood transmitting beam can be tailored to illuminate only the region covered by the receiving beam when it is at a given pointing angle. Furthermore, if a minimum signal-to-noise ratio (SNR) required for detection is acceptable at all bistatic ranges, the flood transmitting beam can also be tailored to satisfy this requirement. The term $\sqrt{T_0/T_s}$ can then be eliminated.

9.2.4   Flood Transmitting Beam and Multiple Receiving Beams

The flood transmitting beam is assumed to cover the search region, Ω, and T_a/T_0 multiple receiving beams are designed to cover Ω, with each beam having a gain that satisfies Eq. (9.20). When the returns in each beam are integrated for T_0 seconds, $(R_T \cdot R_R)_{\max} = \kappa \sqrt{T_0/T_a}$. However, the integration time can be extended to the surveillance frame time T_a. With long integration time, targets can migrate through multiple range, Doppler and angle cells, which requires complex target association algorithms in the signal processor. Furthermore, integration efficiency is usually reduced because target returns decorrelate, which, in turn, requires noncoherent integration.

9.2.5 Flood Transmitting Beam and Flood Receiving Beam

Both the transmitting and receiving beams are assumed to cover the search region Ω. When the returns in each beam are integrated for T_0 seconds, we have

$$(R_T \cdot R_R)_{\max} = \kappa \sqrt{\frac{T_0}{T_a}} \qquad (9.25)$$

When the returns in each beam are integrated for T_a seconds, we can get

$$(R_T \cdot R_R)_{\max} = \kappa \sqrt{\frac{T_0}{T_a} \frac{E_{T_a}}{E_{T_0}}} \qquad (9.26)$$

where E_{T_a} is the integration efficiency over time T_a, and E_{T_0} is the integration efficiency over time T_0.

In this case, the multibeam receiving cost penalty is avoided, but at an expense of significantly reduced range performance, increased mainbeam clutter, and decreased angle measurement accuracy and target resolution.

9.3 Pulse Chasing Technique

Pulse chasing was proposed as a means to reduce the complexity and cost of multibeam bistatic receivers [184, 339, 350]. The fundamental requirements for pulse chasing have been derived in [184]. The simplest pulse chasing concept replaces the multibeam receiving system with a single beam. The single receiving beam rapidly scans the volume covered by the transmitting beam, essentially chasing the pulse as it propagates from the transmitter. In addition to the usual requirements for solving the bistatic triangle, pulse chasing requires knowledge of the transmitter look angle θ_T and pulse transmission time, which can be provided to the receiving site by a data link or be estimated by the receiver. Because the single beam chases one pulse at a time, a maximum allowable transmitting PRF exists such that only one pulse traverses the bistatic coverage area at one time.

Due to pulse propagation delays from the target to the receiver, the pointing angle of the receiving beam, θ_R, must lag the actual pulse position. For an instantaneous pulse position that generates a bistatic angle $\beta/2$, the required receiving beam poiting angle is [184]

$$\theta_R = \theta_T - 2\tan^{-1}\left(\frac{L\cos\theta_T}{R_T + R_R - L\sin\theta_T}\right) \tag{9.27}$$

Suppose the transmitting beamwidth is $\Delta\theta_T$. The minimum receiving beamwidth required to capture all returns from a range cell intersecting the common beam area is approximated by

$$\Delta\theta_{R_{\min}} \approx \frac{c_0 T_p \tan(\beta/2) + \Delta\theta_T R_T}{R_R} = \frac{\sin\theta_R}{\sin\theta_T} \tag{9.28}$$

It can be noticed that the required minimum receiving beamwidth changes as the receiving beam scans out the transmitting beam.

Other pulse chasing concepts are also possible. In one concept, the fixed beam receiving antenna is retained with $\Delta\theta_R = \Delta\theta_T$ for each receiving beam. Two receiver-signal processors are time-multiplexed across the beams. One processor steps across the even-numbered beams and the other processor steps across the odd-numbered beams, so that returns in beam pairs are processed simultaneously. A second concept uses two beams to scan over the required volume. It uses an identical leapfrog sequence and relaxes the fractional beam scan requirements by either sampling or stepping the beams in units of a beamwidth. A third concept uses one relatively broad beam, step scanning in increments of a half beamwidth over the required volume. The beamwidth must be larger than $2\Delta\theta_{R_{\min}}$ in order to capture all returns.

9.4 Sliding Spotlight and Footprint Chasing

The sliding spotlight and footprint chasing geometry was proposed in [282], as shown in Figure 9.5. The illumination start and end of a point target is dependent on the point target position in azimuth and slant range, $(t_{a,0}, r_0)$. For the sliding spotlight geometry, in order to obtain a maximal illumination time for the scene of interest, the full azimuth steering angle range is used for the sliding/staring spotlight mode. In contrast, in the footprint chasing geometry the direction of the receiver antenna pointing depends on the azimuth velocity and time. An important parameter is the time, during which a point target lies within the receiver footprint. The shorter this time interval is, the lower will be the azimuth resolution as it is seen by the receiver.

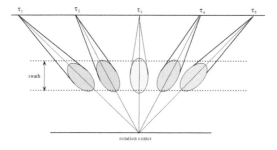

FIGURE 9.5: Sliding spotlight geometry definition.

9.4.1 Transmitter Sliding Spotlight and Receiver Footprint Chasing

The dimension of the receiver footprint in azimuth is calculated as

$$D_{az,r} = 2 \cdot \tan\left(\frac{\Delta\theta_{az,t}}{2}\right) R_c \qquad (9.29)$$

where $\Delta\theta_{az,t}$ is the receiver azimuth beamwidth. The receiver azimuth footprint velocity can be obtained by

$$v_{f,r} = \frac{D_{az,r}}{T_{illu, r}} \qquad (9.30)$$

where $T_{illu, r}$ is the time during which the point target lies within the receiver footprint.

The azimuth dimension of the transmitter footprint $D_{az,t}$ can be calculated according to Eq. (9.29). The azimuth footprint velocity of the transmitter can then be expressed as

$$v_{f,t} = v_{f,r}\left(1 + \frac{D_{az,t} - D_{az,r}}{Y_{sce} + D_{az,r}}\right) \qquad (9.31)$$

where Y_{sce} denotes the scene extension in azimuth. The whole time T_{sce}, during which the transmitter can perform illumination, depends also on the transmitter steering angle range $\Psi_{max,t}$, transmitter altitude h_t and incidence angle η_t

$$T_{sce} = \frac{2h_t}{\tan\left(\frac{\pi - \Psi_{max,t}}{2}\right) \cdot (v_t - v_{f,t})\cos(\eta_t)} \qquad (9.32)$$

9.4.2 Transmitter Staring Spotlight and Receiver Footprint Chasing

When the transmitter operates in staring spotlight mode that the transmitter footprint velocity in azimuth is zero, there are two possibilities to set up the receiver geometry. This is limited by the transmitter steering angle range $\Psi_{\max,t}$, transmitter altitude h_t, transmitter velocity v_t and incidence angle η_t

$$T_{\text{sce}} = \frac{2h_t \cos\left(\frac{\pi - \Psi_{\max,t}}{2}\right)}{v_t \cos(\eta_t)} \tag{9.33}$$

The receiver footprint velocity in azimuth can either be optimized with respect to the receiving time of a single point target or with respect to a desired scene extension in azimuth, Y_{sce}.

9.4.3   Azimuth Resolution

Azimuth resolution will be impacted by the illumination time length of a single target, i.e., $t_{\text{a,start}}$ and $t_{\text{a,end}}$. The illumination start and end azimuth frequencies $f_{\text{a,start}}$ and $f_{\text{a,end}}$ for a point target at position $(t_{a,0}, r_0)$ can be calculated from the instantaneous frequencies [284]

$$f_{\text{a,start}} = -\frac{2v_t^2}{\lambda} \cdot \frac{t_{\text{a,start}} - t_{a,0}}{\sqrt{r_0^2 + v_t^2 \left(t_{\text{a,start}} - t_{a,0}\right)^2}} \tag{9.34}$$

$$f_{\text{a,end}} = -\frac{2v_t^2}{\lambda} \cdot \frac{t_{\text{a,end}} - t_{a,0}}{\sqrt{r_0^2 + v_t^2 \left(t_{\text{a,end}} - t_{a,0}\right)^2}} \tag{9.35}$$

The azimuth resolution for each target position can then be calculated by

$$\rho_a(t_{a,0}, r_0) = \frac{\lambda}{2v_t} \sqrt{\frac{t_{\text{a,start}} - t_{a,0}}{\sqrt{r_0^2 + v_t^2 \left(t_{\text{a,start}} - t_{a,0}\right)^2}} - \frac{t_{\text{a,start}} - t_{a,0}}{\sqrt{r_0^2 + v_t^2 \left(t_{\text{a,start}} - t_{a,0}\right)^2}}} \tag{9.36}$$

It is easily understood that the azimuth resolution is better in far range than in near range since the illumination is longer in far range.

9.5   Multibeam Forming on Receiver

To avoid significant performance degradation, in one synthetic aperture time the range drift in azimuth should be satisfactory with

$$X_{\text{ad}} < \frac{1}{2}\left[\theta_T R_{T_c} - \theta_R R_{R_c}\right] \tag{9.37}$$

where R_{Tc} and R_{Rc} are the nearest range from the target to the transmitter and receiver, respectively. Up to now, the crucial parameters such as baseline length and antenna orientation are usually determined using GPS sensors. The

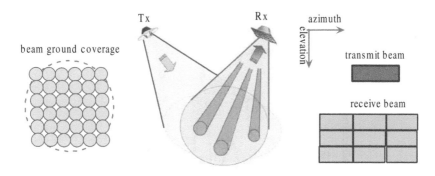

FIGURE 9.6: Illustration of the antenna directing synchronization.

achievable accuracy of a differential GNSS receiver in real-time is about $10cm$ and several millimeters after post-processing. For a high-resolution distributed SAR imaging the demanded real-time accuracy is in the range of millimeters.

As the antenna direction has to be determined in real-time to align it, an approach by installing a dedicated navigation unit on the transmitter and one on the receiver platform is proposed in [436]. The transmitter navigation unit generates a signal from the stable local oscillator and splits it into four individual navigation signals. These four channels are modulated, each one with an unique binary pseudo-random noise sequence. Each of them is transmitted over separate antennas, which are distributed over the transmitter platform. On the receiver platform there are likewise four navigation antennas installed. This idea shows the possible combination of different travel paths from the transmitter to the receiver platform for the navigation signals. However, this approach has an inherent drawbacks that the calibration signals have to be generated on all participating platforms and have to be transferred to the other platforms. As a consequence, no passive receiver exists and the whole system can easily be detected.

To overcome this disadvantages, a passive antenna directing synchronization approach, as shown in Figure 9.6, was proposed in [417] for bistatic SAR antenna synchronization. Suppose 2×2 (two-dimensional) beams are formed at the receiver, where two azimuth (or range) beams are separated by Δ. We take the return signals from the two antennas in range direction (or azimuth direction) and perform range (azimuth) compression processing. Next, their magnitude difference (see Figure 9.7) can be derived from

$$e(t, \Delta) = |A_{\text{left}}(t, \Delta)| - |A_{\text{right}}(t, \Delta)|, \text{subject to } e(t, \Delta) = 0 \qquad (9.38)$$

which gives the estimate of antenna directing errors required to realign the antenna direction. In this way, antenna direction can then be adjusted in real-time. Equally, the antenna directing synchronization can be ensured.

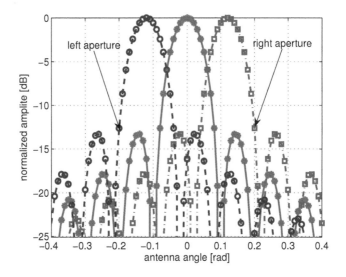

FIGURE 9.7: Passive antenna synchronization using two antennas on receive only.

9.6   Determination of Baseline and Orientation

If moving platforms are involved, as in the case of SAR, high precision navigation is an issue for motion compensation and for alignment of the antenna footprints on ground. For a high degree of automation at the processing of distributed SAR data the baseline between transmitter and receiver and the orientation of the platform towards each other must be known precisely during the recording. These parameters directly affect the achievable geometrical resolution and the geometry of the radar system. If these parameters already known at the measurement, it would be possible to adjust the receiving time gate and align the antenna footprints. Consequential time consuming optimization steps can be avoided at the data.

9.6.1   Four-Antenna-Based Method

Weiβ [436] proposed a method to solve this problem by installing a dedicated navigation unit on the transmitter and one on the receiver platform. The Tx navigation unit generates a signal from the stable local oscillator and splits it into four individual navigation signals. These four channels are modulated, each one with a unique binary pseudo-random noise (PRN) sequence for ranging and an additional navigation (broadcast) message. Each of these navigation signals is transmitted over separate antennas, which are distributed over the transmitter platform. These navigation antennas are pointing towards the Rx platform. The arrangement of the antennas on the Tx platform must be done in such a manner that a maximum face is spanned toward the Rx platform.

For ranging information the navigation signal is modulated by one of several codes. The navigation message includes information about the navigation antenna configuration on the transmitter airborne platform (offset from the primary Tx antenna, GPS coordinates and moving direction of the Tx platform). The PRN codes are chosen from a set of Gold codes and are unique for each navigation signal. Hence, the correlation between any pair of codes is very low. This allows using the same carrier frequency for all navigation signals. The four navigation signals are transmitted using separate navigation antennas and therefore travel on different paths towards the receiver. Then the four navigation signals arrive with unequal time delays at the same point on the receiver side. On the Rx platform there are likewise four navigation antennas installed. These navigation antennas are distributed over the receiver platform in a manner that a maximum area is spanned toward the other platform.

The distance between the Tx navigation antennas and the Rx navigation antennas can be determined by the time of the signal propagations. The pseudo-ranges are

$$R_{i,j} = c_0 \Delta t_{i,j} \tag{9.39}$$

where the indices i denotes the i-th transmitter antenna and j denotes the j-th receiver antenna. The measured time differences $\Delta t_{i,j}$ are biased by the time difference of the local time ΔT. The position of the receiving antenna can be determined from the different pseudo-ranges and the knowledge of the position of the transmitting antennas as follows

$$(x_1 - u_{xj})^2 + (y_1 - u_{yj})^2 + (z_1 - u_{zj})^2 = (R_{1,j} - c_0 \Delta T)^2 \tag{9.40a}$$

$$(x_2 - u_{xj})^2 + (y_2 - u_{yj})^2 + (z_2 - u_{zj})^2 = (R_{2,j} - c_0 \Delta T)^2 \tag{9.40b}$$

$$(x_3 - u_{xj})^2 + (y_3 - u_{yj})^2 + (z_3 - u_{zj})^2 = (R_{3,j} - c_0 \Delta T)^2 \tag{9.40c}$$

$$(x_4 - u_{xj})^2 + (y_4 - u_{yj})^2 + (z_4 - u_{zj})^2 = (R_{4,j} - c_0 \Delta T)^2 \tag{9.40d}$$

where (x_i, y_i, z_i) and (u_{xj}, u_{yj}, u_{zj}) are the position of the navigation antenna on the Tx platform and the navigation antenna on the receiver platform. The four unknown parameters can be determined from these four equations.

Suppose the vector from the receiver antenna j to the point of origin on the transmitter is $\mathbf{R}_j = [u_{xj}, u_{yj}, u_{zj}]$. The baseline between the two platforms can be calculated by

$$L = \frac{|\mathbf{R}_1 + \mathbf{R}_{1e}| + |\mathbf{R}_2 + \mathbf{R}_{2e}| + |\mathbf{R}_3 + \mathbf{R}_{3e}| + |\mathbf{R}_4 + \mathbf{R}_{4e}|}{4} \tag{9.41}$$

where $\mathbf{R}_{je}$ is the offset vector of the navigation antenna j on the receiver platform from the point of origin on the receiver platform. Using three of the four determined position of the navigation antennas on the receiving platform (u_{xj}, u_{yj}, u_{zj}), the orientation between both platforms can be calculated.

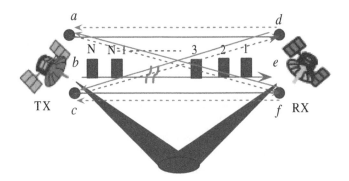

FIGURE 9.8: Two three-antenna radar platforms and the possible combinations of navigation signals.

9.6.2   Three-Antenna-Based Method

In the above four-antenna-based method, the influence of several contributions such as ionospheric propagation and oscillator synchronization errors. To overcome these disadvantages, Figure 9.8 gives a two-way ranging method with 3×3 antennas. At time T_{Tx}, three navigation signals are transmitted from Tx using separate antennas and therefore travel on different paths towards the Rx platform. They are received by RX antenna d at time $T_{Rx_{ad}}$, $T_{Rx_{bd}}$ and $T_{Rx_{cd}}$, respectively.

The pseudo-range between the Tx navigation antennas and each Rx navigation antenna can be determined by the signal propagation time. However, the measured time may be biased due to lack of time synchronization between the clocks inside the TX and RX platforms. Additionally, the time may be biased due to the effects of ionosphere and receiver noise. Hence, for RX antenna d, the measured time equation can be expressed as

$$\Delta T_{ad} = (T_{Rx_{ad}} - T_{Tx}) + \Delta T_{syn} + \Delta t_{ion} + \Delta t_{noi} \tag{9.42}$$

$$\Delta T_{bd} = (T_{Rx_{bd}} - T_{Tx}) + \Delta T_{syn} + \Delta t_{ion} + \Delta t_{noi} \tag{9.43}$$

$$\Delta T_{cd} = (T_{Rx_{cd}} - T_{Tx}) + \Delta T_{syn} + \Delta t_{ion} + \Delta t_{noi} \tag{9.44}$$

where ΔT_{syn}, Δt_{ion} and Δt_{noi} denote the time offsets caused by time synchronization errors, ionosphere effect and receiver noise, respectively.

As the effect of ionosphere can be compensated to satisfactory accuracy by using experiential radio propagation models, e.g., [183]

$$\Delta t_{ion} = \frac{A}{c_0 f_c^2} \tag{9.45}$$

where A is one constant. Eqs. (9.42) to (9.44) can then be further simplified into

$$\Delta T'_{ad} = (T_{Rx_{ad}} - T_{Tx}) + \Delta T_{syn} + \Delta t_{noi} \tag{9.46}$$

$$\Delta T'_{bd} = (T_{Rx_{bd}} - T_{Tx}) + \Delta T_{\text{syn}} + \Delta t_{\text{noi}} \tag{9.47}$$

$$\Delta T'_{cd} = (T_{Rx_{cd}} - T_{Tx}) + \Delta T_{\text{syn}} + \Delta t_{\text{noi}} \tag{9.48}$$

Similarly, for Rx antennas e and f, we also have

$$\Delta T'_{ie} = (T_{Rx_{ie}} - T_{Tx}) + \Delta T_{\text{syn}} + \Delta t_{\text{noi}} \tag{9.49}$$

$$\Delta T'_{if} = (T_{Rx_{if}} - T_{Tx}) + \Delta T_{\text{syn}} + \Delta t_{\text{noi}} \tag{9.50}$$

where $i \in (a, b, c)$.

To cancel out the effect of oscillator synchronization errors, three navigation signals are transmitted from the RX platform to the TX platform at time T_{Rx}. In a like manner, we can get

$$\Delta T'_{ji} = (T_{Tx_{ji}} - T_{Rx}) - \Delta T_{\text{syn}} + \Delta t_{\text{noi}} \tag{9.51}$$

with $j \in (d, e, f)$. Note that the $T_{Tx_{ji}}$ are defined as previously. As there is

$$T_{Rx_{ij}} - T_{Tx} \simeq T_{Tx_{ji}} - T_{Rx} \tag{9.52}$$

We can obtain the time of signal propagation between the navigation Tx antennas and Rx antennas

$$\begin{cases} \tau_{ij} = \tau_{ji} = \frac{\Delta T'_{ij} + \Delta T'_{ji}}{2} - \Delta t_{\text{noi}} \approx \frac{\Delta T'_{ij} + \Delta T'_{ji}}{2} \\ \Delta T_{\text{syn}} = \frac{\Delta T'_{ij} - \Delta T'_{ji}}{2} \end{cases} \tag{9.53}$$

Hence the pseudo-range between the Tx navigation antennas ($i \in (a, b, c)$) and Rx navigation antennas ($i \in (d, e, f)$) is the time offset τ_{ij} (or τ_{ji}) multiplied by the speed of light c_0 and is biased by the effect of receiver noise investigated in subsequent section

$$R_{ij} = R_{ji} = c_0 \cdot \tau_{ij} \approx c_0 \cdot \frac{\Delta T'_{ij} + \Delta T'_{ji}}{2} \tag{9.54}$$

Thus, the relative distance between Tx platform and Rx platform can be determined from the calculated pseudo-range. As an example, for the Rx antenna d we have

$$\begin{cases} (x_i - x_d)^2 + (y_i - y_d)^2 + (z_i - z_d)^2 = R_{id}^2 \\ (x_i - x_e)^2 + (y_i - y_e)^2 + (z_i - z_e)^2 = R_{ie}^2 \\ (x_i - x_f)^2 + (y_i - y_f)^2 + (z_i - z_f)^2 = R_{if}^2 \end{cases} \tag{9.55}$$

In this way, using the three determined positions of the navigation antennas on the Rx platform (x_j, y_j, z_j), the spatial baseline and orientation between both platforms can be determined. More importantly, this method can cancel the effect of oscillator frequency synchronization errors.

9.7   Conclusion

Antenna directing synchronization is also a specific requirement for distributed multi-antenna SAR systems. In this chapter, we analyzed the impacts of antenna directing errors on distributed multi-antenna SAR imaging. Several antenna direction synchronization techniques such as scan-on-scan, pulse chasing, sliding spotlight and multibeam forming on receiver were introduced. The multibeam forming on receiver is the most promising synchronization technique. Additionally, determination of baseline and orientation in distributed multi-antenna SAR systems was also discussed and two measurement methods were introduced.

10

Azimuth-Variant Multi-Antenna SAR Image Formulation

Processing

Distributed multi-antenna SAR, especially bistatic SAR, will play a great role in future remote sensing because of its special advantages [53, 218, 407]. Although bistatic SAR has received much recognition in recent years [37, 67, 276, 406], bistatic configurations considered in open literature are primarily azimuth-invariant bistatic SAR [327, 345, 396, 476]. Typical examples are the "Tandem", where the transmitter and receiver moving one after another with the same trajectory, and "translational invariant" SAR, where the transmitter and receiver move along parallel trajectories with the same velocity. In contrast, azimuth-variant bistatic SAR has received little recognition. One typical azimuth-variant configuration employs a spaceborne transmitter and a receiver located on or near Earth's surface; the receiver may, or may not, be stationary [11, 59]. Thus, azimuth-variant bistatic SAR image formation processing is the topic of this chapter.

When compared with azimuth-invariant bistatic SAR, azimuth-variant bistatic SAR has more complex signal models and resolution characteristics. Correspondingly, azimuth-variant bistatic SAR data processing and image formation focusing will be much more difficult than azimuth-invariant bistatic SAR. This situation will become more complicated for unflat digital-earth model (DEM) topography. In monostatic or azimuth-invariant bistatic SAR systems, scene topography can be ignored in developing image formation algorithms because the measured range delay is related with the double target distance and the observed range curvature; hence, topography is only used to project the compressed image, which is in slant range to the ground range. However, for azimuth-variant bistatic SAR systems, it is mandatory to know both the transmitter-to-target distance and target-to-receiver distance to properly focus the radar data, but they clearly depend on the target height.

Practically, bistatic SAR data processing represents an original scientific task because novel procedures must be developed. As an example, the following problems must be addressed: bistatic Doppler shift and spread; bistatic imagery forming geometry; matched filtering, focusing and motion compensation; synchronization compensation; coregistration with monostatic data.

Several bistatic SAR data processing related techniques have been proposed, e.g., bistatic SAR ambiguity function and its application to signal design was studied in [380, 471], showing the significant effects of bistatic geometry on the resolution capabilities of the transmitted waveform and the need for novel approaches. Procedures for bistatic pulse compression were proposed in [82, 307] and a two-dimensional range-azimuth domain ambiguity function was presented in [352]. A procedure for onboard matched filtering of a spaceborne bistatic SAR which accounts for linear range migration of moving targets was reported in [97]. A processing technique of airborne and spaceborne radar was presented in [377]. A range-Doppler processor accounting for range migration and its validation performed by using airborne bistatic SAR data was proposed in [111] and a procedure for attitude determination of bistatic radar based on single-look complex data processing was presented in [330]. The bistatic SAR point target response in space-time and frequency domains was derived in [260], giving evidence of some peculiar aspects in bistatic imagery geometric characteristics.

In this chapter, we concentrate mainly on azimuth-variant bistatic SAR data processing aspects. The system configuration, signal models, and resolution characteristics are investigated. General two-dimensional spectrum model for azimuth-variant bistatic SAR is derived. Furthermore, a nonlinear chirp scaling (NCS)-based image focusing algorithm is developed to address the problem of azimuth-variant Doppler characteristics. This chapter is organized as follows. In Section 10.1, bistatic SAR system configuration geometry and signal models are given, followed by the system performance analysis in Section 10.2. In Section 10.3, the azimuth-variant Doppler characteristics are investigated, along with two-dimensional spectrum characteristics. Next, motion compensation is discussed in Section 10.4. Finally, one NCS-based image focusing algorithm is developed in Section 10.5.

10.1   Introduction

Bistatic SAR systems can be divided into semi-active and fully active configurations, as shown in Figure 10.1. Semi-active configurations combine passive receiver with an active radar illuminator; fully active bistatic SAR uses two conventional radars flying in close formation to acquire bistatic SAR data during a single pass. Both of the radars have a fully equipped radar payload with transmit and receive capabilities. TanDEM-X is just an example fully active bistatic SAR system [289].

But, in this chapter, we classify bistatic SAR into azimuth-invariant and azimuth-variant configurations. Two typical azimuth-invariant configurations are the pursuit monostatic mode (see Figure 10.2 (left)), where the transmitter and receiver are flying along the same trajectory at an equal velocity, and

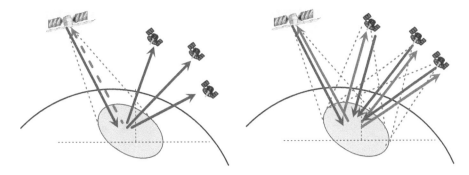

FIGURE 10.1: Semi-active (left) and fully active (right) bistatic SAR.

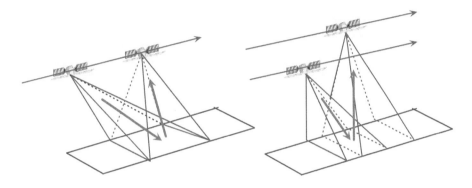

FIGURE 10.2: Two typical azimuth-invariant bistatic SAR configurations.

translational invariant mode (see Figure 10.2 (right)), where the transmitter and receiver are flying along parallel trajectories with the same velocity. Azimuth-invariant configuration can be seen as an equivalent monostatic configuration. In contrast, azimuth-variant configuration cannot be seen as an equivalent monostatic configuration because the geometry between transmitter and receiver will change with azimuth time. Space-surface bistatic SAR with spaceborne transmitter and airborne receiver is just an example azimuth-variant configuration [10, 398]. Before discussing azimuth-variant bistatic SAR data processing, we should first appreciate its differences with the azimuth-invariant case. In an azimuth-invariant situation with straight orbit, the acquisition has rotational symmetry-the axis being the orbit itself-and it may be considered as two-dimensional, rather than three-dimensional. This compaction from a three-dimensional space to a two-dimensional space cannot be happen for azimuth-variant configuration where transmitter and receiver follow different paths. Thus, the three-dimensional geometry along azimuth has to be characterized to simplify subsequent image formation algorithms.

Let us consider a general bistatic SAR geometry as shown in Figure 10.3,

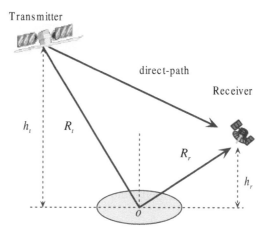

FIGURE 10.3: General bistatic SAR configuration with two-channel receiver.

where the transmitter and receiver may have unequal velocities. Although the transmitter or receiver may be stationary, an aperture synthesis can still be obtained by the transmitter or receiver only. Without loss of generality, we suppose the receiver consists of two channels. One channel is fixed to collect the direct-path signals coming from the transmitter antenna sidelobes, which can be used as the reference function for matched filtering or for bistatic SAR synchronization compensation [422]. The second channel is configured to gather the reflected signals with which radar imaging is attempted. Note that another promising configuration is the digital beamforming on receiver technique [148, 463]. In this case, the receiver antenna is split into multiple subapertures, and each subaperture signal is separately amplified, downconverted and digitized. The digital signals are then combined in a dedicated processor to form multiple sub-beams with arbitrary shapes. If passive receiver is employed, this bistatic SAR has robust survivability in military reconnaissances because detecting the passive receiver is a technical challenge. Moreover, rather than emitting signals, it relies on opportunistic transmitters which allows for the receiver to operate without emitting energy; hence, it has no impacts on existing communication and navigation systems.

For a given imaging target, transmitter, and receiver altitudes, the target must be simultaneously within line-of-sight (LOS) to both the transmitter and receiver sites. That is to say, the antenna directing synchronization should be ensured for bistatic SAR imaging [463]. If an airborne transmitter is used, antenna directing synchronization can be implemented though a cooperative flying plan. However, if a spaceborne transmitter is employed, it will be a challenge due to the fact that a satellite usually has a high flying speed. Moreover, the satellite antenna direction is often uncontrollable for us. The pulse chasing technique proposed in [93] cannot be easily implemented for noncooperative bistatic SAR configurations, particularly when spaceborne transmitter is employed. To reach antenna directing synchronization for spaceborne transmitter, we use a wide antenna beamwidth on receiver due to its advantage of high signal-to-noise ratio (SNR). The radar equation of bistatic radar system is given by [318]

$$P_r = \frac{R_t \lambda^2 G_t G_r \sigma_b^o A_{\mathrm{res}}}{(2\pi)^3 R_t^2 R_r^2} \tag{10.1}$$

where P_t and P_r are the average transmitting and receiving power, G_t and G_r are the gain of the transmitter and receiver, R_t and R_r are the distance from the transmitter and receiver to the imaged scene, σ_b^o is the bistatic scattering coefficient and A_{res} is the size of the resolution cell. The SNR in receiver can then be represented by [392, 391]

$$\mathrm{SNR}_b = \frac{P_t \lambda^2 G_t G_r \sigma_b^o A_{\mathrm{res}} \xi_{\mathrm{int}} \eta}{(4\pi)^3 R_t^2 R_r^2 K_b T_0 F_n} \tag{10.2}$$

where ξ_{int} is the coherent integration time, η is the duty cycle, K_b is the Boltzman constant, T_0 is the system noise temperature and F_n is the noise figure.

Simplify, we suppose the transmitter and receiver are flying in a parallel trajectory (but their flying velocities are not equal), then the size of the resolution cell is expressed as

$$A_{\mathrm{res}} = \frac{\lambda}{v_t/R_t + v_r/R_r} \cdot \frac{1}{\xi_{\mathrm{int}}} \frac{c_0}{2 B_r \cos(\beta/2) \sin(\gamma_b)} \tag{10.3}$$

where v_t (v_r) is the transmitter (receiver) velocity, R_t (R_r) is the transmitter-to-target (target-to-receiver) distance, B_r is the transmitted signal bandwidth, β is the bistatic angle and γ_b is the incidence angle of the bistatic angle bisector. Eq. (10.2) can then be changed into

$$\mathrm{SNR}_b = \frac{P_t \lambda^2 G_t G_r \sigma_b^o \eta}{(4\pi)^3 R_t^2 R_r^2 K_b T_0 F_n} \cdot \frac{\lambda}{v_t/R_t + v_r/R_r} \cdot \frac{c_0}{2 B_r \cos(\beta/2) \sin(\gamma_b)} \tag{10.4}$$

Similarly, for the corresponding monostatic SAR, there is

$$\mathrm{SNR}_m = \frac{P_t \lambda^2 G_t^2 \sigma_m^0 \eta}{(4\pi)^3 R_t^4 K_b T_0 F_n} \cdot \frac{\lambda}{2 v_t/R_t} \cdot \frac{c_0}{2 B_r \sin(\gamma_m)} \tag{10.5}$$

where σ_m^0 and γ_m are the monostatic scattering coefficient and radar incidence angle, respectively. Note that, here, equal system noise temperature and noise figure are assumed. We then have

$$K_\mu = \frac{\mathrm{SNR}_b}{\mathrm{SNR}_m} = \frac{G_r}{G_t} \cdot \left(\frac{R_t}{R_r}\right)^2 \cdot \frac{2 v_t/R_t}{v_t/R_t + v_r/R_r} \cdot \frac{\sigma_b^0}{\sigma_m^0} \cdot \frac{\sin(\gamma_m)}{\cos(\beta/2) \sin(\gamma_b)} \tag{10.6}$$

As an example, supposing the following parameters: $R_t = 800$ km,

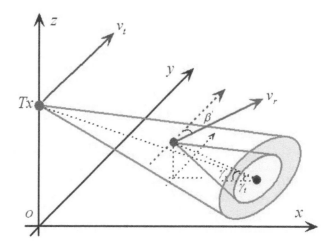

FIGURE 10.4: General geometry of azimuth-variant bistatic SAR systems.

$v_t = 7000$ m/s, $R_r = 30$ km, $v_r = 5$ m/s, $\gamma_b = 45°$, $\gamma_m = 60°$, $\beta = 30°$ and $\sigma_b^0 = \sigma_m^0$, the K_μ is found to be $1973.70 G_r/G_t$. This points out the receiver antenna beamwidth can be significantly extended to provide the same SNR in comparison to the monostatic case; hence, an effective antenna directing synchronization and an extended bistatic SAR imaging coverage can be obtained by extending the receiver antenna's beamwidth.

10.2 Imaging Performance Analysis

We consider a general azimuth-variant bistatic SAR imaging geometry, as shown in Figure 10.4. Both the transmitter and receiver are flying on a straight but nonparallel path with a constant but unequal velocity. The azimuth time is chosen to be zero at the composite beam center crossing time of the reference target. Spatial resolution analysis is an effective way to evaluate its imaging performance [285, 471], but most published methods are only suitable for azimuth-invariant bistatic SAR configurations [286]. Here, we investigate the spatial resolution in a rather general way.

10.2.1 Imaging Time and Imaging Coverage

For bistatic SAR imaging, both of the antennas are steered to obtain an overlapping beam on the ground, and the exposure of a target is governed by this composite beam pattern. An analysis of imaging time and imaging coverage is thus necessary. A space-time diagram-based approach was investigated in [300], but a rectangular approximation is employed.

In fact, the whole bistatic SAR imaging time is determined by

$$T_{\text{image}} = \frac{D_{\text{a,t}} + D_{\text{a,r}}}{|v_t - v_r|} \tag{10.7}$$

where v_t and v_r are the transmitter velocity and receiver velocity, $D_{\text{a,t}}$ and $D_{\text{a,r}}$ are the illuminated ground coverage in azimuth for the transmitter and receiver, respectively.

They are determined, respectively, by

$$D_{\text{a,t}} \approx 2 \frac{h_t}{\sin(\gamma_t)} \tan(\frac{\theta_{\text{a,t}}}{2}) \tag{10.8a}$$

$$D_{\text{a,r}} \approx 2 \frac{h_r}{\sin(\gamma_r)} \tan(\frac{\theta_{\text{a,r}}}{2}) \tag{10.8b}$$

where h_t and h_r are the altitude, γ_t and γ_r are the incidence angle, and $\theta_{\text{a,t}}$ and $\theta_{\text{a,r}}$ are the antenna beamwidth in azimuth direction for the transmitter and receiver, respectively. The imaging coverage in azimuth can then be represented by

$$L_{\text{az}} = D_{\text{a,r}} - v_r \cdot T_{\text{image}} \tag{10.9}$$

Similarly, the imaging coverage in range direction is determined by

$$L_{\text{ra}} = D_{\text{r,r}} \approx 2 \frac{h_r}{\sin(\gamma_r)} \tan(\frac{\theta_{\text{r,r}}}{2}) \tag{10.10}$$

where $D_{\text{r,r}}$ and $\theta_{\text{r,r}}$ are the imaging ground coverage in range direction and antenna beamwidth in range direction for the receiver, respectively.

10.2.2 Range Resolution

From Figure 10.4 we know that the instantaneous range history of the transmitter and receiver to an arbitrary point target $(x, y, 0)$ is

$$R = \sqrt{(x - x_t)^2 + (y - y_t)^2 + h_t^2} + \sqrt{(x - x_r)^2 + (y - y_r)^2 + h_r^2} \tag{10.11}$$

where (x_t, y_t, h_t) and (x_r, y_r, h_r) are the coordinates of the transmitter and the receiver, respectively. We then have

$$\nabla R = \frac{\partial R}{\partial x} \vec{i}_x + \frac{\partial R}{\partial y} \vec{i}_y$$
$$= [\sin(\alpha_t) \cos(\zeta_t) + \sin(\alpha_r) \cos(\zeta_r)] \vec{i}_x + [\sin(\zeta_t) + \sin(\zeta_r)] \vec{i}_y \tag{10.12}$$

where $\alpha_t = \alpha_t(x)$ and $\alpha_r = \alpha_r(x)$ are the instantaneous looking-down angles, $\zeta_t = \zeta_t(x, y; y_t)$ and $\zeta_r = \zeta_r(x, y; y_r)$ (y_t, y_r is the instantaneous location in the y-direction) are the instantaneous squint angles.

Eq. (10.12) can be rewritten as

$$|\nabla R| = \sqrt{[\sin(\alpha_t)\cos(\zeta_t) + \sin(\alpha_r)\cos(\zeta_r)]^2 + [\sin(\zeta_t) + \sin(\zeta_r)]^2} \quad (10.13)$$

As the range resolution of a monostatic SAR is $c_0/2B_r$, the range resolution (in x direction) can then be derived as

$$\rho_r = \frac{c_0/B_r}{|\nabla R|} \cdot \frac{1}{\sin(\zeta_{xy})} \quad (10.14)$$

with

$$\zeta_{xy} = \arctan\left[\frac{\sin(\zeta_t) + \sin(\zeta_r)}{\sin(\alpha_t)\cos(\zeta_t) + \sin(\alpha_r)\cos(\zeta_r)}\right] \quad (10.15)$$

Hence, we have

$$\rho_r = \frac{c_0/B_r}{[\sin(\alpha_t)\cos(\zeta_t) + \sin(\alpha_r)\cos(\zeta_r)]} \quad (10.16)$$

It can be concluded that the range resolution is determined by not only the transmitted signal bandwidth, but also the specific bistatic SAR configuration geometry. When $\alpha_t = \alpha_r = \alpha$ and $\zeta_t = \zeta_r = \zeta$, we get

$$\rho_r = \frac{c_0/2B_r}{[\sin(\alpha)\cos(\zeta)]} \quad (10.17)$$

This is just the range resolution for general monostatic SAR systems.

10.2.3 Azimuth Resolution

Consider again Figure 10.4, at an azimuth time τ, the range sum to an arbitrary reference point is

$$R(\tau) = \sqrt{R_{t0}^2 + (v_t\tau)^2} + \sqrt{R_{r0}^2 + (v_r\tau)^2} \quad (10.18)$$

where R_{t0} and R_{r0} are the closest ranges to a given point target when the transmitter and receiver move along their trajectories, respectively. As the propagation speed of electromagnetic signal is much faster than the speed of platforms, the *stop-and-go* hypothesis is still reasonable.

The instantaneous Doppler frequency can then be derived as

$$f_d(\tau) = -\frac{1}{\lambda} \cdot \frac{\partial R(\tau)}{\partial \tau} = -\frac{1}{\lambda} \cdot \left[\frac{v_t^2\tau}{\sqrt{R_{t0}^2 + (v_t\tau)^2}} + \frac{v_r^2\tau}{\sqrt{R_{r0}^2 + (v_r\tau)^2}}\right] \quad (10.19)$$

Correspondingly the Doppler chirp rate is

$$k_d(\tau) = -\frac{1}{\lambda} \cdot \frac{\partial^2 R(\tau)}{\partial^2 \tau} \approx -\frac{1}{\lambda} \cdot \left[\frac{v_t^2}{R_{t0}}\cos\left(\frac{v_t\tau}{R_{t0}}\right) + \frac{v_r^2}{R_{r0}}\cos\left(\frac{v_r\tau}{R_{r0}}\right)\right] \quad (10.20)$$

The synthetic aperture time is determined by

$$T_s = \min\left\{\frac{\lambda R_{t0}}{L_t v_t}, \frac{\lambda R_{r0}}{L_r v_r}\right\} \tag{10.21}$$

where L_t and L_r are the transmitting antenna length and receiving antenna length, respectively.

If

$$\frac{\lambda R_{t0}}{L_t v_t} > \frac{\lambda R_{r0}}{L_r v_r} \tag{10.22}$$

the corresponding Doppler bandwidth is

$$B_d(\tau) = \int_{-T_s/2}^{T_s/2} |k_d(\tau)|\, d\tau = 2\left[\frac{v_t}{\lambda}\sin\left(\frac{\lambda R_{r0} v_t}{2D_r R_{t0} v_r}\right) + \frac{v_r}{\lambda}\sin\left(\frac{\lambda}{2D_r}\right)\right] \tag{10.23}$$

The azimuth resolution can then be expressed as

$$\rho_a = \frac{v_{eq}}{B_d(\tau)} = \frac{v_{eq}}{2\left[\frac{v_t}{\lambda}\sin\left(\frac{\lambda R_{r0} v_t}{2L_r R_{t0} v_r}\right) + \frac{v_r}{\lambda}\sin\left(\frac{\lambda}{2L_r}\right)\right]} \tag{10.24}$$

where

$$v_{eq} = \sqrt{v_t^2 + v_r^2 - 2v_t v_r \cos(\pi - \beta')} \tag{10.25}$$

is the equivalent velocity, β' is defined as the angle between the transmitter velocity vector and the receiver velocity vector.

Similarly, if

$$\frac{\lambda R_{t0}}{L_t v_t} < \frac{\lambda R_{r0}}{L_r v_r} \tag{10.26}$$

we can get

$$\rho_a = \frac{v_{eq}}{2\left[\frac{v_r}{\lambda}\sin\left(\frac{\lambda R_{t0} v_r}{2L_t R_{r0} v_t}\right) + \frac{v_t}{\lambda}\sin\left(\frac{\lambda}{2L_t}\right)\right]} \tag{10.27}$$

From Eqs. (10.24) and (10.27) we can see that, unlike the azimuth resolution of general monostatic SAR determined only by the azimuth antenna length, the azimuth resolution will be impacted by its bistatic SAR configuration geometry and velocity.

When $R_{t0} = R_{r0}$, $v_t = v_r$ and $L_t = L_r$, there is $\rho_a = L_t/2 = L_r/2$. In this case, it is just a monostatic SAR. If $v_t = 0$ or $v_r = 0$ and $L_t = L_r$, the azimuth resolution is found to be $\rho_a = L_t = L_r$. This case is just a fixed-transmitter or fixed-receiver bistatic SAR [272]. Note that there is a local minimum for the azimuth resolution, which is $L_r/2$ or $L_t/2$ (depending on whether $\frac{\lambda R_{t0}}{L_t v_t} > \frac{\lambda R_{r0}}{L_r v_r}$ or $\frac{\lambda R_{t0}}{L_t v_t} < \frac{\lambda R_{r0}}{L_r v_r}$).

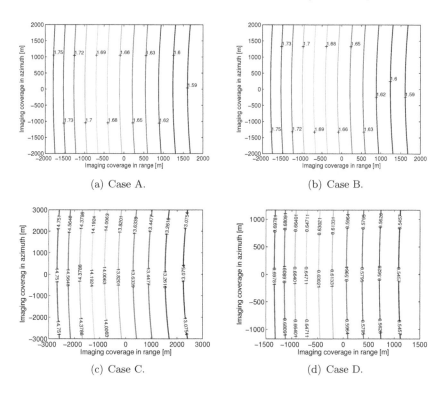

FIGURE 10.5: Range resolution results of the example bistatic SAR configurations.

10.2.4   Simulation Results

To obtain quantitative evaluation, we consider several typical configurations. Table 10.1 gives the simulation parameters. In Cases A and B, the spaceborne TerraSAR-X is used as the transmitter [232]. In Case C, the spaceborne Envisat SAR is used as the transmitter [250]. Note that near-space vehicle is defined as the vehicle flying in near-space altitude, which is from 20 km to 100 km [378, 395, 403, 419]. We can notice that an imaging coverage size of dozens of square kilometers (the size is from 2 km × 3 km to 6 km × 6 km in the simulation examples) can be obtained. This imaging coverage is satisfactory for regional remote sensing applications.

Figure 10.5 gives the range resolution results of the example bistatic SAR configurations. It is noticed that the range resolution has geometry-variant characteristics, which depend on not only slant range but also azimuth range. It degrades with the increase of azimuth range displaced from scene center. To ensure a consistent range resolution, the imaged scene coverage should be limited or a long slant range should be designed.

Figure 10.6 gives the azimuth resolution results of the four example bistatic

TABLE 10.1: Example bistatic SAR configuration parameters.

Parameters	Case A		Case B		Case C		Case D	
	Tx	Rx	Tx	Rx	Tx	Rx	Tx	Rx
Carry frequency (GHz)	9.65	9.65	9.65	9.65	5.33	5.33	1.25	1.25
Flying altitude (km)	515	20	515	20	800	20	10	20
Flying velocity (m/s)	7600	5	7600	0	7450	5	100	5
Signal bandwidth (MHz)	150	150	150	150	16	16	500	500
Beam incidence angle (°)	45	60	45	60	30	60	60	60
Beamwidth in range (°)	2.3	10	2.3	10	2.3	15	15	15
Beamwidth in azimuth (°)	0.33	10	0.33	10	0.28	15	13.75	13.75
Imaging time (s)	1.0844		1.0837		1.8670		85.3248	
Imaging coverage (x, y) (km)	(4.0409, 4.0355)		(4.0409, 4.0409)		(6.0808, 6.0714)		(2.3448, 3.0404)	

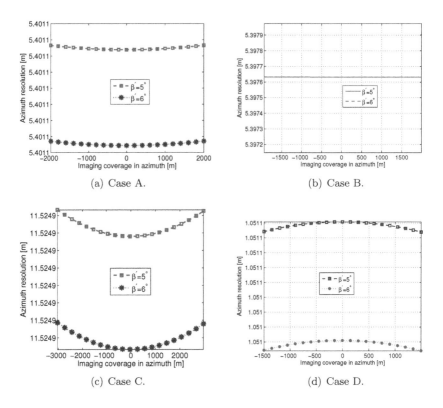

FIGURE 10.6: Azimuth resolution results of the example bistatic SAR configurations.

SAR configurations. It is noticed that the angle (β') between the transmitter velocity vector and the receiver velocity vector has also an impact on the azimuth resolution. Additionally, we notice that the azimuth resolution between scene center and scene edge has a small performance difference. This phenomenon is caused by the change of azimuth time. Further simulation shows that, if the synthetic aperture time is determined primarily by the receiver (e.g., Cases A and B), and suppose the transmitter's velocity is constant, the azimuth resolution will degrade with the increase of receiver's velocity. However, if the synthetic aperture time is determined primarily by the transmitter (e.g., Case D), and suppose the transmitter's velocity is constant, the azimuth resolution will be improved with the increase of the receiver's velocity.

10.3   Azimuth-Variant Characteristics Analysis

As azimuth-variant geometry brings much challenge towards developing high-precision data processing, an analysis of the azimuth-variant characteristics such as Doppler and two-dimensional spectrum is necessary for developing efficient image formation algorithms.

10.3.1   Doppler Characteristics

From Eq. (10.18) we can notice that the range history of a point target does not depend any more on the zero-Doppler distance and the relative distance from the target to the transmitter, but also on the absolute distance to the receiver. According to Eq. (10.20), using the parameters listed in Table 10.1 Figure 10.7 shows the calculated Doppler chirp rate results of the example bistatic SAR configurations. Both Case A and Case B are given in Figure 10.7(a) because they have the only difference in receiver velocity. Note that the synthetic aperture times in Cases A-C are determined primarily by the transmitter, but in Case D it is determined primarily by the receiver. It is noticed that the Doppler chirp rate is azimuth variant. This azimuth-variant phenomenon brings a great challenge towards developing efficient focusing algorithms.

Consequently, the locus of the slant ranges at the beam center crossing times of all targets parallel to the azimuth axis follow an approximate hyperbola. As an example, we consider a typical geometry shown in Figure 10.8 (left), where the (x, y) plane is locally tangent to the surface of the earth. The targets are assumed to lie on this plane, and the transmitter velocity vector is parallel to the y-axis. Figure 10.8 (right) shows the trajectories of three targets A, B and C as well as the hyperbolic locus. Because of the hyperbolic locus, after linear range cell mitigation correction targets such as A, C and D at the same range gate. This is not similar to the monostatic case, which has a linear locus instead of hyperbolic. General range Doppler image formation algorithms cannot handle this problem in a high-precision manner. As a consequence, there is a range ambiguity that does not exist in monostatic cases. It can be observed from Figure 10.8 that two or more targets (e.g., A, C and D) located at different positions can have the same range delay at zero-Doppler but will have different range histories (curvature). Similar phenomena has been investigated in [449], where a ground based stationary receiver is assumed. But, for nonstationary receiver this problem becomes being more complex, because the transmitter follows a rectilinear trajectory, while the receiver follows also a rectilinear trajectory but with a different velocity.

The situation will become more complicated for unflat DEM topography. In monostatic or azimuth-invariant bistatic SAR systems, scene topography can be ignored in developing image formation algorithms because the mea-

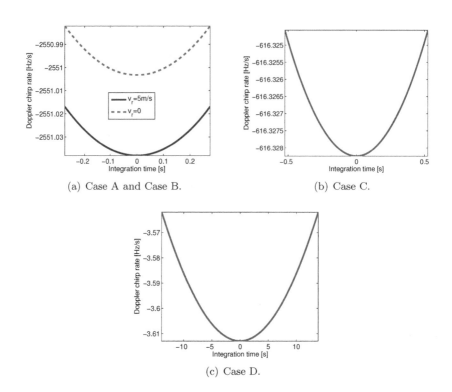

(a) Case A and Case B.   (b) Case C.

(c) Case D.

FIGURE 10.7: Azimuth-variant Doppler characteristics.

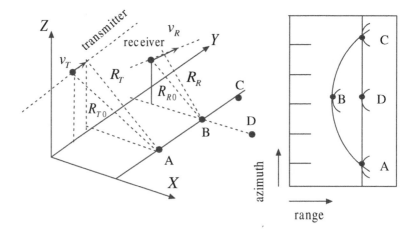

FIGURE 10.8: Targets with the same range delay at zero-Doppler but that will have different range histories.

sured range delay is related with the double target distance and the observed range curvature; hence, topography is only used to project the compressed image which is in slant range to the ground range. However, for azimuth-variant bistatic SAR it is mandatory to know both the transmitter-to-target distance and target-to-receiver distance to properly focus its raw data, but they clearly depend on the target height. In this case, conventional imaging algorithms such as chirp-scaling and wavenumber-domain, may be not suitable to accurately focus the collected data.

It is necessary to develop some image formation algorithms usable for azimuth-variant bistatic SAR configuration. However, most of the published bistatic SAR algorithms are only usable to handle azimuth-invariant bistatic SAR configurations [298, 472], only a few ones usable for azimuth-variant bistatic SAR have been investigated [11, 18]. As the processing efficiency of image formation algorithms depends primarily on the spectrum model of the echo signal, an analysis of two-dimensional spectrum model is necessary for developing an effective image formation algorithm.

10.3.2   Two-Dimensional Spectrum Characteristics

Loffeld et al. [260] divided the bistatic phase history into two quasi-monostatic phase histories and expended them into Taylor series around the individual points of stationary phase. In this way, an approximation of stationary-phase azimuth time is constructed, and a two-dimensional bistatic SAR spectrum is deduced. Several algorithms have been developed based on this model [295]. Another bistatic SAR spectrum model using series expansion to express azimuth time as a function of azimuth frequency during azimuth Fourier transform was proposed in [297], and a chirp scaling algorithm was developed based on this model [450]. However, the accuracy is controlled by keeping enough terms in the power series, but only the first several series can be used. Using the Fresnel approximation, a two-dimensional bistatic SAR spectrum model was derived in [150]. This model can be used only in small squint angle. Moreover, it has no advantages in approximate error because the Fresnel approximation is used. In summary, most bistatic SAR spectrum models have to use some approximations by keeping enough terms in the power series, but cannot be used in the bistatic SAR with large difference between the transmitter-to-target distance and target-to-receiver distance or large squint angle.

To evaluate the applicability of the Loffeld's model for azimuth-variant bistatic SAR, we suppose the transmitted signal is

$$s_t(t) = p(t)\exp(j2\pi f_c t) \tag{10.28}$$

where $p(t)$ is an encoded wide bandwidth signal, for example chirp signal, and upconverted by the transmitter to the carrier frequency f_c. The echo reflected from a point target experiences a delay that is proportional to the range history $R(\tau)$

$$s_r(t, \tau; R_{r0}, \tau_{r0}) = s_t \left(t - \frac{R(\tau)}{c_0} \right) \tag{10.29}$$

where τ_{r0} is the azimuth time when the point target is seen perpendicularly to the receiver track. Note that the antenna pattern and constant amplitude terms in the signal have been omitted. After down-conversion, the demodulated signal is

$$s_r(t, \tau; R_{r0}, \tau_{r0}) = p \left(t - \frac{R(\tau)}{c_0} \right) \exp \left(j2\pi \frac{R(\tau)}{\lambda} \right) \tag{10.30}$$

After Fourier transforming from range-time domain to range-frequency domain, we have

$$s_r(f_r, \tau; R_{r0}, \tau_{r0}) = P(f_r) \exp \left(-j2\pi \frac{f_r + f_c}{c_0} R(\tau) \right) \tag{10.31}$$

where $P(f_r)$ is the Fourier transform of $p(t)$. Transforming from azimuth-time to azimuth-frequency domain, we obtain

$$s_r(f_r, f_a; R_{r0}, \tau_{r0}) = P(f_r) \int_{-\infty}^{\infty} \exp \left(-j2\pi \frac{f_r + f_c}{c_0} R(\tau) - j2\pi f_a \tau \right) d\tau \tag{10.32}$$

where f_a is the azimuth Doppler frequency.

Loffeld et al. expanded the phase histories expressed in Eq. (10.32) into Taylor series around the individual points of stationary phase

$$\Phi_T(\tau) = -2\pi \left[(f_r + f_c) \frac{R_t(\tau)}{c_0} - f_a \tau \right]$$

$$\approx \Phi_T(\tau_T^*) + \Phi_T'(\tau_T^*)(\tau - \tau_T^*) + \frac{1}{2} \Phi_T''(\tau_T^*)(\tau - \tau_T^*)^2 + \cdots \tag{10.33}$$

$$\Phi_R(\tau) = -2\pi \left[(f_r + f_c) \frac{R_r(\tau)}{c_0} - f_a \tau \right]$$

$$\approx \Phi_R(\tau_R^*) + \Phi_R'(\tau_R^*)(\tau - \tau_R^*) + \frac{1}{2} \Phi_R''(\tau_R^*)(\tau - \tau_R^*)^2 + \cdots \tag{10.34}$$

By supposing the second-order range rate contributions are negligible against the linear terms in the common point of stationary phase and that the fourth-order Lagrangian error term (summing up all the higher-order terms) of the Taylor series expansion is negligible against the third-order term, they obtain the common stationary phase point

$$\tau^* = \frac{\Phi_T''(\tau_T^*)\tau_T^* + \Phi_R''(\tau_R^*)\tau_R^*}{\Phi_T''(\tau_T^*) + \Phi_R''(\tau_R^*)} \tag{10.35}$$

However, as noted in [260], the validity of Eq. (10.35) is constrained by

$$|\tau_T^* - \tau_{0T}|^2 = \frac{2R_{0t}^2}{7v_t^2} \tag{10.36a}$$

$$|\tau_R^* - \tau_{0R}|^2 = \frac{2R_{0r}^2}{7v_r^2} \tag{10.36b}$$

Where τ_{0T} is the azimuth time when the point target is seen perpendicularly to the transmitter track. Figure 10.9 shows the constrains expressed in Eq. (10.36) for several typical bistatic SAR velocity configurations. We can see that the Loffeld model can only be used in small squint angle or small difference between the transmitter velocity v_t and receiver velocity v_r. Hence, the Loffeld model is not valid for general azimuth-variant bistatic SAR, particularly for the cases using spaceborne radar as the transmitter.

Considering the phase history in Eq. (10.32) and supposing its stationary phase point is τ^*, from the principle of stationary phase we can get

$$S(f_r, f_a) = R(f_r) \exp\left[-j2\pi \frac{f_0 + f_r}{c_0} R(\tau^*) - j2\pi f_a \tau^*\right] \tag{10.37}$$

Using the instantaneous squint angles ξ_t (transmitter) and ξ_r (receiver), f_a can be represented by

$$f_a = -\frac{f_r + f_a}{c_0} \left.\frac{\partial R(\tau)}{\partial \tau}\right|_{\tau=\tau^*} = -\frac{f_r + f_a}{c_0} [v_t \sin(\xi_t) + v_r \sin(\xi_r)] \tag{10.38}$$

As the instantaneous bistatic range history at azimuth time τ can be presented by

$$\begin{aligned} R(\tau) =& R_{t0} \cos(\xi_t) + [R_{t0} \tan(\xi_{t0}) - v_t \tau] \sin(\xi_t) \\ &+ R_{r0} \cos(\xi_r) + [R_{r0} \tan(\xi_{r0}) - v_r \tau] \sin(\xi_r) \end{aligned} \tag{10.39}$$

where ξ_{t0} and ξ_{r0} are the instantaneous squint angles of the transmitter and receiver at their zero-Doppler frequencies respectively, we then have

$$\begin{aligned} S(f_r, f_a) =& P(f_r) \exp\left\{-j2\pi \frac{f_0 + f_r}{c_0} [R_{t0} \cos(\xi_t) + R_{t0} \tan(\xi_{t0}) \sin(\xi_t)\right. \\ &+ R_{r0} \cos(\xi_r) + R_{r0} \tan(\xi_{r0}) \sin(\xi_r)]\} \end{aligned} \tag{10.40}$$

This is just for general azimuth-variant bistatic SAR spectrum.

For a general monostatic SAR system, a focused SAR image can be obtained from a two-dimensional inverse Fourier transform of Eq. (10.40). However, this method is not effective for azimuth-variant bistatic SAR configurations. One reason is that azimuth Doppler frequency is nonuniformly sampled for azimuth-variant bistatic SAR configurations, and the other reason is that

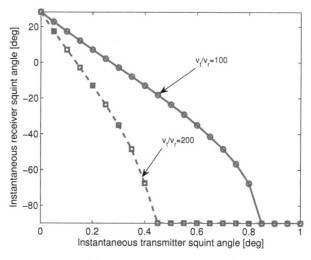

(a) In small squint-angle cases.

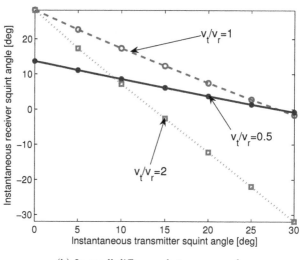

(b) In small difference between v_t and v_t.

FIGURE 10.9: The constraints of the Loffeld's bistatic SAR model.

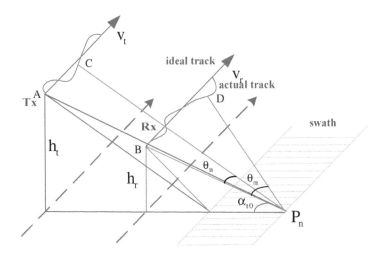

FIGURE 10.10: Illustration of the near-space vehicle-borne bistatic SAR motion errors.

its precision two-dimensional spectrum is difficult to obtain because the instantaneous squint angles can not be obtained in a high-precision manner. This is the reason why azimuth-variant bistatic SAR imaging algorithm has not been thoroughly researched [266, 295, 464], while many azimuth-invariant bistatic SAR imaging algorithms have been proposed [257, 301, 344].

10.4   Motion Compensation

Motion errors are not considered in previous discussions. However, as a matter of fact, problems may arise due to the presence of atmospheric turbulence, which introduce platform trajectory deviations from nominal position as well as altitude (rolls, pitchs, and yaw angles) [126]. To account for such errors, onboard GPS and inertial navigation units (INUs) are widely employed. If motion measurement facilities are not reachable, signal processing-based motion compensation must be applied.

Generally, azimuth-variant bistatic SAR has a short synthetic aperture time. We ignore the acceleration errors in along-track and consider only the motion errors in cross-track in the following discussions. As shown in Figure 10.10, suppose the ideal transmitter and receiver instantaneous positions at azimuth time τ_m are $(v_t \tau_m, y_{t0}, h_t)$ and $(v_r \tau_m, y_{r0}, h_r)$, and their actual positions are $(v_t \tau_m, y_{t0} + \Delta y_t(\tau_m), h_t)$ and $(v_r \tau_m, y_{r0} \Delta y_r(\tau_m), h_r)$.

Suppose the transmitter motion error in cross-track is $\Delta r_t(\tau_m)$, then $\Delta y_t(\tau_m) = -\Delta r_t(\tau_m)\cos(\alpha_{t0})$ (α_{t0} is the instantaneous incidence angle from the transmitter to the point target $P_n(x_n, y_n, 0)$), $\Delta z_t(\tau_m) = -\Delta r_t(\tau_m)\sin(\alpha_{t0})$ and $y_n - y_{t0} = r_{tn}\cos(\alpha_{t0})$ with $r_{tn} = \sqrt{h_t^2 + (y_n - y_{t0})^2}$ and $h_t = r_{tn}\sin(\alpha_{t0})$. The transmitter range history can then be represented by

$$
\begin{aligned}
R_t(\tau_m) &= \sqrt{(x_n - v_t\tau_m)^2 + (y_n - y_{t0} - \Delta y_t(\tau_m))^2 + (h_t - \Delta z_t(\tau_m))^2} \\
&= \sqrt{(x_n - v_t\tau_m)^2 + r_{tn}^2 + 2r_{tn}\cdot\Delta r_t(\tau_m) + \Delta r_t^2(\tau_m)}
\end{aligned}
$$

$$(10.41)$$

Assume the instantaneous transmitter squint angle is θ_{tm} and denote $x_t(\tau_m) = x_n - v_t\tau_m$ and $\tan(\theta_{tm}) = x_t(\tau_m)/r_{tn}$, we can get

$$
\begin{aligned}
R_t(\tau_m) &= \sqrt{r_{tn}^2 + x_t^2(\tau_m)} - \Delta r_t(\tau_m)\cos(\theta_{tm}) \\
&\quad + \sin(\theta_{tm})\sin(2\theta_{tm})\frac{\Delta r_t^2(\tau_m)}{2r_{tn}} + O\left(\frac{\Delta r_t(\tau_m)}{r_{tn}}\right) \\
&\approx \sqrt{r_{tn}^2 + x_t^2(\tau_m)} - \Delta r_t(\tau_m)\cos(\theta_{tm})
\end{aligned}
$$

$$(10.42)$$

Similarly, for the receiver range history we have

$$
R_r(\tau_m) \approx \sqrt{r_{rn}^2 + x_r^2(\tau_m)} - \Delta r_r(\tau_m)\cos(\theta_{rm}) \tag{10.43}
$$

where r_{rn}, $x_r(\tau_m)$, $\Delta r_r(\tau_m)$ and θ_{rm} are defined in an alike manner as the r_{tn}, $x_t(\tau_m)$, $\Delta r_t(\tau_m)$ and θ_{tm}. The bistatic range history can then be represented by

$$
\begin{aligned}
R_b(\tau_m) &\approx \sqrt{r_{tn}^2 + x_t^2(\tau_m)} - \Delta r_t(\tau_m)\cos(\theta_{tm}) \\
&\quad + \sqrt{r_{rn}^2 + x_r^2(\tau_m)} - \Delta r_r(\tau_m)\cos(\theta_{rm})
\end{aligned}
$$

$$(10.44)$$

Since the first and third terms are the ideal range history for subsequent image formation processing, we consider only the second and fourth terms. We have

$$
\begin{aligned}
&\frac{\partial[\Delta r_t(\tau_m)\cos(\theta_{tm}) + \Delta r_r(\tau_m)\cos(\theta_{rm})]}{\partial\tau_m} \\
&\doteq -\frac{\Delta r_t(\tau_{m1})}{r_{tn}}\sin(\theta_{tm})\cdot v_t\cdot(\tau_{m2} - \tau_{m1}) \\
&\quad -\frac{\Delta r_r(\tau_{m1})}{r_{rn}}\sin(\theta_{rm})\cdot v_r\cdot(\tau_{m2} - \tau_{m1})
\end{aligned}
$$

$$(10.45)$$

Because $\Delta r_t(\tau_{m1})/r_{tn} \ll 1$ and $\Delta r_r(\tau_{m1})/r_{rn} \ll 1$, when $\tau_{m2} \approx \tau_{m1}$, Eq. (10.45) will be equal to zero. That is to say, motion errors in a short interval can be seen as a constant. This kind of motion errors can be compensated with subaperture-based motion compensation algorithms [51].

Mapdrift algorithm [75, 36, 154] is the dominant autofocus technique used in SAR processing. Mapdrift directly estimates the quadratic coefficient of the phase error and then, based on this estimate, builds and applies a one-dimensional correction vector to the signal history. Generation and application of the correction vector are simple matters once the quadratic error coefficient is known.

The model for the azimuth signal $g(\tau)$ from a signal range bin is the product of the error-free signal history $s(\tau)$ and a complex exponential with quadratic phase

$$g(\tau) = s(\tau)\exp(jk_a\tau^2), \qquad -\frac{T_s}{2} < \tau < \frac{T_s}{2} \tag{10.46}$$

where k_a is the quadratic phase error coefficient to be estimated and T_s is the synthetic aperture time. The relationship between the center-to-edge phase error Q and the error coefficient is

$$Q = \frac{k_aT_s^2}{4} \tag{10.47}$$

The mapdrift procedure begins by dividing $g(\tau)$ into two subapertures of length $T_s/2$

$$g_1(\tau) = g\left(\tau - \frac{T_s}{4}\right), \qquad -\frac{T_s}{4} < \tau < \frac{T_s}{4} \tag{10.48a}$$

$$g_2(\tau) = g\left(\tau + \frac{T_s}{4}\right), \qquad -\frac{T_s}{4} < \tau < \frac{T_s}{4} \tag{10.48b}$$

Transformation of each subaperture into the image domain via a Fourier transform yields

$$G_1(\omega) = \int_{-T_s/4}^{T_s/4} g_1(\tau)e^{-j\omega\tau}\,d\tau = S_1\left(\omega + \frac{k_aT_s}{2}\right) \tag{10.49a}$$

$$G_2(\omega) = \int_{-T_s/4}^{T_s/4} g_2(\tau)e^{-j\omega\tau}\,d\tau = S_2\left(\omega - \frac{k_aT_s}{2}\right) \tag{10.49b}$$

where

$$S_1(\omega) = \int_{-T_s/4}^{T_s/4} g\left(\tau - \frac{T_s}{4}\right)e^{-j\omega\tau}\,d\tau \tag{10.50a}$$

$$S_2(\omega) = \int_{-T_s/4}^{T_s/4} g\left(\tau + \frac{T_s}{4}\right)e^{-j\omega\tau}\,d\tau \tag{10.50b}$$

The mapdrift algorithm relies on the assumption that

$$|S_1(\omega)|^2 = |S_2(\omega)|^2 \qquad (10.51)$$

In this case, it is clear from Eq. (10.49) that the two intensity images $G_1(\omega)$ and $G_2(\omega)$ are shifted versions of the same function. The relative shift between the two images is directly proportional to the quadratic phase error coefficient k_a over the original, full-length aperture. Various methods are available to estimate this shift. The most common estimation method is to measure the peak location of the cross-correlation of the two intensity images.

The relative shift between two subapertures is $\Delta_\omega = k_a T_s$. A measurement of Δ_ω provides an estimate $\hat{k}_a$ of the quadratic phase error coefficient according to

$$\hat{k}_a = \frac{\Delta_\omega}{T_s} \qquad (10.52)$$

The corresponding estimate of center-to-edge quadratic phase error is

$$\hat{Q} = \frac{\Delta_\omega T_s}{4} \qquad (10.53)$$

A typical mapdrift implementation applies the algorithm to only a small subset of available range bins based on peak energy content. An average of the individual estimates of the error coefficient from each of these range bins provides a final estimate. This procedure naturally reduces the computational burden of the algorithm. It also improves the accuracy of the error estimate. It is also common practice to apply the mapdrift algorithm iteratively. On each iteration, the algorithm forms an estimate of quadrature phase error and applies this estimate to the input signal data. Typically, two to six iterations of mapdrift are sufficient to yield an accurate error estimate.

Extension of the mapdrift algorithm to estimate higher-order phase errors is possible by dividing the signal history aperture into more than two subapertures. In general, N subapertures are adequate to estimate the coefficients of an N-order polynomial error. Figure 10.11 shows a typical implementation of mapdrift algorithm with overlapped subapertures. It can iterate the error estimation and correction procedure in order to improve the final result.

10.5   Azimuth-Variant Bistatic SAR Imaging Algorithm

Taking spaceborne transmitter and slow-moving receiver as an example, which has the following relations

$$v_t \gg v_r \qquad (10.54a)$$

$$R_{t0} \gg v_r \tau, \qquad \tau \in [-T_s/2, T_s/2] \qquad (10.54b)$$

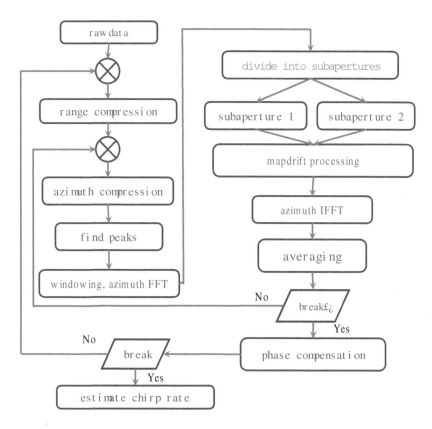

FIGURE 10.11: Illustration of the mapdrift autofocus algorithm.

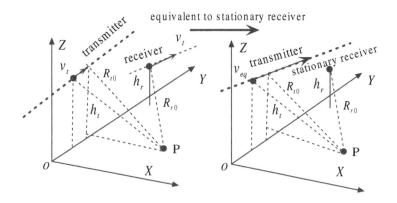

FIGURE 10.12: Azimuth-variant bistatic SAR geometry and its equivalent model with stationary receiver.

we then have

$$R_b(\tau) = \sqrt{R_{t0}^2 + (v_t\tau)^2} + \sqrt{R_{r0}^2 + (v_r\tau)^2}$$
$$\approx \sqrt{R_{t0}^2 + (v_{eq}\tau)^2} + R_{r0} \tag{10.55}$$

where $\vec{v}_{eq} = \vec{v}_t + \vec{v}_r$ is the equivalent velocity between the transmitter and the receiver, as shown in Figure 10.12.

Figure 10.13 gives the range errors caused by the equivalent bistatic SAR system model. As the phase errors that are smaller than $\pi/4$ can be ignored for subsequent imaging algorithms. It is shown that the equivalent model for small v_r is satisfactory for subsequent image formation processing.

The equivalent instantaneous azimuth Doppler frequency f_a can be represented by

$$f_a = \frac{v_{eq}}{\lambda}\sin(\phi_t) \tag{10.56}$$

where ϕ_t is the instantaneous angle between the transmitter and the point target. As there is $\lambda f_a/v_{eq} \ll 1$, for small ϕ_t we have

$$R_b(f_a; R_{t0}) = \frac{R_{t0}}{\cos(\phi_t)} + R_{r0} \approx R_{t0}\left[1 + \frac{\lambda^2 f_a^2}{2v_{eq}^2}\right] + R_{r0} \tag{10.57}$$

For simplicity, we denote $R_b(f_a; R_{t0})$ by

$$R_b(f_a; R_{t0}) = R_{t0}\left[1 + C_s\right] + R_{r0} \tag{10.58}$$

where

$$C_s = \frac{1}{\sqrt{1 - (\lambda f_a/v_{eq})^2}} - 1 \tag{10.59}$$

is the scaling factor of the chirp scaling (CS) algorithm.

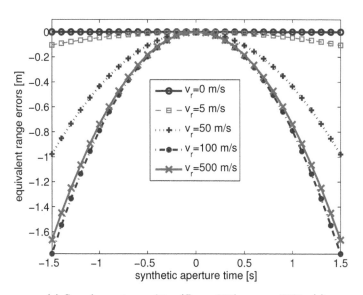

(a) Spaceborne transmitter ($R_{t0} = 800km$, $v_t = 7600m/s$).

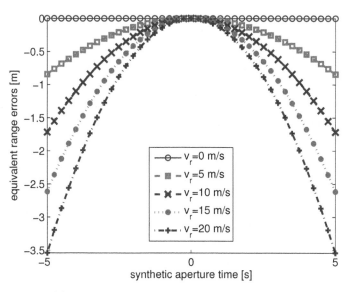

(b) Airborne transmitter ($R_{t0} = 15km$, $v_t = 100m/s$).

FIGURE 10.13: Range errors caused by the equivalent bistatic SAR model.

Next, a NCS algorithm which is similar to the algorithms developed in [449, 450] can then be applied. Suppose the transmitted signal is

$$s_t(t) = \exp\left[j\pi(2f_c t + k_r t^2)\right] \tag{10.60}$$

where k_r is the chirp rate. Note that the amplitude terms are ignored. The demodulated baseband signal in the receiver is

$$s_r(t) = \exp\left\{-j\pi\left[k_r\left(t - \frac{R_b(\tau)}{c_0}\right)^2 + 2\frac{R_b(\tau)}{c_0}\right]\right\} \tag{10.61}$$

Applying Fourier transform to azimuth time τ yields

$$S_r(t, f_a) = \exp\left\{-j\pi k_{eq}\left[t - \frac{R_b(f_a; R_{t0})}{c_0}\right]^2\right\}$$
$$\times \exp\left(-j\pi\frac{\lambda R_{b0} f_a^2}{v_{eq}}\right) \cdot \exp\left(-j2\pi f_a \frac{y_p}{v_{eq}}\right) \tag{10.62}$$

with

$$k_{eq} = \frac{1}{\frac{1}{k_r} - \frac{\lambda R_{t0}}{c_0^2} \cdot \frac{(\lambda f_a/v_{eq})^2}{[\sqrt{1-(\lambda f_a/v_{eq})^2}]^3}} \tag{10.63}$$

where $(x_p, y_p, 0)$ is the point target's coordinate, and R_{b0} is the smallest bistatic range.

Chirp scaling processing with the phase term

$$\Phi_1 = \exp\left\{-j\pi k_{eq}\left[\frac{1}{\sqrt{1 - (\lambda f_a/v_{eq})^2}} - 1\right]\left[t - \frac{R_b(f_a; R_{bref})}{c_0}\right]\right\} \tag{10.64}$$

where the R_{bref} is the reference range. Next, they are transformed into two-dimensional frequency-domain though range FFT. After range compression with the phase term

$$\Phi_2 = \exp\left\{j\pi\frac{k_{eq}}{1 + C_s}f_r^2\right\} \cdot \exp\left\{j\pi f_r \frac{\lambda^2 R_{bref} f_a^2}{c_0 v_{eq}^2}\right\} \tag{10.65}$$

a range inverse FFT (IFFT) is applied. To correct the effects of azimuth-variant Doppler chirp rate, phase filtering is further applied before azimuth IFFT

$$\Phi_3 = \exp\left\{j\frac{\pi\lambda^3 R_{b0}^4 f_a^4}{3L_{az}v_{eq}^6 T_{image}^4}\right\} \tag{10.66}$$

Next, the Doppler chirp rate is corrected by

$$\Phi_4 = \exp\left\{j\pi\left[\frac{2v_{eq}^2}{\lambda L_{az}} - \frac{v_{eq}^2}{\lambda R_{b0}}\right]\tau^2 + j\pi\left[\frac{\left(\frac{2v_{eq}^2}{\lambda L_{az}} - \frac{v_{eq}^2}{\lambda R_{b0}}\right)2R_{b0}}{T_{image}^2 L_{az}}\right]\tau^4\right\} \tag{10.67}$$

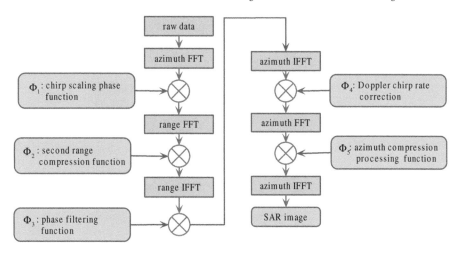

FIGURE 10.14: The NCS-based azimuth-variant bistatic SAR imaging algorithm.

Applying one Fourier transform in azimuth and multiplying a phase compensation term

$$\Phi_5 = \exp\left[-j\pi \frac{2v_{eq}^2}{\lambda L_{az}} f_a^2\right] \cdot \exp\left[\frac{4\pi k_{eq} C_s(1 + C_s)(R_{b0} - R_{\text{bref}})^2}{c_0}\right]$$

$$\times \exp\left\{-j\pi\left[\frac{1}{\left(\frac{2v_{eq}^2}{\lambda L_{az}}\right)^4 + \left(\frac{v_{eq}^2}{\lambda R_{b0}}\right)^4} - \frac{1}{48v_{eq}^6 T_{image}^2}\right] f_a^4\right\} \quad (10.68)$$

The focused bistatic SAR can then be obtained by azimuth inverse Fourier transform. The whole processing steps is illustrated as Figure 10.14. Its processing steps are similar to the NCS image formation algorithm discussed in [57, 449, 450].

To evaluate the performance of the derived imaging algorithm, an example azimuth-variant bistatic SAR data from five point targets are simulated using the parameters of Case C listed in Table 10.1, and further suppose the pulse repetition frequency (PRF) is 2000 Hz and the pulse duration is 25×10^{-6} s. A distance separation of five times of range/azimuth resolution is assumed in the examples. Additionally, $\xi_{t0}(\tau = 0) = 0$ and $\xi_{r0}(\tau = 0) = 0$ are supposed. Figure 10.15 gives the focused images. The point target response quality can be evaluated by the peak sidelobe ratio (PSLR). An additional quality parameter is the integrated sidelobe ratio (ISLR). The processed PSLR and ISLR are about 12 dB and -10 dB, respectively. From the results we conclude that the five point targets are well focused and the imaging performance is acceptable.

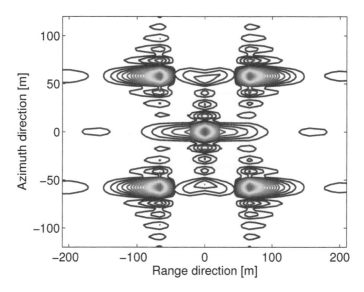

FIGURE 10.15: Imaging simulation results with the NCS-based image formation algorithm.

10.6   Conclusion

Although the feasibility of azimuth-invariant bistatic SAR has been demonstrated by experimental investigations using rather steep incidence angles, resulting in relatively short synthetic aperture time, azimuth-variant bistatic SAR data focusing processing remains a technical challenge. In this chapter, the azimuth-variant bistatic SAR system configurations, system performances, and spatial resolution characteristics were analyzed, along with two-dimensional bistatic SAR phase spectrum characteristics. A NCS-based image formation algorithm for addressing the problem of azimuth-variant Doppler chirp rate was described. Some simulation results were also provided. Azimuth-variant multi-antenna is still a hot research topic in SAR society. There are many open questions, especially for the azimuth-variant with arbitrary geometries.

11

Multi-Antenna SAR Three-Dimensional Imaging

SAR can provide two-dimensional high-resolution radar images in azimuth and elevation dimensions. However, since two-dimensional radar images are obtained by projecting the three-dimensional distributed targets onto the two-dimensional plane, they usually suffer from geometric distortions, such as foreshortening, layover, and so forth. Furthermore, conventional SAR generally operates at the side-looking mode and exhibits strong shadowing caused by buildings, hills, and valleys, which may result in information loss of the explored area. Interferometric SAR (InSAR) with dual antennas can take from the two antennas, and the terrain height can be calculated with the estimated phase difference; however, InSAR has no resolving capability in the elevation dimension. In this chapter, we discuss the potentials of multi-antenna SAR in three-dimensional imaging.

This chapter is organized as follows. Section 11.1 introduces the background and motivation of SAR three-dimensional imaging. Section 11.2 discusses the downward-looking single-input multiple-output (SIMO) linear-array SAR for three-dimensional imaging. Section 11.3 discusses the side-looking linear-array SAR for three-dimensional imaging. Next, Section 11.4 introduces the forward-looking SAR for three-dimensional imaging. Finally, Section 11.5 presents the novel frequency diverse array (FDA) radar which has a potential application to three-dimensional imaging.

11.1 Introduction

Conventional SAR achieves range resolution based on the pulse compression technique and azimuth resolution based on the synthetic aperture principle, therefore two-dimensional high-resolution SAR image of large scene can be obtained. On the zero-Doppler plane, conventional SAR only has the resolving ability in the range direction, and is lack of the resolving ability in the angle direction. Therefore, all scatterers with equal distances to the radar will fall in the same range resolution cell, independently of their angular position. For this reason, conventional SAR usually operates in side-looking mode because it can't distinguish the two scatterers located symmetrically at both sides of the flight axis. Side-looking SAR exhibits the disadvantage of shadowing effects in areas with strong height changes, such as high buildings in urban areas, steep mountains and deep valleys in mountainous areas, which can hide essential information of the explored area[155, 438].

Recently, the development of SAR systems providing full three-dimensional capability has become a field of intensive research [325]. A great extent of the experimental work in this area has been carried out using SAR facilities operating in a controlled environment. Examples of SAR facilities operating a controlled environment can be found in France [13], Sweeden [229], the US [31, 50, 49, 79, 198, 174], Belgium [338], Spain [32], the UK [35], and Germany [120, 122]. Near-field radar cross section (RCS) measurements have also been used to obtain the far-field scattering signature of targets [33, 114, 306].

Multi-baseline SAR tomography can generate a three-dimensional image of the target by multiple passes over the same area [325, 127]. It forms a second aperture in the elevation direction, and has resolving capability. After two-dimensional images from all passes are obtained, it employs array processing technique [168] or inversion technique [127] to get the tomographic image. The signal model and the resolution in the elevation direction of multi-baseline SAR tomography is illustrated under simplified geometry without considering the direction of passes arrangement [325, 128].

Linear array SAR is an extension of SAR interferometry by using linear array antennas. Every antenna in the linear array antennas can get a two-dimensional image of the targets. The array antennas form a second synthetic aperture, so it has a resolving capability in the elevation direction and can generate three-dimensional image according to the principle of multi-baseline SAR tomography [159]. Three-dimensional SAR with linear array combines the theories of real synthetic aperture can provide high-resolution radar images in three dimensions: azimuth, ground range, and elevation.

Linear array SAR has various observation modes including downward-looking, forward-looking, and side-looking. The concept of downward-looking SAR was presented in [155], and two airborne downward-looking SAR systems, DRIVE [302, 303, 304, 305] and ARTINO [205, 206, 207, 208, 209, 210, 438, 439, 440], are being developed for three-dimensional imaging at ON-ERA and FGAN-FHR, respectively. Bistatic configurations are used in these two systems and thus the round-trip range equation is composed of double square roots. Giret et al. [155] made some simplification to the double square roots derived with Fresnel approximation. Klare et al. [208, 205] introduced the concept of virtual antenna elements formed by the mean positions of every single transmitting and receiving antenna element to transform bistatic configuration into monostatic configuration and proposed a three-dimensional imaging algorithm for ARTINO with beamforming operation in the cross-track direction. The concept of forward-looking SAR was proposed in [447]. The first experimental system of forward-looking SAR was developed by Sutor et al. [365]. A forward-looking three-dimensional SAR imaging model using unequally spaced array was proposed in [265] and a three-dimensional imaging algorithm was proposed in [326]. In addition, compressive sensing-based SAR three-dimensional imaging was investigated in [38, 363].

11.2 Downward-Looking SAR Three-Dimensional Imaging

The main advantage of downward-looking SAR is the full ground range coverage without shadowing effects. Image formation processing is a key issue for downward-looking SAR three-dimensional imaging. A three-dimensional range migration algorithm (RMA) was proposed in [324] and a three-dimensional RMA was proposed in [263] for two-dimensional planar scanning aperture near-field imaging. Another three-dimensional RMA with elevation digital spotlight was also proposed in [368] for SAR tomography imaging. All of these three-dimensional RMAs are only suitable for monostatic configuration and cannot be directly applied to bistatic configuration due to the dual square root in the range history. In the following, we introduce a three-dimensional RMA which can be applied to the downward-looking SAR with single-transmission and multiple-reception linear array antennas [100].

11.2.1 Signal and Data Model

Figure 11.1 illustrates a downward-looking SAR in three-dimensional spatial domain $o - x - y - z$. The radar platform is supposed to fly at an altitude h_s along the x-axis with a constant velocity v_s. The illuminated scenario is located at the nadir area of the platform, and the origin of the coordinate system is located at the nadir point.

Suppose a point target P is stationary at (x_p, y_p, z_p). The instantaneous distance from the m-th transmitting antenna element to the target P is

$$R_{T_m}(\tau) = \sqrt{(x_m + v_s\tau - x_p)^2 + (y_m - y_p)^2 + (h_s - z_p)^2} \qquad (11.1)$$

where (x_m, y_m, h_s) is the position of the m-th transmitting array element at azimuth time τ. Similarly, the distance from the target P to the n-th receiving array element is

$$R_{R_m}(\tau) = \sqrt{(x_n + v_s\tau - x_p)^2 + (y_n - y_p)^2 + (h_s - z_p)^2} \qquad (11.2)$$

where (x_n, y_n, h_s) is the n-th receiving array element at azimuth time τ. The round-trip range history can then be written as

$$R_{mn}(\tau) = R_{T_m}(\tau) + R_{R_n}(\tau) \qquad (11.3)$$

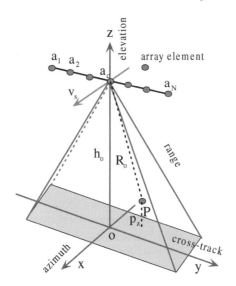

FIGURE 11.1: Illustration of a downward-looking SAR system.

Suppose the transmitted signal vector is $s_0(t)$. The radar returns at the n-th receiving element can be written in a matrix form

$$s_{r_n}\left(t, \tau; \boldsymbol{\xi}_p\right) = \exp\left\{-jk_0 \mathbf{R}_n(t, \tau; \boldsymbol{\xi}_p)\right\} s_0\left(t - \frac{\mathbf{R}_n(t, \tau; \boldsymbol{\xi}_p)}{c_0}\right) \qquad (11.4)$$

where $\mathbf{R}_n(t, \tau; \boldsymbol{\xi}_p)$ is a matrix for the round-trip range history with $k_0 = 2\pi/\lambda$ and $\boldsymbol{\xi}_p = (x_p, y_p, z_p)$.

For distributed targets we then have

$$s_{r_n}\left(t, \tau\right) = \iiint \exp\left\{-jk_0 \mathbf{R}_n(t, \tau; \boldsymbol{\xi}_p)\right\} s_0\left(t - \frac{\mathbf{R}_n(t, \tau; \boldsymbol{\xi}_p)}{c_0}\right) d\boldsymbol{\xi}_p \quad (11.5)$$

After range compression and azimuth compression, we can get

$$s_{r_n}\left(t, \tau\right) = \iiint \rho_r\left(r - \mathbf{R}_{n_0}(\boldsymbol{\xi}_p)\right) \cdot \rho_x\left(x - x_p(\boldsymbol{\xi}_p)\right) \cdot \exp\left\{-jk_0 \mathbf{R}_{n_0}(\boldsymbol{\xi}_p)\right\} d\boldsymbol{\xi}_p$$
$$(11.6)$$

where $\rho_r(r)$ and $\rho_x(x)$ denote the point spread functions in the range and x-axis direction, respectively. $\mathbf{R}_{n_0}(\boldsymbol{\xi}_p)$ is the reference range. Next, the elevation angle compression yields [102]

$$s_{r_n}\left(t, \tau\right) = \iiint \rho_r\left(r - \mathbf{R}_{n_0}(\boldsymbol{\xi}_p)\right) \cdot \rho_x\left(x - x_p(\boldsymbol{\xi}_p)\right) \cdot \rho_\theta\left(\sin\theta - \sin\theta_0\right) d\boldsymbol{\xi}_p$$
$$(11.7)$$

where θ_0 is the elevation angle from the target P to the reference element, ρ_θ

is the point spread function in the elevation angle direction. It can be noticed that the downward-looking SAR has three-dimensional resolution capability in range r, azimuth x, and elevation angle θ.

11.2.2 Imaging Resolution Analysis

Without loss of generality, we suppose there is only a single point target $P(x_p, y_p, z_p)$ in the scene. The whole three-dimensional image formation processing includes range compression, azimuth compression, and angle compression. The range and azimuth compressed returns of the n-th receiving element is given by [101]

$$s_{r,a}(t, \tau, d_n) = \mathrm{sinc}\left[B_r\left(t - \frac{2R_n}{0}\right)\right]\mathrm{sinc}\left[B_a\left(\tau - x_p\right)\right]$$
$$\times \exp\left[-j\frac{2\pi}{\lambda}\frac{d_n^2 - 2d_n y_p}{R_0}\right]\exp\left[-j\frac{4\pi}{\lambda}R_0\right]$$

(11.8)

where B_r is the transmitted signal bandwidth, B_a is the Doppler bandwidth, R_n is the distance between the n-th antenna element and the point target P, R_0 is the distance between the center of the antenna array and the target P on the zero-Doppler plane, d_n is the position of the n-th antenna element in the cross-track direction.

Considering only the first exponential term in Eq. (11.8) which is the only term dependent on the angular position, the angle compressed signal focused to arbitrary angle θ is given by

$$s'_{r,a,\theta}(t, \tau, \theta) = C_{r,a}\sum_{m=-\frac{N-1}{2}}^{\frac{N-1}{2}}\exp\left[-j\frac{2\pi}{\lambda}\frac{d_n^2 - 2d_n y_p}{R_0}\right]\exp\left[-j\frac{2\pi}{\lambda}K(d_n, \theta)\right]$$

(11.9)

where $K(d_n, \theta)$ is the focusing kernel function [102]

$$K(d_n, \theta) = \frac{d_n^2}{R_0} - 2d_n\sin\theta$$

(11.10)

The $C_{r,a}$ is expressed as

$$C_{r,a} = \exp\left(-j\frac{4\pi}{\lambda}R_0\right)\mathrm{sinc}\left[B_r\left(t - \frac{2R_n}{c_0}\right)\right]\mathrm{sinc}\left[B_a(\tau - x_p)\right]$$

(11.11)

The signal focused to the target angle θ is

$$s_{r,a,\theta}(t, \tau, \theta) = C_{r,a}\sum_{m=-\frac{N-1}{2}}^{\frac{N-1}{2}}\exp\left[-j\frac{4\pi}{\lambda}d_n(\sin\theta - \sin\theta_0)\right]$$
$$\approx NC_{r,a}\mathrm{sinc}\left[\frac{(2N-1)d}{\lambda}(\sin\theta - \sin\theta_0)\right]$$

(11.12)

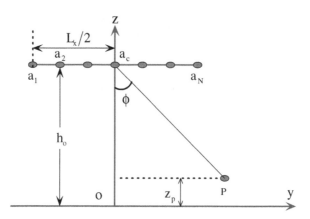

FIGURE 11.2: Geometry of downward-looking SAR cross-track resolution.

where d is the distance between individual antenna elements. Thus, the angle resolution is inversely proportional to $(2N - 1)d/\lambda$. The corresponding angle resolution is

$$\rho_\theta = \frac{\lambda}{(2N - 1)d} \tag{11.13}$$

The cross-track imaging resolution can be further derived as

$$\rho_y = \frac{\lambda}{4} \sqrt{\frac{\sqrt{\left(C_1 + \frac{L_x^2}{4}\right)^2 - C_2} + C_1 - \frac{L_x^2}{4}}{\sqrt{\left(C_1 + \frac{L_x^2}{4}\right)^2 - C_2} - C_1 - \frac{L_x^2}{4}}} \tag{11.14}$$

where L_x is the length of the linear array (see Figure 11.2), and

$$C_1 = (h_s - z_p)^2 \cdot (1 + \tan^2 \phi) \tag{11.15a}$$
$$C_2 = [(h_s - z_p)L_x \tan \phi]^2 \tag{11.15b}$$

11.2.3   Three-Dimensional Range Migration Algorithm

Suppose the transmitted signal is

$$s_0(t) = \text{rect}\left[\frac{t}{T_p}\right] \exp\left[j2\pi\left(f_c t + \frac{1}{2}k_r t^2\right)\right] \tag{11.16}$$

The demodulated radar returns can be represented by [100]

$$s_1(t, \tau, v) = \text{rect}\left\{\left[t - \frac{R(\tau, v)}{c_0}\right] / T_p\right\} \exp\left[j\pi k_r \left(t - \frac{R(\tau, v)}{c_0}\right)^2\right]$$
$$\times \exp\left[-j2\pi f_c \frac{R(\tau, v)}{c_0}\right] \tag{11.17}$$

Fourier transforming $s_1(t, \tau, v)$ with respect to the fast-time t yields

$$
\begin{aligned}
S_1(f_t, \tau, v) =& \text{rect}\left[\frac{f_t}{k_r T_p}\right] \exp\left[-j\pi\frac{f_t^2}{k_r}\right] \exp\left[-j2\pi f_t\frac{R(\tau, v)}{c_0}\right] \\
& \times \exp\left[-j2\pi f_c\frac{R(\tau, v)}{c_0}\right]
\end{aligned}
\tag{11.18}
$$

where f_t is the slant-range frequency associated with t, and $|f_t| \le B_r/2$.

After range compression with the reference function

$$
H_1(f_t) = \text{rect}\left[\frac{f_t}{k_r T_p}\right] \exp\left[j\pi\frac{f_t^2}{k_r}\right]
\tag{11.19}
$$

we can get

$$
\begin{aligned}
S_2(f_t, \tau, v) =& S_1(f_t, \tau, v) \cdot H_1(f_t) \\
=& \exp\left[-j2\pi\frac{f_t + f_c}{c_0}R(\tau, v)\right]
\end{aligned}
\tag{11.20}
$$

Let $k_t = 2\pi(f_t + f_c)/c_0$ denote the wavenumber domain of the transmitted signal, Eq. (11.20) can be rewritten as

$$
S_2(k_t, \tau, v) = \exp\left[-jk_t R(\tau, v)\right]
\tag{11.21}
$$

Two-dimensional Fourier transforming $S_2(k_t, \tau, v)$ with respect to (τ, v) yields

$$
\begin{aligned}
S_3(k_t, k_\tau, k_v) =& \iint S_2(k_t, \tau, v) \exp\left[-j(k_\tau\tau + k_v v)\right] d\tau dv \\
=& \iint \exp\left\{-j\left[k_t R(\tau, v) + k_\tau\tau + k_v v\right]\right\} d\tau dv
\end{aligned}
\tag{11.22}
$$

where k_τ and k_v denote the azimuth wavenumber domain for τ and cross-track wavenumber domain for v, respectively. Let $U = \tau - x_p$ and $V = v - y_p$, Eq. (11.22) can then be written as [100]

$$
\begin{aligned}
S_3(k_t, k_\tau, k_v) =& \exp\left[-j\left(k_\tau x_p + k_v y_p\right)\right] \\
& \times \iint \exp\left\{-j\left[k_t R(U, V) + k_\tau U + k_v V\right]\right\} dU dV
\end{aligned}
\tag{11.23}
$$

where

$$
R(U, V) = \sqrt{U^2 + y_p^2 + (h_s - z_p)^2} + \sqrt{U^2 + V^2 + (h_s - z_p)^2}
\tag{11.24}
$$

Let

$$
\Phi(U, V) = k_t R(U, V) + k_\tau U + k_v V
\tag{11.25}
$$

According to the principle of stationary phase, the stationary points U_0 and V_0 are

$$\frac{\partial \Phi(U,V)}{\partial U}\bigg|_{U=U_0,V=V_0} = 0 \tag{11.26a}$$

$$\frac{\partial \Phi(U,V)}{\partial V}\bigg|_{U=U_0,V=V_0} = 0 \tag{11.26b}$$

Since both U_0 and V_0 are the function of k_t, k_τ, k_v, y_p and z_p, $\Phi(U_0,V_0)$ can be factorized into a combination of linear phase terms with respect to y_p and z_p

$$\Phi(U_0,V_0) = a(k_t,k_\tau,k_v)y_p + b(k_t,k_\tau,k_v)z_p \tag{11.27}$$

Expanding it to the Taylor series at $y_p = 0$, $z_p = 0$ and neglecting the quartic and higher-order terms, we then have [100]

$$
\begin{aligned}
\Phi(U_0,V_0) \approx & \,\Phi_{\text{const}} + \left(y_p\frac{\partial}{\partial y_p} + z_p\frac{\partial}{\partial z_p} \right) \Phi(U_0,V_0)\bigg|_{y_p=0,z_p=0} \\
& + \frac{1}{2} \left(y_p\frac{\partial}{\partial y_p} + z_p\frac{\partial}{\partial z_p} \right)^2 \Phi(U_0,V_0)\bigg|_{y_p=0,z_p=0} \\
& + \frac{1}{6} \left(y_p\frac{\partial}{\partial y_p} + z_p\frac{\partial}{\partial z_p} \right)^3 \Phi(U_0,V_0)\bigg|_{y_p=0,z_p=0}
\end{aligned}
\tag{11.28}
$$

The first term is constant and independent of the target location, the second term is the linear phase term which contributes to SAR image focusing, and the third and forth terms are the main linear phase terms which can be compensated in three-dimensional wavenumber domain. Then Eq. (11.23) can be rewritten as

$$S_4(k_x,k_y,k_z) = \exp\left[-j\left(x_p k_x + y_p k_y + z_p k_z\right)\right] \tag{11.29}$$

with

$$k_x = k_\tau \tag{11.30a}$$
$$k_y = k_v + a(k_t,k_\tau,k_v) \tag{11.30b}$$
$$k_z = b(k_t,k_\tau,k_v) \tag{11.30c}$$

Finally, the three-dimensional image can be obtained by wavenumber interpolation and three-dimensional IFFT (inverse fast Fourier transform).

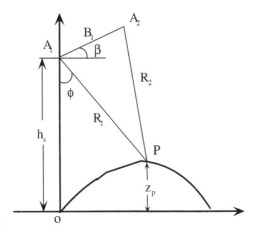

FIGURE 11.3: Geometry of InSAR for terrain elevation mapping.

11.3 Side-Looking SAR Three-Dimensional Imaging

11.3.1 InSAR for Terrain Elevation Mapping

Previously we saw that a SAR produces a two-dimensional image of the ground being illuminated. The two SAR receivers shown in Figure 11.3 produce two such images of the ground. Because the SAR is coherent, these two images possess phase as well as magnitude. In fact the image functions are comprised of a complex reflectivity modulated by a phase term due to the round trip path length as follows

$$g_1(x,y) = a_1(x,y) \exp\left(\frac{j4\pi R_1(x,y)}{\lambda}\right) \tag{11.31}$$

$$g_2(x,y) = a_2(x,y) \exp\left(\frac{j4\pi R_2(x,y)}{\lambda}\right) \tag{11.32}$$

where a_i is the complex terrain reflectivity, R_i is the range from antenna i to the point (x,y).

If we arrange the imaging geometry precisely and perform the signal processing correctly, the reflectivity terms in the image equations will be highly correlated. Then the two images can be interfered with each other by conjugate multiplication [163]

$$g_1(x,y) \cdot g_2^*(x,y) \cong |a(x,y)|^2 \exp\left\{\frac{j4\pi}{\lambda}[R_1(x,y) - R_2(x,y)]\right\} \tag{11.33}$$

The interferometric phase pattern

$$\Phi_{in} = \arctan[g_1(x,y) \cdot g_2^*(x,y)] \tag{11.34}$$

contains the effects of geometry and wavelength for flat terrain as well as the influence of terrain height. Armed with precise knowledge of the imaging geometry, it is possible to remove the pattern due to flat terrain, leaving only the height-induced information.

From the geometry given in Figure 11.3 we have

$$R_2^2 = R_1^2 + B_l^2 - 2R_1 B_l \cos\left(\frac{\pi}{2} - \phi + \beta\right) \tag{11.35}$$

Since $R_2 - R_1 \ll R_1$ and $B_l \ll R_1$, we have the following relationship

$$\phi = \arcsin\left(\frac{\lambda \Phi_{in}}{4\pi B_l}\right) + \beta \tag{11.36}$$

The target altitude can then be calculated by

$$z_p = h_s - R_1 \cos\phi \tag{11.37}$$

11.3.2 Side-Looking Linear Array SAR

Figure 11.4 illustrates a side-looking linear array SAR for three-dimensional imaging. The linear array elements are placed orthogonal to the azimuth direction. In the slant range-azimuth plane, the two-dimensional focused image of the target from the n-th antenna can be expressed as

$$s_n(r', x') = \iint \exp\left[-j4\pi \frac{R_n(v_0)}{\lambda}\right] \text{sinc}\left[\frac{\pi}{\rho_x}(x' - x)\right] \text{sinc}\left[\frac{\pi}{\rho_r}(r' - r)\right] dr dx \tag{11.38}$$

It can be easily understood that the slant range resolution and azimuth resolution are determined, respectively, by

$$\rho_r = \frac{c_0}{2B_r} \tag{11.39a}$$

$$\rho_x = \frac{D_a}{2} \tag{11.39b}$$

with D_a being the antenna length in the azimuth direction.

Assuming the two-dimensional sinc function can be approximated as a two-dimensional Dirac function, the focused signal of the target in the n-th antenna element can then be expressed as

$$s_n(v_0) = \exp\left[-j4\pi \frac{R_n(v_0)}{\lambda}\right] \tag{11.40}$$

According to the Fresnel approximation the range $R_n(v_0)$ can be approximated as [429]

$$R_n(v_0) = \sqrt{(R_0 \sin\theta + v_0 \cos\theta - d_n \cos\phi)^2 + (R_0 \cos\theta - v_0 \sin\theta + d_n \sin\phi)^2}$$
$$\approx R_0 - d_n \sin(\theta - \phi) + \frac{[v_0 - d_n \cos(\theta - \phi)]^2}{2[R_0 - d_n \sin(\theta - \phi)]} \tag{11.41}$$

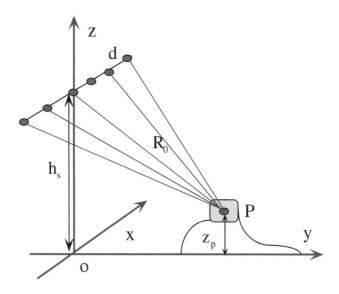

FIGURE 11.4: Geometry of a side-looking linear array SAR for three-dimensional imaging.

Multiplying the deramping factor

$$h_1 = \exp\left[j\frac{4\pi}{\lambda}R_n(0)\right] \tag{11.42}$$

we can get the focused signal in the n-th antenna

$$g_n(v_0) = \exp\left\{-j\frac{2\pi}{\lambda R_0}R\left[v_0^2 - 2v_0 d_n \cos(\theta - \phi)\right]\right\} \tag{11.43}$$

As the wavenumber along the linear array direction is only related with the elevation position of the scatters, the three-dimensional point spread function (PSF) can be derived as [429]

$$\text{PSF} \sim \text{sinc}\left[\frac{\pi}{\rho_x}(x' - x)\right]\text{sinc}\left[\frac{\pi}{\rho_r}(r' - r)\right]\text{sinc}\left[\frac{\pi}{\rho_v}(v' - v)\right] \tag{11.44}$$

The corresponding resolution in the elevation direction is

$$\rho_v = \frac{h_s}{2L\cos\theta\cos(\theta - \phi)} \tag{11.45}$$

The specific resolution for different θ and ϕ combinations can be found in [429].

11.4   Forward-Looking SAR Three-Dimensional Imaging

The forward-looking SAR discussed in [447] can obtain only range resolution and cross-track direction resolution. Two apertures are formed in the system, one is realized by the movement of aircrafts, and the other is achieved by a physically existent linear antenna array perpendicular to the flight direction. Forward-looking SAR has three-dimensional resolution capabilities; however, the real-aperture antenna arrays cannot allow an interelement space larger than half of the wavelength. This restriction makes the array system costly and leads to severe coupling problem. A more cost-effective realization may be achieved by using unequally spaced array antennas. In this section we introduce the unequally spaced array forward-looking SAR proposed in [265].

An array is sparse when the interelement space is larger than $\lambda/2$. Sparse arrays potentially reduce the array cost and control complexity. Sparse arrays can be further classified into two types: one is periodic with equally spaced elements, and the other is aperiodic with unequally spaced elements. The former type exhibit grating lobes. This problem can be resolved by using the latter type of arrays.

Figure 11.5 illustrates the geometry of a forward-looking linear array SAR for three-dimensional imaging. Each antenna of the M unequally spaced antennas can operate as both transmitter and receiver. At azimuth slow time τ the range history of the n-th antenna to a given point target $P(x_p, y_p, z_p)$ can be expressed as

$$R(\tau, n) = \sqrt{(v_s\tau - x_p)^2 + (d_n - y_p)^2 + (h_s - z_p)^2} \tag{11.46}$$

where d_n is the position of the n-th antenna along the cross-track direction. Let

$$R_{n_0} = \sqrt{x_p^2 + (d_n - y_p)^2 + (h_s - z_p)^2} \tag{11.47}$$

$R(\tau, n)$ can be expanded to its third-order Taylor series as [265]

$$R(\tau, n) \approx R_{n_0} - \frac{v_s x_p}{R_{n_0}}\tau + \frac{1}{2}\left(\frac{v_s^2}{R_{n_0}} - \frac{v_s^2 x_p^2}{R_{n_0}^3}\right) + \frac{1}{6}\left(\frac{v_s^3 x_p}{R_{n_0}^3} - \frac{v_s^3 x_p^3}{R_{n_0}^3}\right) \tag{11.48}$$

As N channels can be formed in the forward-looking SAR receiver, the total radar returns can be seen as a three-dimensional matrix comprised of each channel's two-dimensional data. The subsequent image formation processing can be performed in a manner like the algorithm discussed in Section 11.2.

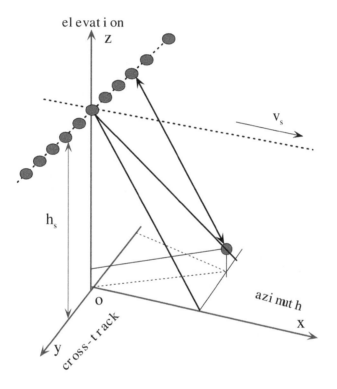

FIGURE 11.5: Geometry of a forward-looking linear array SAR for three-dimensional imaging.

11.5 Frequency Diverse Array SAR Three-Dimensional Imaging

Recently, a flexible beam-scanning array named a frequency diverse array (FDA) was introduced in [8, 9, 179, 294]. Two patents have been issued [442, 443], which discussed the increased degrees-of-freedom of one FDA system when compared to conventional phased-array systems. Note that the FDA discussed in this book is different from conventional multiple-input multiple-output (MIMO) arrays [411, 267, 71, 103], where orthogonal waveforms are employed in each antenna element, whereas for the FDA the same waveform is employed in each antenna element. The most prominent distinction from a conventional phased array is that a small amount of frequency increment, compared to the carrier frequency, is used across the array elements.

FDA radar has a range-dependent beampattern, which can be used to suppress range-dependent clutter and high-resolution radar imaging. Specially, the application of FDA in SAR was investigated in [16, 116, 117]. The concept is to increase the angular extent of the measured scene by exploiting the bending-beam phenomenology, so as to decrease the observation time of the radar platform in the scene; however, they reject the received signals that are not frequency matched to the transmitted signal. A FDA bistatic radar system was introduced in [336, 337], which could have applications in low probability of intercept (LPI) radars or suppressing the echoes from undesired range cells. Additional study was presented in [15, 30, 181, 192, 322, 343] to analyze the range-dependent beampattern characteristics. Although not completely associated with FDA, a concept by merging the phased-MIMO and FDA was proposed in [427] and a technique by merging frequency diversity and MIMO radar for suppressing grating lobes was proposed in [480].

FDA provides increased degrees-of-freedom to efficiently gather and therefore potentially use scene information. But, current research undermines potential FDA capability and much further work should be carried out. An open question is how to precisely steer the beam direction for a desired range cell. Since FDA apparent scan angle is not equal to the nominal scanning angle, which means that precisely beam steering depends on both the range and angle parameters. Consequently it is insufficient to precisely steer the beam like conventional phased array; nevertheless, FDA provides new degrees-of-freedom in range, angle, and time to design and control the antenna beampattern. This provides a potential to beam scan without the need for phase shifters or mechanical steering. Moreover, FDA allows for new system operating modes, e.g., simultaneously ground moving target indication (GMTI) and SAR imaging.

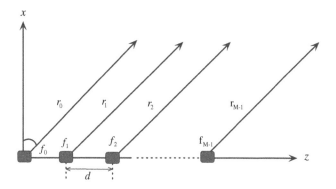

FIGURE 11.6: A linear FDA with an identical frequency increment.

11.5.1 FDA System and Signal Model

In conventional phased arrays, it is assumed that the same waveform is radiated by each array element. Different from conventional phased arrays, FDA elements can be excited by either the same waveform [180] or different waveforms [427]. For simplicity and without loss of generality, we assume that the waveform radiated from each array element is identical with a frequency offset of Δf Hz, as shown in Figure 11.6. That is, the radiated frequency from each element is [8, 9, 294]

$$f_m = f_0 + m\Delta f, \qquad m = 0, 1, 2, \ldots, M - 1 \tag{11.49}$$

where f_m is the FDA operating carrier frequency and M is the number of the antenna elements.

Consider a given far-field point, the phase of the signal transmitted by the first element can be represented by

$$\psi_1 = \frac{2\pi f_0 r}{c_0} \tag{11.50}$$

where c_0 and r are the speed of light and the distance between the first element and the observed point, respectively. Similarly, the phase of the signal transmitted by the second element can be represented by

$$\psi_2 = \frac{2\pi f_2 r_2}{c_0} = \frac{2\pi (f_0 + \Delta f) r_2}{c_0} \tag{11.51}$$

The phase difference between the signals arriving at the first and second elements can be expressed as

$$
\begin{aligned}
\Delta \psi_1 &= \psi_2 - \psi_1 \\
&= -\frac{2\pi f_0 d \sin\theta}{c_0} - \frac{2\pi \Delta f d \sin\theta}{c_0} + \frac{2\pi \Delta f r}{c_0}
\end{aligned} \tag{11.52}
$$

where θ and d are the desired direction angle and element spacing, respectively. Note that $r_2 = r - d\sin\theta$ is used in the above equation.

Similarly, the phase difference between the signals coming from the first and the $(m-1)$-th element can be expressed as

$$
\begin{aligned}
\Delta\psi_{m-1} &= \psi_m - \psi_1 \\
&= -\frac{2\pi f_0(m-1)d\sin\theta}{c_0} + \frac{2\pi(m-1)\Delta fr}{c_0} \\
&\quad - \frac{2\pi(m-1)^2\Delta f d\sin\theta}{c_0}
\end{aligned}
\tag{11.53}
$$

where the range difference between individual elements is approximated by

$$
r_m \approx r - md\sin\theta, \qquad m = 0, 1, 2, \ldots, M - 1 \tag{11.54}
$$

Considering Eq. (11.53), the first term is simply for the conventional array factor seen frequently in array theory. The second term is important because it shows that the array radiation pattern depends on the range and frequency increment. Taking the first element as the reference for the array, the steering vector can be expressed as

$$
\mathbf{a}(\theta, r) = \begin{bmatrix} 1 & \cdots & e^{-j\left(\frac{2\pi f_0(M-1)d\sin\theta}{c_0} + \frac{2\pi(M-1)^2\Delta f d\sin\theta}{c_0} - \frac{2\pi(M-1)\Delta fr}{c_0}\right)} \end{bmatrix}^T
\tag{11.55}
$$

where $[\cdot]^T$ denotes the transpose. Note that the FDA is different from conventional frequency scanning arrays. Frequency scanning arrays use the frequency increments as a function of time for all elements, while the FDA use the frequency increment as a function of the element index. In the following, we make a summary on the FDA characteristics:

1) If the frequency offset Δf is fixed, the beam direction will vary as a function of the range r (see Figures 11.7 and 11.8), i.e., it is a range-dependent beam. It can be noticed from Figure 11.8 that, each beampattern is repeated two times within the spatial domain due to spatial aliasing. This is because the interelement spacing is two times half a wavelength. Range-dependent beampattern can also be noticed from the frequency diverse array.

2) If the range r is fixed, the beam direction will vary as a function of Δf, as shown in Figure 11.9. This means that the FDA also is a frequency-offset dependent beam.

3) If the frequency increment across the array is not applied (i.e., $\Delta f = 0$), the corresponding FDA is just a conventional uniform linear phased-array.

In summary, the FDA is different from conventional phased arrays and frequency scanning arrays. The range-angle dependent beampattern is of great

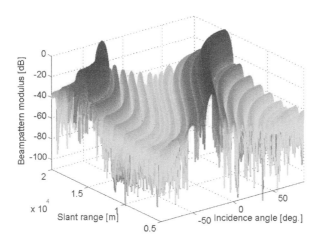

(a) FDA beampattern.

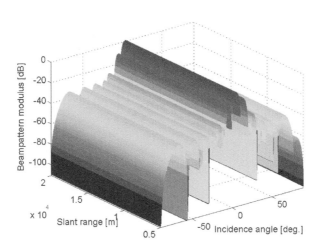

(b) Phased array beampattern.

FIGURE 11.7: Comparative beampattern between FDA and conventional phased array, where $\Delta f = 30kHz$, $M = 10, N = 12$, $f_0 = 10$ GHz and $d = \lambda/2$ are assumed.

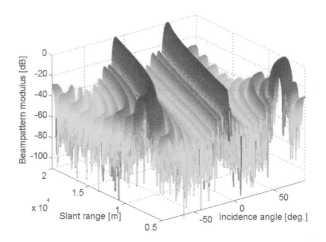

(a) FDA beampattern.

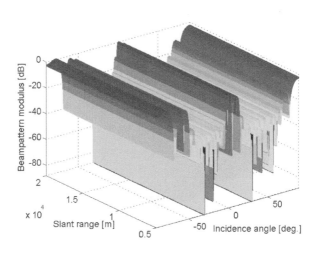

(b) Phased array beampattern.

FIGURE 11.8: Comparative beampattern between FDA and conventional phased array, where $\Delta f = 30 kHz$, $M = 10, N = 12$, $f_0 = 10$ GHz and $d = \lambda$ are assumed.

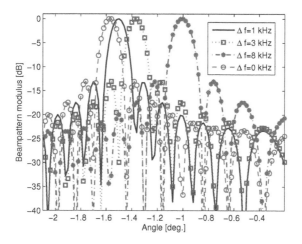

FIGURE 11.9: A linear FDA with an identical frequency increment.

importance because it can provide local maxima at different range cells, and this may provide many novel application potentials, including three-dimensional imaging.

11.5.2 Application Potentials in Target Imaging

The FDA antenna pattern changes the direction of focus with the range parameter. This can provide more flexible multibeam scan options than a phased array. Moreover, coding technique can be employed at the beam level to enable separation of multiple beams at the receiver [191]. When coding technique is employed, the signal transmitted by the m-th element for the k-th beam can be modelled as

$$\phi_{km}(t) = c_k(t) \exp\left(j2\pi f_{km}t\right) \tag{11.56}$$

where $c_k(t)$ is the coding sequences and $f_{km} = f_c + m\Delta f_k$ with Δf_k being the frequency increment used in the k-th beam is the carrier frequency.

If we steer the k-th beam to the angle $\hat{\theta}_k$ and $\hat{r}_k$ with the transmitting beam weighting function

$$\mathbf{w}_{km,t}(t) = \exp\left\{ j2\pi f_{km}\left(\frac{\hat{r}_k}{c_0} - \frac{md\sin\hat{\theta}_k}{c_0} \right) \right\} \tag{11.57}$$

The composite multibeam (i.e, K) signal propagating to the point target located at angle θ_k and range r_k is given by

$$s_T(t) = \sum_{k=1}^{K} \sum_{m=0}^{M-1} c_k \left(t - \frac{r_k}{c_0} + \frac{md \sin \theta_k}{c_0} \right)$$

$$\times \exp \left\{ j2\pi f_{km} \left(t - \frac{r_k}{c_0} + \frac{md \sin \theta_k}{c_0} \right) \right\} \qquad (11.58)$$

$$\times \exp \left\{ j2\pi f_{km} \left(\frac{\hat{r}_k}{c_0} - \frac{md \sin \hat{\theta}_k}{c_0} \right) \right\}$$

Similarly, using the receive beam weighting function

$$\mathbf{w}_{kn,r}(t) = \exp \left\{ j2\pi f_{kn} \left(\frac{\hat{r}_k}{c_0} - \frac{nd \sin \hat{\theta}_k}{c_0} \right) \right\} \qquad (11.59)$$

we can get the matched filtering result [191]

$$y_k(t) = \int_{t=0}^{T_0} \sum_{n=0}^{N-1} \sum_{m=0}^{M-1} c_k^* \left(t - \frac{2\hat{r}_k}{c_0} + \frac{md \sin \hat{\theta}_k + nd \sin \hat{\theta}_k}{c_0} \right)$$

$$\times c_k \left(t - \frac{2r_k}{c_0} + \frac{md \sin \theta_k + nd \sin \theta_k}{c_0} \right)$$

$$\times \exp \left\{ j2\pi f_{km} \left(t - \frac{2(r_k - \hat{r}_k)}{c_0} \right. \right.$$

$$\left. \left. + \frac{md(\sin \theta_k - \sin \hat{\theta}_k)}{c_0} + \frac{nd(\sin \theta_k - \sin \hat{\theta}_k)}{c_0} \right) \right\} dt \qquad (11.60)$$

$$\cong C_k \exp (j\Phi_0) \frac{\sin \left[2\pi f_0 M \left(\sin \theta_k - \sin \hat{\theta}_k \right) \right]}{\sin \left[2\pi f_0 \left(\sin \theta_k - \sin \hat{\theta}_k \right) \right]}$$

$$\times \frac{\sin \left(\pi \Delta f M \left(t - \frac{2(R_0 - \hat{R}_0)}{c_0} \right) + 2\phi_0 M \left(\sin \theta_0 - \sin \hat{\theta}_0 \right) \right)}{\sin \left(\pi \Delta f \left(t - \frac{2(R_0 - \hat{R}_0)}{c_0} \right) + 2\phi_0 \left(\sin \theta_0 - \sin \hat{\theta}_0 \right) \right)}$$

where T_0 and C_k are the duration and auto-correlation amplitude of the code sequence, respectively. Note that $M = N$ is assumed in the above equation.

Figure 11.10 shows the transmit-receive beampatterns for each beam separately, where $\Delta f_1 = 40$ kHz and $\Delta f_2 = -3$ kHz are assumed. Note that only two beams are considered in the simulation. Multiple beams over two beams can be obtained in a similar way.

Three-dimensional imaging is possible by applying the synthetic aperture principle to FDA radar. We present a frequency-coding FDA radar with stationary platform for two-dimensional imaging of targets (if moving platform is employed, three-dimensional imaging is possible for the system). Figure

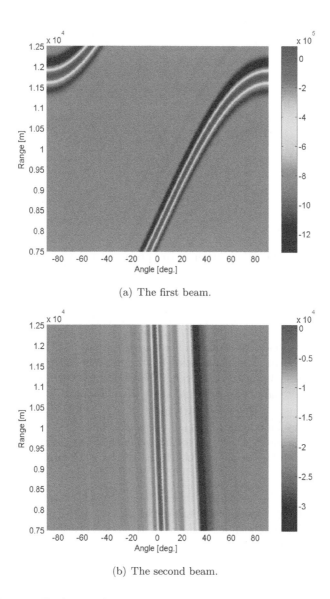

(a) The first beam.

(b) The second beam.

FIGURE 11.10: FDA multibeam forming with coding technique.

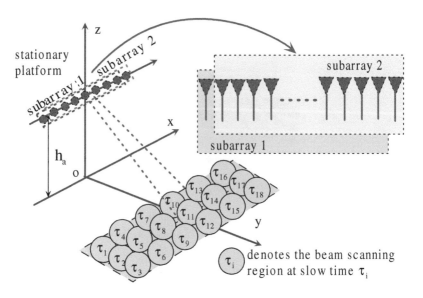

FIGURE 11.11: Illustration of the proposed FDA radar. Here, only 18 scanning beams are shown. It is for illustration purposes only.

11.11 illustrates the proposed FDA radar, where τ_i denotes the beam scanning region at the slow time τ_i. Only 18 scanning beams are shown. It is for illustration purpose only. The array antenna is placed inside a stationary radar platform. The synthesized beam scans the desired observation area like a conventional phased-array radar. Note that the two subarrays are allowed to overlap, but non-overlapped subarrays are assumed in this chapter. The first subarray uses a frequency increment of Δf_1, while the second subarray uses a frequency increment of Δf_2. Note that, here, $\Delta f_1 \neq \Delta f_2$ is assumed.

Suppose $s_0(t)$ is the waveform used in the FDA radar. The baseband equivalent model, in complex-valued form, of the FDA transmitted signals can be expressed as

$$\mathbf{a}_t^*(\theta, r)s_0(t) \tag{11.61}$$

where $(\cdot)^*$ denotes the conjugate and $\mathbf{a}_t(\theta, r)$ is the new transmitter steering vector. Taking the first element as the reference for the array, the steering vector can be represented by

$$\mathbf{a}_t(\theta, r) = \begin{bmatrix} 1 & \cdots & e^{-j(N-1)\omega_1} & e^{-j\omega_3} & \cdots & e^{-j[\omega_3+(N-1)\omega_2]} \end{bmatrix}^T \tag{11.62}$$

where

$$\omega_1 = \frac{2\pi f_0 d \sin \theta_0}{c_0} + \frac{2\pi \Delta f_1 d \sin \theta_0}{c_0} - \frac{2\pi \Delta f_1 r_0}{c_0} \tag{11.63a}$$

$$\omega_2 = \frac{2\pi f_0 d \sin \theta_0}{c_0} + \frac{2\pi \Delta f_2 d \sin \theta_0}{c_0} - \frac{2\pi \Delta f_2 r_0}{c_0} \tag{11.63b}$$

$$\omega_3 = \frac{2\pi N f_0 d \sin \theta_0}{c_0} \tag{11.63c}$$

Note that, since $2\pi n \Delta f_1 d \sin \theta \ll c_0$ and $2\pi n \Delta f_1 d \sin \theta \ll c_0$, $\frac{2\pi n^2 \Delta f_1 d \sin \theta}{c_0}$ is approximated in the above equation by $\frac{2\pi n \Delta f_1 d \sin \theta}{c_0}$, where $m = 1, 2, \ldots, 2N - 1$.

The signal seen at a specific far-field location with angle θ_i and range r is a superposition of the delayed and attenuated version of the transmitted signals

$$\mathbf{a}_t^T (\theta_i, r_i) \mathbf{a}_t^* (\theta_0, r_0) s_0(t) \triangleq \beta_i s_0(t) \tag{11.64}$$

where the first $\mathbf{a}_t(\theta_i, r_i)$ on the left side is a propagation vector due to propagation effects, and takes the same form of the steering vector. For notional simplicity and following the convention of many existing works in array signal processing, both are simply referred to as the steering vector hereafter. Specially, the signal seen at the look angle θ_0 and range r_0 is given by

$$\mathbf{a}_t^T (\theta_0, r_0) \mathbf{a}_t^* (\theta_0, r_0) s_0(t) \triangleq \beta_0 s_0(t) = 2N s_0(t) \tag{11.65}$$

It has a directional gain $2N$ (the size of the transmit aperture) at the looking direction, which is similar to the property of conventional phased-array radars. It is easily understood that $|\beta_i| \leq 2N$ for $\theta_i \neq \theta_0$ or $r_i \neq r_0$.

Suppose that the transmitting beam is steered to (θ_0, r_0) and there are L targets, where one target is located at (θ_0, r_0) and $L - 1$ targets are located at (θ_i, r_i), $i \in [1, 2, \ldots, L - 1]$. The baseband equivalent of the signals at the receiving array is given by

$$\mathbf{x}(t) = \sigma_0 \beta_0 \mathbf{a}_r (\theta_0, r_0) s_0(t) + \sum_{i=1}^{L-1} \sigma_i \beta_i \mathbf{a}_r (\theta_i, r_i) s_0(t) + \mathbf{n}(t) \tag{11.66}$$

where σ_i denotes the complex amplitude of the i-th source, $\mathbf{a}_r(\theta, r)$ similarly defined as the transmitter steering vector $\mathbf{a}_t(\theta, r)$ is an $2N \times 1$ propagation vector due to the propagation delays from a source to the receive elements (i.e., the receiver steering vector), and $\mathbf{n}(t)$ is the $2N \times 1$ additive white Gaussian noise vector with zero mean and covariance matrix $\sigma_n^2 \mathbf{I}$ with $\mathbf{I}$ being an identity matrix.

By matched filtering the received signal with the transmitted signal $s_0(t)$, we can get

$$
\mathbf{y} \triangleq \frac{\int_{T_p} \mathbf{x}(t) s_0^*(t) dt}{\int_{T_p} |s_0(t)|^2}
$$

$$
= \sigma_0 \beta_0 \mathbf{a}_r(\theta_0, r_0) + \sum_{i=1}^{L-1} \sigma_i \beta_i \mathbf{a}_r(\theta_i, r_i) + \mathbf{n} \tag{11.67}
$$

$$
= \sigma_0 \mathbf{u}_r(\theta_0, r_0) + \sum_{i=1}^{L-1} \sigma_i \mathbf{u}_r(\theta_i, r_i) + \mathbf{n}
$$

where T_p is the transmitted pulse signal duration, and the virtual steering vector is

$$
\mathbf{u}(\theta_i, r_i) \triangleq \beta_i \mathbf{a}_r(\theta_i, r_i) \tag{11.68}
$$

It can be easily proved that $\mathbf{n}$ also has zero mean and covariance matrix $\sigma_n^2 \mathbf{I}$. Two-dimensional imaging of targets can be achieved by beamforming at the receiver

$$
\mathbf{z}(\theta, r) = \left| \mathbf{w}^H(\theta, r) \mathbf{y} \right|^2 \tag{11.69}
$$

where $(\cdot)^H$ is the conjugate transpose and $\mathbf{w}(\theta, r)$ is the weight vector of the beamformer. We consider a simple data-independent beamformer for simplicity. Furthermore, the two-dimensional location of the targets can be estimated from the beamforming output peaks.

$$
(\hat{\boldsymbol{\theta}}, \hat{\mathbf{r}}) = \max_{\theta, r} \{ \mathbf{z}(\theta, r) \} \tag{11.70}
$$

This is not directly accessible for conventional phased-array radars.

As an example, we consider a linear FDA array with 64 array elements used for transmitting the baseband waveform and 64 array elements at the receiver. one point target of interest is supposed to reflect a plane-wave that impinges on the array from direction of angle $\theta_0 = 10°$ and range $r = 10$ km. Suppose $\Delta f_1 = 100$ kHz and $\Delta f_2 = -100$ kHz, Figure 11.12 shows the comparative two-dimensional target response of the beamformer. Different from Figure 11.12(a) where $\Delta f_1 = \Delta f_2 = 100$ kHz is assumed for a conventional FDA radar, the target is unambiguously imaged. The peak shown in Figure 11.12(b) can be used to estimate the target angle and range parameters, but this is not accessible for Figure 11.12(a) which has a coupling angle-range response.

11.5.3   Several Discussions

FDA provides many promising potentials including three-dimensional imaging (if moving platform is employed), but there are several open issues.

11.5.3.1   Waveform Optimization

Beyond the basic range-dependent beamforming, more sophisticated transmit waveforms may enable the development of multiple modes that support simul-

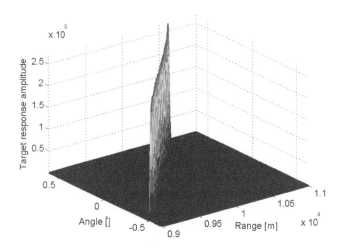

(a) Conventional FDA radar.

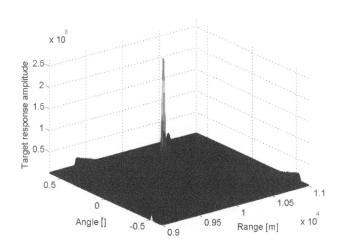

(b) The proposed FDA radar.

FIGURE 11.12: Comparative single-point target imaging results.

taneous SAR and GMTI through a single aperture. It is possible for the FDA to transmit a chirp signal towards a desired direction at the transmitter, while use a single carrier to demodulate the chirp signal at the receiver. This concept combines the flexible beamforming feature of FDA and improved target detection ability of chirp signal, at the same time eliminates the use of phase shifters. Figure 11.13(a) shows an implementation of the chirp modulation across the FDA elements, where six elements are assumed. Signals received from all spatial channels are processed as in a conventional MTI processor, and the frequency offset between elements provides range-dependent beamforming for flexible beam scanning and for multipath mitigation. A potential further extension may be implemented by utilizing contiguous and nonoverlapping spectra on adjacent channels, as shown in Figure 11.13(b). Using the contiguous bandwidth implementation, it may be possible to construct very large bandwidth signals, with each element radiating a non-overlapping segment of the entire frequency extent. In this way, a higher resolution may be obtained.

FDA provides more degrees-of-freedom at a large cost in terms of complexity and analyticity. Waveform optimization is required to further understand how the signal parameters impact the system performance. The ambiguity function may provide a useful optimization metric. Using knowledge of the ambiguity function's primary sidelobe locations, an optimization algorithm could be designed to optimize the ambiguity function at those locations. The waveform is determined by the optimization constraints including total bandwidth, number of subcarriers, number of transmitting and receiving elements, maximum chirp rate, maximum transmit signal amplitude and maximum peak sidelobe level. This constrained optimization problem requires much further work.

11.5.3.2 Array Configuration

Linear array geometry is exclusively used in literature because it allows the relationship between the temporal, spatial and spectral aspects of FDA to be clearly visualized. In fact, M transmitting antennas and N receiving antennas can be used to synthesizes apertures of various geometries comprising $M \times N$ phantom elements. Two-dimensional and three-dimensional array can also be synthesized midway between each transmitter-receiver pair. Figure 11.14 shows an example two-dimensional planar array and a three-dimensional cylindrical array. Moreover, the FDA with constant element spacing may be not the ideal configuration due to the frequency diversity. It is thus necessary to optimally design the FDA array geometric configuration.

11.5.3.3 Optimal Array Processing

The receiver discussed in Eq. (11.60) is quite computationally expensive. It is thus necessary to reduce the number of computations required to process the received signals. For example, the complex transmitter/receiver weighting

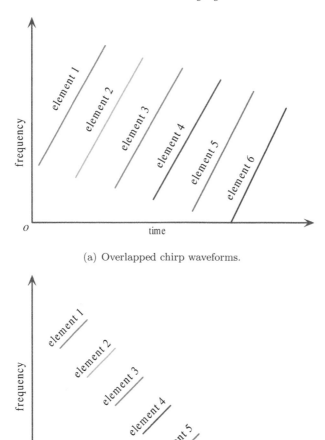

(a) Overlapped chirp waveforms.

(b) Continuous bandwidth chirp waveform.

FIGURE 11.13: Chirp waveform implementation of an FDA system.

functions depend on the baseband frequency as well as the frequency offset. Relaxing the weighting function's frequency offset dependence would allow the spatial weighter to be factored out from the summation and applied once to the entire signal. This can significantly reduce the computation complexity, but we found no related literature. On the other hand, the FDA's target detection and estimation performance should be evaluated in noise and clutter. Although a number of optimal processing techniques exist in literature for narrow-band,

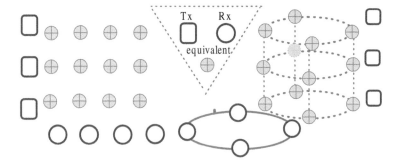

FIGURE 11.14: Example two-dimensional planar virtual array and three-dimensional cylindrical virtual array.

monochromatic signals, existing optimal techniques to process FDA signals have not been studied.

11.6 Conclusion

This chapter investigated the role of multi-antenna SAR in three-dimensional imaging from a top-level system description, with an aim for further research. Several representative three-dimensional imaging techniques including downward-looking linear array SAR, side-looking linear array SAR, and forward-looking linear array SAR were introduced. Their signal models and imaging resolutions were derived. Particularly, we introduced the FDA radar which provides promising potentials for future three-dimensional imaging (if a moving platform is employed, otherwise, two-dimensional imaging is possible for stationary platforms); however, much further research work should explore the waveform optimization, array configuration, optimal array processing, and three-dimensional imaging algorithm. Additionally, MIMO SAR also provides potential three-dimensional imaging.

Bibliography

[1] http://igscb.jpl.nasa.gov.

[2] M.M. Abousetta. Noise analysis of digitised FMCW radar waveforms. *IEE Proceedings Radar, Sonar and Navigation*, 145(4):209–215, August 1998.

[3] R.S. Adve, R.A. Schneible, and R. McMillan. Adaptive space/frequency processing for distributed apertures. In *Proceedings of the IEEE Radar Conference*, pages 160–164, Huntsville, AL, May 2003.

[4] S.M. Alamouti. A simple transmit diversity technique for wireless communications. *IEEE Journal on Selected Areas in Communications*, 16(8):1451–1458, October 1998.

[5] D.W. Allan. Statistics of atomic frequency standards. *Proceedings of the IEEE*, 54(2):221–230, February 1966.

[6] D.W. Allan and J.A. Barnes. A modified "Allan variance" with increased oscillator characterization ability. In *Proceedings of the 35th Annual Frequency Control Symposium*, pages 470–475, Pennsywania, May 1981.

[7] T. Amiot, F. Douchin, and E. Thouveno. The interferometric cartwheel: a multi-purpose formation of passive radar microsatellites. In *Proceedings of the IEEE Geoscience and Remote Sensing Symposium*, pages 435–437, Torono, Canada, June 2002.

[8] P. Antonik, M.C. Wicks, H.D. Griffiths, and C.J. Baker. Frequency diverse array radars. In *Proceedings of the IEEE IEEE Radar Conference*, pages 215–217, Verona, NY, April 2006.

[9] P. Antonik, M.C. Wicks, H.D. Griffiths, and C.J. Baker. Range dependent beamforming using element level waveform diversity. In *Proceedings of the International Waveform Diversity and Design Conference*, pages 1–4, Las Vegas, January 2006.

[10] M. Antoniou, M. Cherniakov, and C. Hu. Space-surface bistatic SAR image formation algorithms. *IEEE Transactions on Geoscience and Remote Sensing*, 47(6):1827–1843, June 2009.

[11] M. Antoniou, R. Saini, and M. Cherniakov. Results of a space-surface bistatic SAR image formation algorithm. *IEEE Transactions on Geoscience and Remote Sensing*, 45(11):3359–3371, November 2007.

[12] M. Ares. Optimum burst waveforms for detection of targets in uniform range-extended clutter. *IEEE Transactions on Aerospace and Electronics Systems*, 3(1):138–141, January 1967.

[13] S.E. Assad, I. Lakkis, and J. Saillard. Holographic SAR image formation by coherent summation of impulse response derivatives. *IEEE Transactions on Antennas and Propagation*, 41(5):620–624, May 1993.

[14] J.L. Auterman. Phase stability requirements for a bistatic SAR. In *Proceedings of the IEEE National Radar Conference*, pages 48–52, Atlanta, GA, March 1984.

[15] A. Aytun. Frequency Diverse Array Radar. Master's thesis, Naval Postgraduate School, Monterey, CA, 2010.

[16] P. Baizert, T.B. Hale, M.A. Temple, and M.C. Wicks. Forward-looking radar GMTI benefits using a linear frequency diverse array. *Electronic Letters*, 42(22):1311–1312, October 2006.

[17] R. Bamler. A comparison of range-Doppler and wavenumber domain SAR focusing algorithms. *IEEE Transactions on Geoscience and Remote Sensing*, 30(4):706–713, July 1992.

[18] R. Bamler, F. Meyer, and W. Liebhart. Processing of bistatic SAR data from quasi-stationary configurations. *IEEE Transactions on Geoscience and Remote Sensing*, 45(11):3350–3357, November 2007.

[19] J.A. Barnes, A.R. Chi, L.S. Cutler, D.J. Healey, D.B. Leeson, T.E. McGunigal, J.A. Mullen, W.L. Smith, R.L. Sydnor, R. Vessot, and G.M.R. Winkler. Characterization of frequency stability. *IEEE Transactions on Instrument and Measurement*, 20(1):105–120, May 1971.

[20] R.A. Baugh. Frequency modulation analysis with Hadamard variance. In *Proceedings of the 25th Annual Frequency Control Symposium*, pages 222–225, Atlantic City, NJ, April 1971.

[21] R.W. Bayma and P.A. McInnes. Aperture size and ambiguity constraints for a synthetic aperture radar. In *Proceedings of the IEEE Radar Conference*, pages 499–504, Arlington, VA, April 1975.

[22] M.R. Bell. Information theory and radar waveform design. *IEEE Transactions on Information Theory*, 39(5):1578–1597, September 1993.

[23] A. Bellettini and M. Pinto. *Accuracy of Synthetic Aperture Sonar Micronavigation Using a Displaced Phase Centre Antenna: Theory and Experimental Validation*. North Atlantic Treaty Organization, La Spezia, Italy, 2002.

[24] A. Bellettini and M.A. Pinto. Theoretical accuracy of synthetic aperture sonar micronavigation using a displaced phase-center antenna. *IEEE Journal of Oceanic Engineering*, 27(4):780–789, October 2002.

[25] W.R. Bennett. Spectra of quantized signals. *Bell System Technology Journal*, 27:467–472, 1948.

[26] C.R. Berger, B. Demissie, J. Heckenbach, P. Willett, and S.L. Zhou. Signal processing for passive radar using OFDM waveforms. *IEEE Journal of Oceanic Engineering*, 4(1):226–238, February 2010.

[27] U. Berthold, F.K. Jondral, S. Brandes, and M. Schnell. OFDM-based overlay systems: a promising approach for enhancing spectral efficiency. *IEEE Communication Magazine*, 45(12):52–58, December 2007.

[28] D.W. Bliss and K.W. Forsythe. Multiple-input and multiple-output (MIMO) radar and imaging: degrees of freedom and resolution. In *Proceedings of the 37th Asilomar Signals, Systemss Computers Conference*, pages 54–59, Pacific Grove, USA, November 2003.

[29] F. Bordoni, M. Younis, and G. Krieger. Ambiguity suppression by azimuth phase coding in multichannel SAR systems. *IEEE Transactions on Geoscience and Remote Sensing*, 50(2):617–629, February 2012.

[30] S. Brady. Frequency Diverse Array Radar: Signal Characterization and Measurement Accuracy. Master's thesis, Air Force Institute of Technology, Wright Ptrsn AFB, OH, 2010.

[31] J. Bredow, K. Xie, R. Porco, and M. Shah. An experimental study on the use of multistatic imaging for investigating wave-object interaction. *Journal of Electromagnetic Waves and Applications*, 7:811–831, June 1993.

[32] A. Broquetas, L. Jofre, and A. Cardama. A near field spherical wave inverse synthetic aperture radar technique. In *Proceedings of the IEEE International Antennas and Propagation Symposium*, pages 114–117, Chicago, IL, July 1992.

[33] A. Broquetas, J. Palau, L. Jofre, and A. Cardama. Spherical wave near-field imaging and radar cross-section measurement. *IEEE Transactions on Antennas and Propagation*, 46(5):730–735, May 1998.

[34] J.L. Brown. Multi-channel sampling of low-pass signals. *IEEE Transactions on Circuits Systems*, 28(2):101–106, February 1981.

[35] S.C.M. Brown and J.C. Bennett. High-resolution microwave polarimetric imaging of small trees. *IEEE Transactions on Geoscience and Remote Sensing*, 37(1):48–53, January 1999.

[36] W.D. Brown and D.C. Ghiglia. Some methods for reducing propagation-induced phase errors in coherent imaging systems. I: formalism. *Journal of the Optical Society of America*, 5(6):924–942, June 1988.

[37] D. Bruyere and N. Goodman. Adaptive detection and diversity order in multistatic radar. *IEEE Transactions on Aerospace and Electronic Systems*, 44(4):1615–1623, October 2008.

[38] A. Buddillon, A. Evangelista, and G. Schirinzi. Three-dimensional SAR focusing from multipass signals using compressive sampling. *IEEE Transactions on Geoscience and Remote Sensing*, 49(1):488–499, January 2011.

[39] A. Buddillon, V. Pascazio, and G. Schirinzi. Estimation of radial velocity of moving targets by along-track interferometric SAR systems. *IEEE Geoscience and Remote Sensing Letters*, 5(3):349–353, July 2008.

[40] R.J. Burkholder, I.J. Gupta, and J.T. Johnson. Comparison of monostatic and bistatic radar images. *IEEE Antennas Propagation Magazine*, 45(3):41–50, June 2003.

[41] C. Cafforio, C. Prati, and F. Rocca. SAR focusing using seismic mitigation techniques. *IEEE Transactions on Aerospace and Electronic Systems*, 27(2):194–207, April 1991.

[42] R. Calderbank, S.D. Howard, and B. Moran. Waveform diversity in radar signal processing. *IEEE Signal Processing Magazine*, 32(1):32–41, January 2009.

[43] G.D. Callaghan. *Wide-Swath Space-borne SAR: Overcome the trade-off between Swath-width and Resolution*. PhD thesis, University of Queensland, Queensland, Australia, 1999.

[44] G.D. Callaghan and I.D. Longstaff. Wide-swath spaceborne SAR and range ambiguity. In *Proceedings of IEEE Radar Conference*, pages 248–252, Syracuse, NY, October 1997.

[45] G.D. Callaghan and I.D. Longstaff. Wide swath spaceborne SAR using a quad element array. *IEE Proceedings-Radar Sonar and Navigation*, 146(3):159–165, June 1999.

[46] T.M. Calloway and G.W. Donohoe. Subaperture autofocus for synthetic aperture radar. *IEEE Transactions on Aerospace and Electronic Systems*, 30(2):617–621, April 1994.

[47] P. Capaldo, M. Crespi, F. Fratarcangeli, A. Nascetti, and F. Pieralice. High-resolution SAR radargrammetry: A first application with COSMO-SkyMed spotlight imagery. *IEEE Geoscience and Remote Sensing Letters*, 8(6):1100–1104, December 2011.

[48] G.T. Capraro, I. Bradaric, D.D. Weiner, R. Day, J. Perretta, and M.C. Wicks. Waveform diversity in multistatic radar. In *Proceedings of International Waveform Diversity and Design Conference*, Lihue, HI, January 2006.

[49] L. Carin, N. Geng, M. McClure, J. Sichina, and L. Nguyen. Ultra-wideband synthetic-aperture radar for mine-field detection. *IEEE Antennas and Propagation Magazine*, 41(1):18–33, February 1999.

[50] L. Carin, R. Kapoor, and C.E. Baum. Polarimetric SAR imaging of buried lanmines. *IEEE Transactions on Geoscience and Remote Sensing*, 36(11):1985–1988, November 1998.

[51] W.G. Carrara, R.S. Goodman, and R.M. Majewski. *Spotlight Synthetic Aperture Radar: Signal Processing Algorithms*. Artech House, Boston, MA, 1995.

[52] M.R. Cassola, S.V. Baumgartner, G. Krieger, and A. Moreira. Bistatic TerraSAR-X/F-SAR spaceborne-airborne SAR experiment: description, data processing, and results. *IEEE Transactions on Geoscience and Remote Sensing*, 48(2):781–794, February 2010.

[53] M.R. Cassola, P. Prats, D. Schulze, N.T. Ramon, U. Steinbrecher, L. Marotti, M. Nannini, M. Younis, P.L. Dekker, M. Zink, A. Reigber, G. Krieger, and A. Moreira. First bistatic spaceborne SAR experiments with TanDEM-X. *IEEE Geoscience and Remote Sensing Letters*, 9(1):33–37, January 2012.

[54] Franceschetti G. Ceraldi, E., A. Iodice, and D. Roccio. Estimating the soild dielectric constant via scattering measurements along the specular direction. *IEEE Transactions on Geoscience and Remote Sensing*, 43(2):295–305, February 2005.

[55] E. Chapin and C.W. Chen. Along-track interferometry for ground moving target indication. *IEEE Aerospace and Electronic Systems Magazine*, 23(6):19–24, June 2008.

[56] C.C. Chen and H.C. Andrews. Target-motion-induced radar imaging. *IEEE Transactions on Aerospae and Electronic Systems*, 16(1):2–14, January 1980.

[57] X.L. Chen, C.B. Ding, X.D. Liang, and Y.R. Wu. An improved NLCS imaging algorithm of bistatic SAR with stationary receiver. *Journal of Electronic and Information Technology*, 30(5):1041–1046, May 2008. in Chinese.

[58] C.L. Cheng, F.R. Chang, and K.Y. Tu. Highly accurate real-time GPS carrier phase-disciplined oscillator. *IEEE Transactions on Instrument and Measurement*, 54(2):819–824, April 2005.

[59] M. Cherniakov. Space-surface bistatic synthetic radar-prospective and problems. In *Proceedings of IEEE International Radar Conference*, pages 22–26, Edinburgh, UK, 2002.

[60] M. Cherniakov. *Bistatic Radars: Principles and Practice*. Wiley, Hoboken, NJ, 2007.

[61] M. Cherniakov. *Bistatic Radars: Emerging Technology*. Wiley, Hoboken, NJ, 2008.

[62] S. Chiu. Clutter effects on ground moving target velocity estimation with SAR along-track interferometry. In *Proceedings of the IEEE International Geoscience and Remote Sensing Symposium*, pages 1314–1319, Toulouse, France, July 2003.

[63] S. Chiu and C. Livingstone. A comparison of displaced phase centre antenna and along-track interferometry technique for RADARSAT-2 ground moving target indication. *Canadian Journal of Remote Sensing*, 34(1):27–51, February 2005.

[64] Jr. L. Cimini. Analysis and simulation of a digital mobile channel using orthogonal frequency division multiplexing. *IEEE Transactions on Communications*, 33(7):665–675, July 1985.

[65] J.P. Claassen and J. Eckerman. A system concept for wide swath constant incident angle coverage. In *Proceedings of the Synthetic Aperture Radar Technology Conference*, volume 4, pages 1–19, Las Cruces, New Mexico, March 1978.

[66] C. Coleman and H. Yardley. Passive bistatic radar based on target illuminations by digital audio broadcasting. *IET Radar Sonar and Navigation*, 2(5):366–375, October 2008.

[67] F. Colone, D.W. O'Hagan, P. Lombardo, and C.J. Baker. A multistage processing algorithm for disturbance removal and target detection in passive bistatic radar. *IEEE Transactions on Aerospace and Electronic Systems*, 45(3):698–722, April 2009.

[68] Jr.B. Correll. Efficient spotlight SAR MIMO linear collection geometries. In *Proceedings of the European Microwave Conference*, pages 21–24, Amsterdam, The Netherlands, October 2008.

[69] J.P. Costas. A study of a class of detection waveforms having nearly ideal range-Doppler ambiguity properties. *Proceedings of IEEE*, 72(8):996–1099, August 1984.

[70] L.W. Couch. *Digital and Analog Communication Systems*. Prentice Hall, Upper Saddle River, NJ, 2001.

[71] D. Cristallini, D. Pastina, and P. Lombardo. Exploiting MIMO SAR potentialities with efficient cross-track constellation configurations for improved range resolution. *IEEE Transactions on Geoscience and Remote Sensing*, 49(1):38–52, January 2011.

[72] D. Cristallini, M. Sedehi, and P. Lombardo. Imaging solution based on azimuth phase coding. In *Proceedings of the European Conference on Synthetic Aperture Radar*, pages 57–60, Friedrichshafen, Germany, June 2008.

[73] I.G. Cumming and W.F. Wong. *Digital Processing of Synthetic Aperture Radar Data: Algorithms And Implementation*. Artech House, Boston, MA, 2005.

[74] J.C. Curlander. Location of spaceborne SAR imagery. *IEEE Transactions on Geoscience and Remote Sensing*, 20(3):359–364, July 1982.

[75] J.C. Curlander, Wu. C., and A. Pang. Automated processing of spaceborne SAR data. In *Proceedings of the IEEE International Geoscience and Remote Sensing Symposium*, pages FA–1:3.1–3.6, Munich, Germany, June 1982.

[76] J.C. Curlander and R.N. McDonough. *Synthetic Aperture Radar: Systems and Signal Processing*. Wiley-Interscience, New York, 1991.

[77] A. Currie and M.A. Brown. Wide-swath SAR. *IEE Proceedings-Radar Sonar and Navigation*, 139(2):122–135, April 1992.

[78] A. Currie and C.D. Hall. A synthetic aperture radar technique for the simultaneous provision of high-resolution wide-swath coverage. In *Proceedings of the Military Microwave Conference*, pages 539–544, West Point, NY, June 1990.

[79] N.C. Currie. *Radar Reflectivity Measurement: Techniques and Applications*. Artech House, Norwood, MA, 1989.

[80] L.S. Cutler and C.L. Searle. Some aspects of the theory and measurement of frequency fluctuation in frequency standards. *Proceedings of The IEEE*, 54(2):89–100, February 1966.

[81] J. Dall and A. Kusk. Azimuth phase coding for range ambiguity suppression in SAR. In *Proceedings of the IEEE International Geoscience and Remote Sensing Symposium*, volume 3, pages 1734–1737, Anchorage, AK, September 2004.

[82] D. D'Aria, M.A. Guarnieri, and F. Rocca. Focusing bistatic synthetic aperture radar using dip move out. *IEEE Transactions on Geoscience and Remote Sensing*, 42(7):1362–1376, July 2004.

[83] G.W. Davidson, I.G. Cumming, and M.R. Ito. A chirp scaling approach for processing squint mode SAR data. *IEEE Transactions on Aerospace and Electronic Systems*, 32(1):121–133, January 1996.

[84] B Dawidowicz, K.S. Kulpa, and M. Malanowski. Suppression of the ground clutter in airborne PCL radar using DPCA technique. In *Proceedings of the European Radar Conference*, pages 306–309, Rome, September 2009.

[85] P. Defraigne, Q. Baire, and N. Guyennon. GLONASS and GPS PPP for time and frequency transfer. In *Proceedings of the Frequency Control Symposium*, pages 306–309, Geneva, Switzerland, May 2007.

[86] P.L. Dekker, J.J. Mallorqui, P.S. Morales, and J.S. Marcos. Phase synchronization and Doppler centroid estimation in fixed receiver bistatic SAR systems. *IEEE Transactions on Geoscience and Remote Sensing*, 46(11):3459–3471, November 2008.

[87] C.Y. Delisle and H.Q. Wu. Moving target imaging and trajectory computation using ISAR. *IEEE Transactions on Aerospace and Electronic Systems*, 30(3):887–899, July 1994.

[88] D. DeLong and E. Hofstetter. On the design of optimum radar waveforms for clutter rejection. *IEEE Transactions on Information Theory*, 13(3):454–463, July 1967.

[89] J. Denes and A.D. Keedwell. *Latin Squares and Their Applications*. Academic Press, Salt Lake City, UT, 1991.

[90] H. Deng. Discrete frequency-coding waveform design for netted radar systems. *IEEE Signal Processing Letters*, 11(2):179–182, February 2004.

[91] H. Deng. Polyphase code design for orthogonal netted radar systems. *IEEE Transactions on Signal Processing*, 52(11):3126–3135, November 2004.

[92] H. Deng and B. Himed. A virtual antenna beamforming (VAB) approach for radar systems by using orthogonal coding waveform. *IEEE Transactions on Antennas and Propagation*, 57(2):425–435, February 2009.

[93] M. D'Errico and A. Moccia. Attitude and antenna pointing design of bistatic radar formations. *IEEE Transactions on Aerospace and Electronic Systems*, 39(4):949–959, July 2003.

[94] R.H. Dicke. Object detection systems, January 6 2009. US patent, 2,624,876.

[95] F.R. Dickey, Jr.M. Labitt, and F.M. Staudaher. Development of airborne moving target radar for long range surveillance. *IEEE Transactions on Aerospace and Electronic Systems*, 27(6):959–972, December 1991.

[96] J.V. Difranco and W.L. Rubin. *Radar Detection*. SciTech Publishing, New York, 2004.

[97] R.C. DiPietro, R.L. Fante, and R.P. Perry. Space-based bistatic GMTI using low resolution SAR. In *Proceedings of the IEEE Aerospace Conference*, pages 181–192, Big Sky, MT, March 1997.

[98] Z. Dong, B. Cai, and D.N. Liang. Detection of ground moving targets for two-channel spaceborne SAR-ATI. *EURASIP Journal on Advances in Signal Processing*, 2010. Article ID 230785.

[99] D.M. Drumheller and E.L. Titlebaum. Cross-correlation properties of algebraically constucted Costas arrays. *IEEE Transanctions on Aerospace and Electronic Systems*, 72(1):2–10, January 1991.

[100] L. Du, Y.P. Wang, W. Hong, W.X. Tan, and Y.R. Wu. A three-dimensional range migration algorithm for downward-looking 3D-SAR with single-transmitting and multiple-receiving linear array antennas. *EURASIP Journal on Advances in Signal Processing*, 2010:1–15, 2010. Article ID 957916.

[101] L. Du, Y.P. Wang, W. Hong, and Y.R. Wu. Analytic modeling and three-dimensional imaging of downward-looking SAR using bistatic uniform linear array antennas. In *Proceedings of the Asia-Pacific Synthetic Aperture Radar Conference*, pages 49–53, Huangshan, China, November 2007.

[102] L. Du, Y.P. Wang, W. Hong, and Y.R. Wu. Analysis of 3D-SAR based on angle compression principle. In *Proceedings of the IEEE International Geoscience and Remote Sensing Symposium*, volume IV, pages 1324–1327, Boston, MA, July 2008.

[103] G.Q. Duan, D.W. Wang, X.Y. Ma, and Y. Su. Three-dimensional imaging via wideband MIMO radar systems. *IEEE Geoscience and Remote Sensing Letters*, 7(3):445–449, July 2010.

[104] M. D'Urso, M.G. Labate, and A. Buonanno. Reducing the number of amplitude controls in radar phased arrays. *IEEE Transanctions on Antennas and Propagation*, 58(9):3060–3064, September 2010.

[105] R. Eigel and P. Collins. Phase synchronization and Doppler centroid estimation in fixed receiver bistatic SAR systems. *IEEE Transactions on Geoscience and Remote Sensing*, 38(5):2078–2092, May 2000.

[106] M. Eineder. Oscillator drift compensation in bistatic interferometric SAR. In *Proceedings of the IEEE Geoscience and Remote Sensing Symposium*, pages 1449–1451, Toulouse, France, July 2003.

[107] B. Eissfeller, T. Zink, R. Wolf, J. Hammesfahr, A. Hornbostel, J. Hahn, and P. Tavella. Autonomous satellite state determination by use of two-directional links. *International Journal of Satellite Communications*, 18(4-5):325–346, August 2000.

[108] A. Eloualkadi, J.M. Paillot, and D. Flandre. Clock jitter effect on switched-capacitor. *IEEE Transactions on Geoscience and Remote Sensing*, 38(5):2078–2092, May 2000.

[109] J.H.G. Ender, C.H. Gierull, and D.C. Maori. Improved space-based moving target indication via alternate transmission and receiver switching. *IEEE Transactions on Geoscience and Remote Sensing*, 46(12):3960–3974, December 2008.

[110] J.H.G. Ender, J. Klare, and S. Perna. Bistatic exploration using spaceborne and airborne SAR sensors: a close collaboration between FGAN, ZESS, and FOMAAS. In *Proceedings of the IEEE Geoscience and Remote Sensing Symposium*, pages 1828–1831, Denver, CO, July 2006.

[111] J.H.G. Ender, I. Walterscheid, and A.R. Brenner. New aspects of bistatic SAR: processing and experiments. In *Proceedings of the IEEE Geoscience and Remote Sensing Symposium*, pages 1758–1762, Anchorage, Alaska, September 2004.

[112] T. Espeter, I. Walterscheid, and J. Klare. Synchronization techniques for the bistatic spaceborne/airborne SAR experiment with TerraSAR-X and PARMIR. In *Proceedings of the IEEE Geoscience and Remote Sensing Symposium*, pages 2160–2163, Barcelona, Spain, July 2007.

[113] T. Espeter, I. Walterscheid, and J. Klare. Process of hybrid bistatic SAR: synchronization experiments and first imaging results. In *Proceedings of the European Synthetic Aperture Radar Conference*, pages 217–220, Friedrichshafen, Germany, May 2008.

[114] D.G. Falconer. Extrapolation of near-field RCS measurements to the far zone. *IEEE Transactions on Antennas and Propagation*, 36(6):822–829, June 1988.

[115] L. Falk. *P. M. Woodward and the ambiguity function*, chapter in Principles of Waveform Diversity and Design, pages xiii–xx. Scitech Publishing Inc., Raleigh, NC, 2010.

[116] J. Farooq, M.A. Temple, and M.A. Saville. Application of frequency diverse arrays to synthetic aperture radar imaging. In *Proceedings of the International Electromagetics in Advances in Applications Conference*, pages 447–449, Torino, September 2007.

[117] J. Farooq, M.A. Temple, and M.A. Saville. Exploiting frequency diverse array processing to improve SAR imaging resolution. In *Proceedings of the the IEEE Radar Conference*, pages 1–5, Rome, Italy, May 2008.

[118] F. Feng, S.Q. Li, and W.D. Yu. Study on the processing scheme for space-time waveform encoding SAR system based on two-dimensional digital beamforming. *IEEE Transactions on Geoscience and Remote Sensing*, 50(3):910–932, March 2012.

[119] R. Filler. The acceleration sensitivity of quartz crystal oscillators: a review. *IEEE Transactions on Ultrasonic Ferrolectronic, Frequency Control*, 35(3):297–305, May 1988.

[120] C. Fischer, J. Fortuny, and W. Wiesbeck. 3-D imaging for near-range ground penatrating radar based on $\omega - K$ migration. In *Proceedings of European Conference on Synthetic Aperture Radar*, pages 841–844, Munich, Germany, May 2000.

[121] C. Fischer, C. Heer, G. Krieger, and R. Werninghaus. A high resolution wide swath SAR. In *Proceedings of European Conference on Synthetic Aperture Radar*, Dresden, Germany, 2006.

[122] C. Fischer and W. Wiesbeck. Laboratory verification for a forward-looking multi-receiver mine-detection GPR. In *Proceedings of the IEEE International Geoscience and Remote Sensing Symposium*, volume IV, pages 1643–1645, Honolulu, HI, July 2000.

[123] E. Fishler, A. Haimovich, and R. Blum. Spatial diversity in radars-models and detection performace. *IEEE Transactions on Signal Processing*, 54(3):823–838, March 2006.

[124] R. Fitzgerald. Effects of range-Doppler coupling on chirp radar tracking accuracy. *IEEE Transactions on Aerospace and Electronic Systems*, 10(3):528–532, July 1974.

[125] F.L. Fleming and N.J. Willis. Sanctuary radar. In *Proceedings of Millitary Microwave Conference*, pages 103–108, London, October 1980.

[126] G. Fornaro, G. Franceschetti, and S. Perna. Motion compensation errors: effects on the accuracy of airborne SAR images. *IEEE Transactions on Aerospace and Electronic Systems*, 41(4):1338–1352, October 2005.

[127] G. Fornaro, F. Lombardini, and F. Serafino. Three-dimensional multi-pass SAR focusing: experiments with long-term spaceborne data. *IEEE Transactions on Geoscience and Remote Sensing*, 43(4):702–714, April 2005.

[128] G. Fornaro, F. Serafino, and F. Soldovieri. Three-dimensional focusing with multipass SAR data. *IEEE Transactions on Geoscience and Remote Sensing*, 41(3), March 2003.

[129] K.W. Forsythe and D.W. Bliss. MIMO radar waveform constraints for GMTI. *IEEE Journal of Selected Topics in Signal Processing*, 4(1):21–32, February 2010.

[130] G.J. Foschini and G. Vannucci. Characterizing filtered light waves corrupted by phase noise. *IEEE Transactions on Information Theory*, 34(6):1437–1448, November 1988.

[131] R.W. Fox, S.A. Diddams, A. Bartel, and L. Hollberg. Optical frequency measurements with the global positioning system: tests with an iodine-stabilized He-Ne laser. *Applied Optics*, 44(1):113–120, January 2005.

[132] G. Franceschetti and G. Schirinzi. A SAR processor based on two-dimensional FFT codes. *IEEE Transactions on Aerospace and Electronic Systems*, 26(2):356–366, March 1990.

[133] R.L. Frank. Polyphase codes with good nonperiodic correlation properties. *IEEE Transactions on Information Theory*, 9(1):43–45, January 1963.

[134] G. Franken, H. Nikookar, and P. van Genderen. Doppler tolerance of OFDM-coded radar signals. In *Proceedings of the European Radar Conference*, pages 108–111, Manchester, UK, September 2006.

[135] A. Freeman, W.T.K. Johnson, B. Huneycutt, R. Jordan, S. Hensley, P. Siqueira, and J. Curlander. The "myth" of the minimum SAR antenna area constain. *IEEE Transactions on Geoscience and Remote Sensing*, 38(1):320–324, January 2000.

[136] P. Friedlander. Waveform design for MIMO radars. *IEEE Transactions on Aerospace and Electronics Systems*, 43(3):1227–1238, July 2007.

[137] W. Friedlander. A subspace framework for adaptive radar waveform design. In *Proceedings of the 39th Asilomar Conference on Signals, Systems and Computers*, pages 1135–1139, Pacific Grove, CA, 2005.

[138] D. Fuhrmann, P. Browning, and M. Rangaswamy. Signal strategies for the hybrid MIMO phased-array radar. *IEEE Journal of Selected Topics in Signal Processing*, 4(1):66–78, February 2010.

[139] E.A.Y. Gadallah. *Global Positioning System (GPS) Receiver Design for Multipath Mitigation*. PhD thesis, Air Force Institute of Technology, Wright Ptrsn AFB, OH, 1998.

[140] W.A. Gardner. *Statistical Spectral Analysis: A Nonprobabilistic Theory*. Prentice-Hall, Englewood Cliffs, NJ, 1988.

[141] D. Garmatyuk. Simulated imaging performance of UWB SAR based on OFDM. In *Proceedings of the IEEE International Conference on Ulta-Wideband*, pages 237–242, Waltham, MA, September 2006.

[142] D. Garmatyuk, J. Schuerger, Y. Morton, K. Binns, M. Durbin, and J. Kimani. Feasibility study of a multi-carrier dual-use imaging radar and communication system. In *Proceedings of the European Radar Conference*, pages 194–197, Munich, Germany, October 2007.

[143] D.A. Garren, M.K. Osborn, A.C. Odom, J.S. Goldstein, S.U. Pillai, and J.R. Guerci. Enhanced target detection and identification via optimized radar transmission pulse shape. *Proceedings of the IEEE*, 148(3):130–138, June 2001.

[144] L. Gasparini, O. Zadedyurina, G. Fontana, D. Macii, A. Boni, and Y. Ofek. A digital circuit for jitter reduction of GPS-disciplined 1-pps synchronization signals. In *Proceedings of International Workshop Advanced Methods for Uncertainty Estimation in Measurement*, pages 1–5, Trento, Italy, July 2007.

[145] N. Gebert. *Multi-channel Azimuth Processing for High-resolution Wide-swath SAR Imaging*. PhD thesis, Der University Fridericiana Karlsruhe, Karlsruhe, Baden-Wurttemberg, Germany, 2009.

[146] N. Gebert, G. Krieger, and A. Moreira. SAR signal reconstruction from non-uniform displaced phase centre sampling in the presence of perturbations. In *Proceedings of the IEEE International Geoscience and Remote Sensing Symposium*, pages 1034–1037, Seoul, Korea, July 2005.

[147] N. Gebert, G. Krieger, and A. Moreira. High resolution wide swath SAR imaging with digital beamforming-performance analysis, optimization and system design. In *Proceedings of the European Conference on Synthetic Aperture Radar*, pages 1–4, Dresden, Germany, May 2006.

[148] N. Gebert, G. Krieger, and A. Moreira. Digital beamforming on receive: techniques and optimization strategies for high-resolution wide-swath SAR imaging. *IEEE Transactions on Aeropsace and Electronic Systems*, 45(2):564–592, April 2009.

[149] U. Gebhardt, O. Loffeld, H. Nies, K. Natroshvili, and S. Knedlik. Bistatic space borne / airborne experiment: geometrical modeling and simulation. In *Proceedings of the IEEE International Geoscience and Remote Sensing Symposium*, pages 1832–1835, Denver, CO, July 2006.

[150] X.P. Geng, H.H. Yan, and Y.F. Wang. A two-dimensional spectrum model for general bistatic SAR. *IEEE Transactions on Geoscience and Remote Sensing*, 46(8):2216–2223, August 2008.

[151] J. George, K.V. Mishra, C.M. Nguyen, and V. Chandrasekar. Implementation of blind zone and range-velocity ambiguity mitigation for solid-state weather radar. In *Proceedings of the IEEE Radar Conference*, pages 1434–1438, Washington, DC, May 2010.

[152] A.B. Gershman and N.D. Sidiropoulos. *Space-Time Processing for MIMO Communications*. John Wiley & Sons, Inc., New York, 2005.

[153] D. Geudtner, M. Zink, and C. Gierull. Interferometric alignment of the X-SAR antenna system on the space shuttle radar topography mission. *IEEE Transactions on Geoscience and Remote Sensing*, 40(5):995–1005, May 2002.

[154] D.C. Ghiglia and W.D. Brown. Some methods for reducing propagation-induced phase errors in coherent imaging systems. II: numerical results. *Journal of the Optical Society of America*, 5(6):943–957, June 1988.

[155] C. Gierull. On a concept for an airborne downward-looking imaging radar. *AEU*, 53(6):295–304, June 1999.

[156] C. Gierull. Mitigation of phase noise in bistatic SAR systems with extremely large synthetic apertures. In *Proceedings of the European Synthetic Aperture Radsr Symposium*, pages 1251–1254, Dresden, Germany, May 2006.

[157] C.H. Gierull. Ground moving target parameter estimation for two-channel SAR. *IEE Proceedings on Radar, Sonar and Navigation*, 153(3):224–233, June 2006.

[158] C.H. Gierull, D.C. Maori, and J. Ender. Ground moving target indication with Tandem satellite constellation. *IEE Geoscience and Remote Sensing Letters*, 5(4):710–714, October 2008.

[159] C. Gini and F. Lombardini. Multibaseline cross-track SAR interferometry: a signal processing perspective. *IEEE Aerospace and Electronic Systems Magazine*, 20(8), August 2005.

[160] S.W. Golomb and M.Z. Win. Recent results on polyphase sequences. *IEEE Transactions on Information Theory*, 44(2):817–824, March 1988.

[161] N. Goodman, S. Lin, D. Rajakrisha, and J. Stiles. Processing of multiple-receiver spaceborne arrays for wide-area SAR. *IEEE Transactions on Geoscience and Remote Sensing*, 40(4):841–852, April 2002.

[162] I.S. Gradsheyn and I.M. Rtzhik. *Tables of Integrals, Series, and Products*. Academic Press, Salt Lake City, UT, 1965.

[163] L.C. Graham. Synthetic interferometric radar for topographic mapping. *Proceedings of the IEEE*, 62(6):763–768, June 1974.

[164] H. Griffiths and P. Mancini. Ambiguity suppression in SARs using adaptive array techniques. In *Proceedings of the IEEE International Geoscience and Remote Sensing Symposium*, pages 1015–1018, Espoo, Finland, July 1991.

[165] H.D. Griffiths. From a different prospective: principles, practice and potential of bistatic radar. In *Proceedings of the IEEE Radar Conference*, pages 1–7, Huntsville Marriott, July 2003.

[166] A.M. Guardieri and S. Tebaldini. On the exploitation of target statistics for SAR interferometry applications. *IEEE Transactions on Geoscience and Remote Sensing*, 46(11):3436–3443, November 2008.

[167] A.M. Guarnieri and C. Prati. ScanSAR focusing and interferometry. *IEEE Transactions on Geoscience and Remote Sensing*, 34(4):1029–1038, July 1996.

[168] S. Guillaso and A. Reighber. Scatterer characterisation using polarimetric SAR tomography. In *Proceedings of the IEEE International Geoscience and Remote Sensing Symposium*, pages 2685–2688, Seoul, Korea, July 2005.

[169] C. Hackman, J. Levine, T.E. Parker, D. Piester, and J. Becker. A straightforward frequency-estimation technique for GPS carrier-phase time transfer. *IEEE Transactions on Untra Ferr. Frequency Control*, 53(9):1570–1583, September 2006.

[170] A.M. Haimovich, R.S. Blum, and L.J. Cimini. MIMO radar with widely separated antennas. *IEEE Signal Processing Magazine*, 25(1):116–129, January 2008.

[171] D. Halford, J.H. Shoaf, and A.S. Risley. Spectral density analysis: frequency domain specifications and measurement of signal stability. In *Proceedings of the 27th Annual Frequency Control Symposium*, pages 421–431, Fort Monmouth, NJ, June 1973.

[172] D. Halford, A.E. Wainwright, and J.A. Barns. Flicker noise in RF amplifiers and frequency multipliers: characterization, cause, and cure. In *Proceedings of the 22nd Annual Frequency Control Symposium*, pages 340–341, Atlantic, USA, April 1968.

[173] L. Hanzo, W. Webb, and T. Keller. *Single- and Multi-carrier Quadrature Amplitude Modulation - Principles and Applications for Personal Communications, WLANs and Broadcasting*. John Wiley & Sons, Hoboken, NJ, 2000.

[174] R.L. Harris, B.E. Freburger, M.E. Lewis, and C.F. Zappala. Three-dimensional radar cross section imaging. In *Proceedings of the AMTA - Sixteenth Meeting and Symposium*, pages 443–448, October 1994.

[175] A. Hassanien and S.A. Vorobyov. Phased-MIMO radar: a tradeoff between phased-array and MIMO radars. *IEEE Transactions on Signal Processing*, 58(6):3137–3151, June 2010.

[176] C. Heer, F. Soualle, R. Zahn, and R. Reber. Investigations on a new high resolution wide swath SAR concept. In *Proceedings of the IEEE International Geoscience and Remote Sensing Symposium*, pages 521–523, Toulouse, France, July 2003.

[177] N.C. Helsby. GPS disciplined offset-frequency quartz oscillator. In *Proceedings of the IEEE Frequency Control Symposium*, pages 435–439, Tampa, FL, May 2003.

[178] H.F. Huang and D.N. Liang. The comparison of altitude and antenna directing design stategies of noncooperative spaceborne bistatic radar. In *Proceedings of the IEEE Radar Conference*, pages 568–571, Alabama, May 2005.

[179] H.F. Huang, K.F. Tong, and C.J. Baker. Frequency diverse array with beam scanning feature. In *Proceedings of the IEEE Antennas and Propagation Conference*, pages 1–4, San Diego, California, July 2008.

[180] H.F. Huang, K.F. Tong, and C.J. Baker. Frequency diverse array: simulation and design. In *Proceedings of the LAPS Antennas and Propagation Conference*, pages 253–256, Loughborough, UK, May 2009.

[181] J.J. Huang. *Frequency Diverse Array: Theory and Design*. PhD thesis, University College London, London, England, 2010.

[182] A.W. Hyatt. Doppler Aliasing Reduction in Wide-angle Synthetic Aperture Radar Using Phase Modulation Random Stepped-frequency Waveforms. Master's thesis, Air University, Maxwell Gunter AFB, AL, 2005.

[183] A. Ishimaru, Y.Y. Kuga, and J. Liu. Ionospheric effects on synthetic aperture radar at 100MHz to 2GHz. *Radio Science*, 34(1):257–268, January 1999.

[184] M.C. Jackson. The geometry of bistatic radar systems. *IEE Proceedings on Radar and Signal Processing*, 133(7):604–612, December 1986.

[185] H. Jafarkhani. A quasi-orthogonal space-time block code. *IEEE Transactions on Communications*, 49(1):1–4, January 2001.

[186] A. Jain. Multibeam synthetic aperture radar for global oceangraphy. *IEEE Transactions on Antennas and Propagation*, 27(4):535–538, July 1979.

[187] B.R. Jean and J.W. Rouse. A multiple beam synthetic aperture radar design concept for geoscience applications. *IEEE Transactions on Geoscience and Remote Sensing*, 21(2):201–207, April 1983.

[188] M.Y. Jin and C. Wu. A SAR correlation algorithm which accomodates large range mitigation. *IEEE Transactions on Geoscience and Remote Sensing*, 22(6):592–597, June 1984.

[189] T. Johnsen. Time and frequency synchronization in multistatic radar: consequences to usage of GPS disciplined references with and without GPS signals. In *Proceedings of the IEEE National Radar Conference*, pages 141–147, Long Beach, CA, April 2002.

[190] W.T. Johnson. Magellan imaging radar mission to Vensus. *Proceedings of the IEEE*, 79.

[191] A.M. Jones. Frequency Diverse Array Receiver Architecture. Master's thesis, Wright State University, Dayton, OH, 2011.

[192] B.W. Jung, R.S. Adve, and J. Chun. Frequency diversity in multistatic radars. In *Proceedings of the IEEE National Radar Conference*, pages 1–6, Rome, Italy, May 2008.

[193] J.H. Jung, J.S. Jung, C.H. Jung, and Y.K. Kwag. Ground moving target displacement compensation in the DPCA based SAR GMTI system. In *Proceedings of the IEEE Radar Conference*, pages 1–4, Pasadena, California, May 2009.

[194] D.W. Kammler. *A First Course in Fourier Analysis*. Prentice-Hall, Upper Saddle River, NJ, 2000.

[195] S. Kay. Optimal signal design for detection of gaussian point targets in stationary Gaussian clutter/reverberation. *IEEE Journal of Selected Topics in Signal Processing*, 1(1):31–41, June 2007.

[196] B.M. Keel, J.M. Baden, and T.H. Heath. A comprehensive review of quasi-orthogonal waveforms. In *Proceedings of the IEEE Radar Conference*, pages 122–127, Boston, MA, April 2007.

[197] H.A. Khan, Y.Y. Zhang, and C.L. Jin. Optimizing polyphase sequences for orthogonal netted radar. *IEEE Signal Processing Letters*, 13(10):589–592, October 2006.

[198] H. Kim, J.T. Johnson, and B.A. Baertlein. High resolution Ka-band images of a small tree: measurements and models. *IEEE Transactions on Geoscience and Remote Sensing*, 38(3):899–910, March 2000.

[199] J. Kim, J.T. Park, W.Y. Song, S.H. Rho, and Y.K. Kwag. Ground moving target displacement compensation for DPCA based SAR GMTI system. In *Proceedings of the Asia-Pacific Synthetic Aperture Radar Conference*, pages 177–180, Xi'an, China, October 2009.

[200] J. Kim and B.D. Tapley. Simulation of dual one-way ranging measurements. *Journal of Spacecr. Rockets*, 40(3):419–425, May 2003.

[201] J.H. Kim, T. Fugen, and W. Wiesbeck. A new approach for reconfigurable SAR system using space-time coding. In *Proceedings of the European Conference on Synthetic Aperture Radar*, pages 1–4, Friedrichshafen, Germany, June 2008.

[202] J.H. Kim, A. Ossowska, and W. Wiesbeck. Investigation of MIMO SAR for interferometry. In *Proceedings of the European Radar Conference*, pages 51–54, Munich, Germany, October 2007.

[203] J.H. Kim and W. Wiesbeck. Investigation of a new multifunctional high performance SAR system concept exploiting MIMO technology. In *Proceedings of the IEEE International Geoscience and Remote Sensing Symposium*, volume 2, pages 221–224, Boston, MA, July 2008.

[204] P.B. Kistner. Enhanced Detection of Orthogonal Radar Waveforms Using Time-Frequency and Bi-frequency Signal Processing Techniques. Master's thesis, Naval Postgraduate School, Monterey, CA, 2008.

[205] J. Klare. A new airborne radar for 3D imaging-simulation of ARTINO. In *Proceedings of the European Conference on Synthetic Aperture Radar*, pages 1–4, Dresden, Germany, May 2006.

[206] J. Klare. Digital beamforming for a 3D MIMO SAR-improvements through frequency and waveform diversity. In *Proceedings of the IEEE International Geoscience and Remote Sensing Symposium*, pages 17–20, Boston, MA, July 2008.

[207] J. Klare, A. Brenner, and J. Ender. Imapct of platform altitude disturbance on the 3D imaging quality of the UAV ARTINO. In *Proceedings of the European Conference on Synthetic Aperture Radar*, pages 1–4, Friedrichshafen, Germany, June 2008.

[208] J. Klare, A.R. Brenner, and J.H.G. Ender. A new airborne radar for 3D imaging - image formation using the ARTNO principle. In *Proceedings of the European Conference on Synthetic Aperture Radar*, pages 1–4, Dresden, Germany, May 2006.

[209] J. Klare, D.C. Maori, A.R. Brenner, and J.H.G. Ender. Image quality analysis of the vibrating sparse MIMO antenna array of the airborne 3D imaging radar ARTINO. In *Proceedings of the IEEE International Geoscience and Remote Sensing Symposium*, pages 5310–5314, Barcelona, Spain, July 2007.

[210] J. Klare, M. Weiβ, A.R. Brenner, and J.H.G. Ender. ARTNO: a new high resolution 3D imaging radar on an autonomous airborne platform. In *Proceedings of the IEEE International Geoscience and Remote Sensing Symposium*, pages 3842–3845, Denver, CO, August 2006.

[211] J.J. Kovaly. *Synthetic Aperture Radar*. John Wiley & Sons, New York, 1976.

[212] G. Krieger, M.R. Cassola, M. Younis, and R. Metzig. Impact of oscillator noise in bistatic and multistatic SAR. In *Proceedings of the IEEE International Geoscience and Remote Sensing Symposium*, pages 1043–1046, Seoul, Korea, July 2005.

[213] G. Krieger, H. Fiedler, and J. Mittermayer. Analysis of multistatic configurations for space-borne SAR interferometry. *IEE Proc. on Radar, Sonar and Navigation*, 150(3):87–96, June 2003.

[214] G. Krieger, N. Gebert, and A. Moreira. Unambiguous SAR signal reconstruction from nonuniform displaced phase center sampling. *IEEE Geoscience and Remote Sensing Letters*, 1(4):260–264, October 2004.

[215] G. Krieger, N. Gebert, and A. Moreira. Multidimensional waveform encoding for spaceborne synthetic aperture radar systems. In *Proceedings of the International Waveform Diversity and Design Conference*, pages 282–286, Pisa, June 2007.

[216] G. Krieger, N. Gebert, and A. Moreira. Multidimensional waveform encoding: a new digital beamforming technique for synthetic aperture radar remote sensing. *IEEE Transactions on Geoscience and Remote Sensing*, 46(1):31–46, January 2008.

[217] G. Krieger and A. Moreira. Potentials of digital beamforming in bi- and multistatic SAR. In *Proceedings of the IEEE International Geoscience and Remote Sensing Symposium*, pages 527–529, Toulouse, France, July 2003.

[218] G. Krieger and A. Moreira. Spaceborne bi- and multistatic SAR: potential and challenges. *IEE Proceedings-Radar, Sonar and Navigation*, 153(3):184–198, June 2006.

[219] G. Krieger, M. Wendler, and H. Fiedler. Performance analysis for bistatic interferometric SAR configurations. In *Proceedings of the IEEE Geoscience and Remote Sensing Symposium*, pages 650–652, Toronto, Canada, June 2002.

[220] G. Krieger and M. Younis. Impact of oscillator noise in bistatic and multistatic SAR. *IEEE Geoscience and Remote Sensing Letters*, 3(3):424–429, June 2006.

[221] M. Krieger, G. Younis, S. Huber, F. Bordoni, A. Patyuchenko, J. Kim, P. Laskowski, M. Villano, T. Rommel, P. Lopez-Dekker, and A. Moreira. Digital beamforming and MIMO SAR: review and new concepts. In *Proceedings of the 9th European Conference and Synthetic Aperture Radar*, pages 11–14, Nuremberg, Germany, April 2012.

[222] V.F. Kroupa. Noise properties of PLL systems. *IEEE Transactions on Communications*, 30(10):2244–2252, October 1982.

[223] V.F. Kroupa. Low-noise microwave-frequency synthesizers design principles. *IEE Proc. Microwave, Optics & Antennas*, 130(7):239–243, December 1983.

[224] V.F. Kroupa. Close-to-carrier noise in DDFS. In *Proceedings of the IEEE International Frequency Control Symposium*, pages 934–941, Honolulu, USA, June 1996.

[225] V.F. Kroupa. Jitter and phase noise in frequency dividers. *IEEE Transactions on Instrument and Measurement*, 50(5):1241–1243, October 2001.

[226] A.L. Lance, W.D. Seal, and F. Labaar. Phase noise and AM noise measurements in the frequency domain. *Infrared and Millimeter Waves*, 11, 1984.

[227] L. Landi, R.S. Adve, and A.De. Maio. Time-orthogonal-waveform-space-time adaptive processing for distributed aperture radars. In *Proceedings of the 3rd International Waveform Diversity and Design Conference*, pages 13–17, Pisa, Italy, June 2007.

[228] R.O. Lane. Super-resolution and the radar point spread function. In *Proceedings of the London Communications Symposium*, pages 1–4, London, September 2003.

[229] C.U.S. Larsson, R. Erickson, and O. Lunden. 3-D processing and imaging of near-field ISAR data in an arbitrary measurement geometry. In *Proceedings of the the AMTA Seventeenth Meeting and Symposium*, pages 106–110, November 1995.

[230] H. Lee and A.N. Mousa. GPS travelling wave fault locator systems: investigation into the anomalous measurements related to lighting strikes. *IEEE Transactions on Power Dilivery*, 11(3):1214 – 1223, July 1996.

[231] G. Lellouch and H. Nikookar. On the capability of a radar network to support communications. In *Proceedings of the 14th IEEE Symposium on Communications and Vehicular Technology in the Benelux*, pages 1–5, November 2007.

[232] R. Lenz, K. Schuler, M. Younis, and W. Wiesbeck. TerraSAR-X active raddar ground calibration system. *IEEE Aerospace and Electronic Systems Magazine*, 21(5):30–33, May 2006.

[233] P. Lesage and T. Ayi. Characterization of frequency stability: analysis of the modified Allan variance and properties of its estimate. *IEEE Transactions on Instrument and Measurement*, 33(4):332–337, December 1984.

[234] Y. Leshem, O. Naparstek, and A. Nehorai. Information theoretic adaptive radar waveform design for multiple extended targets. *IEEE Journal of Selected Topics in Signal Processing*, 1(1):42–55, June 2009.

[235] N. Levanon. Multifrequency complementary phased-coded radar signal. *IEE Proceedings- Radar, Sonar and Navigation*, 147(6):276–284, December 2000.

[236] N. Levanon. Mitigating range ambiguity in high PRF radar using interpulse binary coding. *IEEE Transactions on Aerospace and Electronic Systems*, 45(2):687–697, April 2009.

[237] N. Levanon and E. Mozeson. *Radar Signals*. Wiley-Interscience, New York, 2004.

[238] W. Lewandowski and C. Thomas. GPS time trnasfer. *Proceedings of The IEEE*, 79(7):991–1000, July 1991.

[239] B.L. Lewis and F.F. Kretschmer. A new class of polyphase pulse compression codes. *IEEE Transactions on Aerospace and Electronic Systems*, 17(3):364–371, May 1981.

[240] B.L. Lewis and F.F. Kretschmer. Linear frequency modulation derived polyphase pulse compression codes. *IEEE Transactions on Aerospace and Electronic Systems*, 18(5):637–641, September 1982.

[241] F.K. Li and W.T.K. Johnson. Ambiguities in spaceborne synthetic aperture radar systems. *IEEE Transactions on Aerospace and Electronic Systems*, 19(3):389–397, May 1983.

[242] G. Li, X.G. Xia, and Y.N. Peng. Doppler keystone transform: an approach suitable for parallel implementation of SAR moving target imaging. *IEEE Geosciene and Remote Sensing Letters*, 5(4):573–577, October 2008.

[243] H.B. Li and B. Himed. Transmit subaperturing for MIMO radars with colocated antennas. *IEEE Journal of Selected Topics in Signal Processing*, 4(1):55–65, February 2010.

[244] J. Li and P. Stoica. MIMO radar with colocated antennas. *IEEE Signal Processing Magazine*, 24(5):106–114, May 2007.

[245] J. Li and P. Stoica. *MIMO Radar Signal Processing*. John Wiley & Sons, New York, 2009.

[246] J. Li, P. Stoica, and W. Roberts. On parameter indentifiability of MIMO radar. *IEEE Signal Processing Letters*, 14(12):968–971, December 2007.

[247] J. Li, P. Stoica, and X.M. Xu. Signal synthesis and receiver design for MIMO radar imaging. *IEEE Transactions on Signal Processing*, 56(8):3959–3968, August 2008.

[248] J. Li, L.Z. Xu, P. Stoica, K.W. Forsythe, and D.W. Bliss. Range compression and waveform optimization for MIMO radar: a Cramer-Rao bound based study. *IEEE Transactions on Signal Processing*, 56(1):218–232, January 2008.

[249] Z.F. Li, H.Y. Wang, T. Su, and Z. Bao. Generation of wide-swath and high-resolution SAR images from multichannel small spaceborne SAR systems. *IEEE Geosciene and Remote Sensing Letters*, 2(1):82–86, January 2005.

[250] J.R. Liebe, N. van de Giesen, M.S. Andreini, T.S. Steenhuis, and M.T. Walter. Suitability and limitations of ENVISAT ASAR for monitoring small reservoirs in a semiarid area. *IEEE Transactions on Geoscience and Remote Sensing*, 47(5):1536–1547, May 2009.

[251] J. Linder. Binary sequences up to length 40 with best possible autocorrelation function. *Electronic Letters*, 11(21):507–507, October 1975.

[252] C. Little and C. Green. GPS disciplined rubidium oscillator. In *Proceedings of the European Frequency Time Forum*, pages 105–110, Brighton, UK, March 1996.

[253] B. Liu. *Research on Generation of Orthogonal Waveform and Signal Processing for MIMO Radar*. PhD thesis, University of Electronic Science and Technology of China, Chengdu, China, 2007.

[254] B. Liu. Orthogonal discrete frequency-coding waveform set design with minizied autocorrelation sidelobe. *IEEE Transactions on Aerospace and Electronic Systems*, 45(4):1650–1657, October 2009.

[255] B. Liu, Z. He, and J. Li. Mitigation of autocorrelation sidelobe peaks of orthogonal discrete frequency-coding waveform for MIMO radar. In *Proceedings of the International Radar Conference*, pages 1–6, Chengdu, China, 2008.

[256] B. Liu and Z.S. He. Orthogonal discrete frequency-coding waveform for MIMO radar. *Journal of Electronics (China)*, 25(4):471–476, July 2008. in Chinese.

[257] B.C. Liu, T. Wang, Q.S. Wu, and Z. Bao. Bistatic SAR data focusing using an omega-K algorithm based on method of series reversion. *IEEE Transactions on Geoscience and Remote Sensing*, 47(8):2899–2912, August 2009.

[258] M. Liuse and R. Reggiannini. Carrier frequency acquisition and tracking for OFDM systems. *IEEE Transactions on Communications*, 44(11):1590–1598, November 1996.

[259] C. Livingstone, I. Sikaneta, C. Gierull, B. Chiu, A. Beaudoin, J. Campbell, J. Beaudoin, S. Gong, and T. Knight. An airborne synthetic aperture radar (SAR) experiment to support RADARSAT-2 ground moving target indication (GMTI). *Canadian Journal of Remote Sensing*, 28(6):794–813, July 2002.

[260] O. Loffeld, H. Nies, V. Peters, and S. Knedlik. Models and useful relations for bistatic SAR processing. *IEEE Transactions on Geoscience and Remote Sensing*, 42(10):2031–2038, October 2004.

[261] M.A. Lombardi, A.N. Novick, and V.S. Zhang. Characterizing the performance of GPS disciplined oscillators with respect to UTC (NIST). In *Proceedings of IEEE International Frequency Control Symposium and Exposition*, pages 677–684, Vancouver, August 2005.

[262] P. Lombardo, F. Colone, and D. Pastina. Monitoring and surveillance potentialities obtained by splitting the antenna of the COSMO-SkyMed SAR into multiple sub-apertures. *IEE Proceedings-on Radar, Sonar and Navigation*, 153(2):104–116, April 2006.

[263] J.M. Lopez-Sanchez and J. Fortuny. 3-D radar imaging using range migration techniques. *IEEE Transactions on Antennas and Propagation*, 48(5):728–737, May 2000.

[264] J.E. Luminati. *Wide-angle Multistatic Synthetic Aperture Radar: Focused Image Formation and Aliasing Artifact Mitigation*. PhD thesis, Air Force Institute of Technology, Wright Ptrsn AFB, OH, March 2009.

[265] L. Lv, X.L. Zhang, and Z.J. Cao. A forward-looking 3-D SAR imaging model using unequally spaced array. In *Proceedings of International Symposium on Intelligent Signal Processing and Communication Systems*, pages 1–4, Chengdu, December 2010.

[266] X.L. Lv, M.D. Xing, Y.K. Deng, S.H. Zhang, and Y.R. Wu. Coherence-improving algorithm for image pairs of bistatic SARs with nonparallel trajectories. *IEEE Transactions on Geoscience and Remote Sensing*, 47(8):2884–2898, August 2009.

[267] C.Z. Ma, T.S. Yeo, C.S. Tan, and Z.F. Liu. Three-dimensional imaging of targets using colocated MIMO radar. *IEEE Transactions on Geoscience and Remote Sensing*, 49(8):3009–3021, August 2011.

[268] J.L. MacArthur and A.S. Posner. Satellite-to-satellite range-gate measurements. *IEEE Transactions on Geoscience and Remote Sensing*, 23(4):517–523, July 1985.

[269] B.R. Mahafza and A.Z. Elsherbeni. *MATLAB Simulations for Radar Systems Design*. CRC Press, New York, 2004.

[270] C.E. Mancill and J.M. Swiger. A map drift autofocus technique for correcting higher order SAR phase errors. In *Proceedings of the 27th Annual Trip-Service Radr Symposium*, pages 391–400, Monterey, CA, June 1981.

[271] D.G. Manolakis, V.K. Ingle, and S.M. Kogon. *Statistical and Adaptive Signal Processing: Spectral Estimation, Signal Modeling, Adaptive Filtering and Array Processing*. McGraw-Hill, New York, 2000.

[272] J.S. Marcos, P.L. Dekker, Mallorqui. J.J., A. Aguasca, and P. Prats. SABRINA: a SAR bistatic receiver for interferometric applications. *IEEE Geoscience and Remote Sensing Letters*, 4(2):307–311, April 2007.

[273] R.J. Marks. *Introduction to Shannon Sampling and Interpolation Theory*. Springer-Verlag, New York, 1991.

[274] S.L. Marple. *Digital Spectral Analysis with Applications*. Prentice-Hall, Englewood Cliffs, NJ, 1987.

[275] P.A.C. Marques and D.J.M. Bioucas. Moving targets processing in SAR spatial domain. *IEEE Transactions on Aerospace and Electronic Systems*, 43(3):864–874, July 2007.

[276] M. Martorella, J. Palm, J. Homer, B. Littleton, and I.D. Longstaff. Bistatic inverse synthetic aperture radar. *IEEE Transactions on Aerospace and Electronic Systems*, 43(3):1125–1134, July 2007.

[277] D. Massonnet. Capabilities and limitations of the interferometric cartwheel. *IEEE Transactions on Geoscience and Remote Sensing*, 39(3):506–520, March 2001.

[278] J. Mcneill. Jitter in ring oscillators. *IEEE Journal of Solidstate Circuits*, 32(6):870–879, June 1997.

[279] D. Mensa and G. Heidbreder. Bistatic synthetic aperture radar imaging of rotating objects. *IEEE Transactions on Aerospace and Electronic Systems*, 18(4):423–431, July 1982.

[280] D.G. Meyer. A test set for the accurate measurement of phase noise on high-quality signal sources. *IEEE Transactions on Instrument and Measurement*, 19(4):215–227, 1970.

[281] J. Mittermayer, G. Krieger, and A. Moreira. Interferometric performance estimation for the interferom cartwheel in combination with a transmitting SAR-satellite. In *Proceedings of the IEEE Geoscience and Remote Sensing Symposium*, pages 2955–2957, Sydney, Australia, July 2001.

[282] J. Mittermayer, R. Lord, and E. Borner. Sliding spotlight SAR processing for TerraSAR-X using a new formulation of the extended chirp scaling algorithm. In *Proceedings of the IEEE International Geosience and Remote Sensing Symposium*, pages 1642–1464, Toulouse, France, July 2003.

[283] J. Mittermayer and J.M. Martinez. Analysis of range ambiguity suppression in SAR by up and down chirp modulation for pint and distributed targets. In *Proceedings of the IEEE International Geosience and Remote Sensing Symposium*, pages 4077–4079, Toulouse, France, July 2003.

[284] J. Mittermayer, A. Moreira, and O. Loffeld. Spotlight SAR data processing using the frequency scaling algorithm. *IEEE Transactions on Geoscience and Remote Sensing*, 37(5):2198–2214, September 1999.

[285] A. Moccia and G. Fasano. Analysis of spaceborne Tandem configurations for complementing Cosmo with SAR interferometry. *EURASIP Journal of Applied Signal Processing*, 20:3304–3315, 2005.

[286] A. Moccia, G. Salzillo, M. D'Errico, G. Rufino, and G. Alberti. Performance of spaceborne bistatic synthetic aperture radar. *IEEE Transactions on Aerospace and Electronic Systems*, 41(4):1383–1395, October 2005.

[287] G.E. Moore. Cramming more components onto integrated circuits. *Electronics*, 38(8), April 1965.

[288] R.K. Moore, J.P. Claassen, and Y.H. Lin. Scanning spaceborne synthetic aperture radar with integrated radiometer. *IEEE Transactions on Aerospace and Electronic Systems*, 17(3):410–420, May 1981.

[289] A. Moreira, G. Krieger, I. Hajnsek, D. Hounam, M. Werner, S. Riegger, and E. Settelmeyer. TanDEM-X: a TerraSAR-X add-on satellite for single-pass SAR interferometry. In *Proceedings of the IEEE International Geosience and Remote Sensing Symposium*, pages 1000–1003, Anchorage, Alaska, September 2004.

[290] L.C. Morena, K.V. James, and J. Beck. An introduction to the RADARSAT-2 mission. *Canada Journal of Remote Sensing*, 30(3):221–234, June 2004.

[291] W.H. Mow and S.Y.R. Li. Aperiodic autocorrelation and crosscorrelation of polyphase sequences. *IEEE Transactions on Information Theory*, 43(3):1000–1007, May 1997.

[292] E. Mozeson. *Multicarrier Radar Signals and New Single Carrier Derivatives with Good Pulse Compression Properties*. PhD thesis, Tel Aviv University, Tel Aviv, Israel, 2003.

[293] C.E. Muehe and M. Labitt. Displaced-phase-center antenna technique. *Lincolin Laboratory Journal*, 12(2):281–295, 2000.

[294] S. Mustafa, D. Simsek, and H.A.E. Taylan. Frequency diverse array antenna with periodic time modulated pattern in range and angle. In *Proceedings of the IEEE Radar Conference*, pages 427–430, Boston, MA, April 2007.

[295] K. Natroshvili, O. Loffeld, H. Nies, A.M. Ortiz, and S. Knedlik. Focusing of general bistatic SAR configuration data with a two-dimensional inverse scaled FFT. *IEEE Transactions on Geoscience and Remote Sensing*, 44(10):2718–2727, October 2006.

[296] R. Navid, T.H. Lee, and R.W. Dutton. An analysis formulation of phase noise of signals with guassian-distributed jitter. *IEEE Transactions on Circuits and Systems-II: Express Briefs*, 52(3):149–153, March 2005.

[297] Y.L. Neo, F.H. Wong, and I.G. Cumming. A two-dimensional spectrum for bistatic SAR processing using series reversion. *IEEE Geoscience and Remote Sensing Letters*, 4(1):93–96, January 2007.

[298] Y.L. Neo, F.H. Wong, and I.G. Cumming. A comparison of point target spectra derived for bistatic SAR processing. *IEEE Transactions on Geoscience and Remote Sensing*, 46(9):2481–2492, September 2008.

[299] L.B. Neronskiy, S.G. Likhanski, I.V. Elizavetin, and D.V. Sysenko. Phase and amplitude histories adapted to spaceborne SAR survey. *IEE Proceedings-on Radar, Sonar and Navigation*, 150(3):184–192, June 2003.

[300] H. Nico and M. Tesauro. On the existence of coverage and integration time regimes in bistatic SAR configurations. *IEEE Geoscience and Remote Sensing Letters*, 4(3):426–430, July 2007.

[301] H. Nies, O. Loffeld, and K. Natroshvili. Analysis and focusing of bistatic SAR data. *IEEE Transactions on Geoscience and Remote Sensing*, 45(11):3342–3349, November 2007.

[302] J.F. Nouvel and O.R. du Plessis. The ONERA compact SAR in Ka band. In *Proceedings of the European Conference on Synthetic Aperture Radar Symposium*, pages 1–4, Friedrichshafen, Germany, June 2008.

[303] J.F. Nouvel, O.R. du Plessis, J. Svedin, and A. Gustafsson. ONERA DRIVE project. In *Proceedings of the European Conference on Synthetic Aperture Radar Symposium*, pages 1–4, Friedrichshafen, Germany, June 2008.

[304] J.F. Nouvel, H. Jeuland, G. Bonin, S. Roques, O. du Plessis, and J. Peyret. A Ka band imaging radar: DRIVE on board ONERA motorglider. In *Proceedings of the IEEE International Geosience and Remote Sensing Symposium*, pages 134–136, Denver, Colo, August 2006.

[305] J.F. Nouvel, S. Roques, and O.R. du Plessis. A low-cost imaging radar: DRIVE on board ONERA motorglider. In *Proceedings of the IEEE International Geosience and Remote Sensing Symposium*, pages 5306–5309, Barcelona, Spain, July 2007.

[306] J.W. Odendaal and J. Joubert. Radar cross section measurements using near-field radar imaging. *IEEE Transactions on Antennas and Propagation*, 45(12):948–958, December 1996.

[307] R.F. Ogrodnik, W.E. Wolf, R. Schneible, and J. McNamara. Bistatic vairants of space-based radar. In *Proceedings of the IEEE Aerospace Conference*, pages 159–169, Big Sky, USA, March 1997.

[308] A.V. Oppenheim, A.S. Willsky, and S. Hamid. *Signals and Systems*. Prentice Hall, Upper Saddle River, NJ, 1996.

[309] A. Papoulis. *The Fourier Integral and Its Applications*. McGraw Hill, New York, 1962.

[310] A. Papoulis. *Systems and Transforms with Applications in Optics*. McGraw Hill, New York, 1968.

[311] A. Papoulis. *Probability, Random Variables, and Stochastic Processes*. McGraw Hill, New York, 1984.

[312] B.W. Parkinson and J.J. Spilker. *Global Positioning System: Theory and Applications*. American Institute of Aeronautics and Asronautics, Washington, DC, 1996.

[313] A. Paulraj, R. Nabar, and D. Gore. *Introduction to Space-Time Wireless Communications*. Cambridge University Press, Cambridge, England, 2003.

[314] B.M. Penrod. A new class of precision UTC and frequency reference using IS-95 CDMA base station transmissions. In *Proceedings of the 33rd Annual Precise Time and Time Interval Meeting Conference*, pages 279–291, Pasadena, CA, December 2001.

[315] M.I. Pettersson. Extraction of moving targets by a bistatic ultra-wideband SAR. *IEE Proceedings on Radar, Sonar and Navigation*, 148(1):35–40, February 2001.

[316] U. Pillai, K.Y. Li, I. Selesnick, and B. Himed. *Waveform Diversity: Theory & Applications*. McGraw Hill, New York, 2011.

[317] R. Prasad. *OFDM for Wireless Communication Systems*. Artech House, Boston, MA, 2004.

[318] D.S. Purdy. Receiver antenna scan rate requirements needed to implement pulse chasing in a bistatic radar receiver. *IEEE Transactionas on Aerospace and Electronic Systems*, 37(1):285–288, January 2001.

[319] J. Qian, X. Lv, M. Xing, L. Li, and Z. Bao. Motion parameter estimation of multiple ground fast-moving targets with a three-channel synthetic aperture radar. *IET Radar, Sonar and Navigation*, 5(5):582–592, June 2011.

[320] R.K. Raney, H. Runge, R. Bamler, I.G. Cumming, and F.H. Wong. Precision SAR processing using chirp scaling. *IEEE Transactions on Geoscience and Remote Sensing*, 32(4):786–799, July 1994.

[321] R.K. Raney and P.W. Vachon. A phase preserving SAR processor. In *Proceedings of the IEEE Geoscience and Remote Sensing Symposium*, pages 2588–2591, Vancouver, Canada, July 1989.

[322] V. Ravenni. Performance evaluations of frequency diversity radar system. In *Proceedings of the 4th European Radar Conference*, pages 436–439, Munich, Germany, October 2007.

[323] B.R. Reddy and M.U. Kumari. *Generation of Orthogonal Discrete Frequency Coded Waveform Using Accelerated Particle Swarm Optimization Algorithm for MIMO Radar*, chapter in Advances in Computer Science, Engineering & Applications, pages 13–23. Springer-Verlag., Berlin, Heidelberg, 2012.

[324] A. Reigber. *Airborne Polarimetric SAR Tomography*. PhD thesis, University Stuttgart, Stuttgart, Germany, 2001.

[325] A. Reigber and A. Moreira. First demonstration of airborne SAR tomography using multibaseline L-band data. *IEEE Transactions on Geoscience and Remote Sensing*, 38(9):2142–2152, September 2000.

[326] X.Z. Ren, J.T. Sun, and R.L. Yang. A new three-dimensional imaging algorithm for airborne forward-looking SAR. *IEEE Geoscience and Remote Sensing Letters*, 8(1):153–157, January 2001.

[327] A. Renga and A. Moccia. Performance of stereor radar grammetric methods applied to spaceborne monostatic-bistatic synthetic aperture radar. *IEEE Transactions on Geoscience and Remote Sensing*, 47(2):544–560, Feburary 1999.

[328] Z. Rodriguez and J.M. Martin. Theory and design of interferometric synthetic aperture radars. *IEE Proceedings on Radar and Signal Processing*, 153(2):147–159, April 1992.

[329] R. Romeiser, H. Runge, J. Suchandt, Sprenger, H. Weilbeer, A. Sohrmann, and D. Stammer. Current measurements in rivers by spaceborne along-track InSAR. *IEEE Transactions on Geoscience and Remote Sensing*, 45(12):4019–4031, December 2008.

[330] G. Rufino and A. Moccia. A procedure for alttitude determination of a bistatic SAR by using raw data. In *Proceedings of the International Astronautical Federation Congress*, pages 1–8, Torino, Italy, October 1997.

[331] W.D. Rummier. Clutter suppression by complex weighting of coherent pulse trains. *IEEE Transactions on Aerospace and Electronics Systems*, 2(6):689–699, November 1966.

[332] H. Runge and R. Bamler. A novel high precision SAR focusing algorithm based on chirp scaling. In *Proceedings of the IEEE International Geoscience and Remote Sensing Symposium*, pages 372–375, Houston, TX, 1992.

[333] J. Rutman. Characterization of phase and frequency stabilities in precision frequency sources: fifteen years of progress. *Proceedings of The IEEE*, 66(9):1048–1073, September 1978.

[334] J. Rutman and J. Groslambert. Characterization and measurement of frequency stability: twenty five years of progress. In *Proceedings of the 4th Annual European Frequency and Time Forum*, pages 145–148, 1990.

[335] J. Rutman and F.L. Walls. Characterization of frequency stability in precision frequency sources. *Proceedings of The IEEE*, 79(6):952–960, June 1991.

[336] P.F. Sammartino and C. Baker. Developments in the frequency diverse bistatic system. In *Proceedings of the IEEE Radar Conference*, pages 1–5, Pasadena, CA, May 2009.

[337] P.F. Sammartino and C. Baker. The frequency diverse bistatic system. In *Proceedings of the 4th Waveform Diversity & Design Conference*, pages 155–159, Orlando, FL, Febuary 2009.

[338] B. Scheers. *Ultra-wideband Ground Penetrating Radar, with Application to the Detection of Anti Personnel Landmines*. PhD thesis, Royal Military Academy, Ontario, Canada, March 2001.

[339] J.G. Schoenenberger and J.R. Forrest. Principles of independent receivers for use with cooperative radar transmitters. *The Radio and Electronic Engineer*, 52(2):93–101, April 1982.

[340] S. Searle and S. Howard. A novel polyphase code for sidelobe suppression. In *Proceedings of the 3rd International Waveform Diversity and Design Conference*, pages 377–381, Pisa, Italy, June 2007.

[341] M.A. Sebt, A. Sheikhi, and M.M. Nayebi. Orthogonal frequency-diversion multiplexing radar signal design with optimized ambiguity function and low peak-to-average power ratio. *IET Radar Sonar and Navigation*, 3(2):122–132, April 2009.

[342] S. Sen and A. Nehorai. Adaptive design of OFDM radar signal with improved wideband ambiguity function. *IEEE Transactions on Signal Processing*, 58(2):928–933, February 2010.

[343] K.V. Shanbhag, D. Deb, and M. Kulkarni. MIMO radar with spatial-frequency diversity for improved detection performance. In *Proceedings of the IEEE International Communication, Control and Computing Technology Conference*, pages 66–70, Nagercoil, India, October 2010.

[344] J. Shi, X.L. Zhang, and J.Y. Yang. Principle and methods on bistatic SAR signal processing via image correction. *IEEE Transactions on Geoscience and Remote Sensing*, 46(10):3163–3178, October 2008.

[345] H.S. Shin and J.T. Lim. Omega-K algorithm for airborne spatial invariant bistatic spotlight SAR imaging. *IEEE Transactions on Geoscience and Remote Sensing*, 47(2):238–250, February 2009.

[346] L. Sibul and L. Ziomek. Generalized wideband crossambiguity function. In *Proceedings of the IEEE International Conference on Acoustics, Speech, and Signal Processing*, pages 1239–1242, Atlanta, Georgia, April 1981.

[347] S. Sira, D. Morell, and A.P. Suppappola. Waveform design and scheduling for agile sensors for target tracking. In *Proceedings of the Asilomar Conference on Signals, Systems and Computers*, pages 820–824, November 2004.

[348] M.I. Skolnik. *Radar Handbook*. McGraw-Hill, Columbus, OH, 1990.

[349] M.I. Skolnik. *Introduction to Radar Systems*. McGraw-Hill, Desoto,TX, 2002.

[350] T.A. Soame and D.M. Gould. Description of an experimental bistatic radar system. In *Proceedings of the IEE Intenational Radar Conference*, pages 7–11, 1987.

[351] U. Somaini. Binary sequences with good autocorrelation and crosscorrelation properties. *IEEE Transactions on Aerospace and Electronic Systems*, 11(6):1226–1231, November 1975.

[352] M. Soumekh. Wide-bandwidth continuous-wave monostatic/bistatic synthetic aperture radar imaging. In *Proceedings of the International Conference on Information Processing*, pages 361–365, Chicago, IL, October 1998.

[353] M. Soumekhm. *Synthetic Aperture Radar Signal Processing Using MAT-LAB algortihms*. Wiley-Interscience, Hoboken, NJ, 1999.

[354] S. Sowelam and A. Tewfik. Waveform selection in radar target classification. *IEEE Transactions on Information Theory*, 46(3):1014–1029, May 2000.

[355] M. Stangl, R. Werninghaus, B. Schweizer, C. Fischer, M. Brandfass, J. Mittermayer, and H. Breit. TerraSAR-X technologies and first results. *IEE Proceedings-on Radar, Sonar and Navigation*, 153(2):86–95, April 2006.

[356] P. Stoica, J. Li, and Y. Xie. On probing signal design for MIMO radar. *IEEE Transactions on Signal Processing*, 55(8):4151–4161, August 2007.

[357] J.P. Stralka. *Applications of Orthogonal Frequency-division Multiplexing (OFDM) to Radar*. PhD thesis, The Johns Hopkins University, Baltimore, MD, March 2008.

[358] F.G. Stremler. *Introduction to Communication Systems*. Addison-Wesley, Boston MA, 1982.

[359] C.A. Stutt and L.J. Spafford. A 'best' mismatched filter response for radar clutter discrimination. *IEEE Transactions on Information Theory*, 14(2):280–287, March 1968.

[360] I. Suberviola, I. Mayordomo, and J. Mendizabal. Experimental results of air target detection with a GPS forward-scattering radar. *IEEE Geoscience and Remote Sensing Letters*, 9(1):47–51, January 2012.

[361] M. Suess, B. Grafmueller, and R. Zahn. A novel high resolution, wide-swath SAR system. In *Proceedings of the IEEE International Geoscience and Remote Sensing Symposium*, pages 1013–1015, Sydney, Australia, July 2001.

[362] M. Suess and W. Wiesbeck. Side-looking synthetic aperture radar system, September 18 2002. European patent, 1241487A1.

[363] X.L. Sun, A.X. Yu, Z. Dong, and D.N. Liang. Three-dimensional SAR focusing via compressive sensing: the case study on angel stadium. *IEEE Geoscience and Remote Sensing Letters*, 9(4):759–763, July 2012.

[364] S. Sussman. Least-square synthesis of radar ambiguity functions. *IEEE Transactions on Information Theory*, 8(3):246–254, April 1962.

[365] T. Sutor, S. Buckreuss, G. Krieger, M. Wendler, and F. Witte. Sector imaging radar for enhanced vision (SIREV): theory and applications. In *Proceedings of the SPIE*, volume 4023, pages 292–297, 2000.

[366] H.B. Sverdlik and L. Levanon. Family of multicarrier bi-phase radar signals represented by ternary arrays. *IEEE Transactions on Aerospace and Electronic Systems*, 42(3):933–953, July 2006.

[367] K. Tajima, Y. Imai, and Y. Kanagawa. Low spurious frequency setting algorithm for a triple type PLL synthesizer driven by a DDS. *IEICE Transactions on Electronics*, E85-C(3):595–598, March 2002.

[368] W. Tan, W. Hong, Y. Wang, Y. Lin, and Y.R. Wu. Synthetic aperture radat tomography sampling critera and three-dimensional range migration algorithm with elevation digital spotligh. *Sciences in China F*, 52(1):100–114, January 2009.

[369] Y. Tang, M.A. Lombardi, and D.A. Howe. Frequency uncertainty analysis for Jpsephson voltage standard. In *Proceedings of the Precision Electromagnetic Measurement Conference*, pages 338–339, London, UK, June 2004.

[370] Z.Y. Tang and S.R. Zhang. *Bistatic Synthetic Aperture Radar System and Principle*. National Defense Industry Press, Beijing, China, 2003. in Chinese.

[371] J. Tao and B.D. Steinber. Reduction of sidelobe and speckle antifacts in microwave imaging. *IEEE Transactions on Antennas and Propagation*, 36(4):543–556, April 1998.

[372] V. Tarokh, H. Jafarkhani, and A.R. Calderbank. Space-time block codes from orthogonal designs. *IEEE Transactions on Information Theory*, 45(5):744–765, May 1999.

[373] V. Tarokh, H. Jafarkhani, and A.R. Calderbank. Space-time block coding for wireless communications: performance results. *IEEE Journal on Selected Areas in Communications*, 17(3):451–460, March 1999.

[374] I.E. Telatar. Capability of multi-antenna gaussian channels. *European Transactions on Telecommunications*, 10(6):585–595, November/December 1999.

[375] S.C. Thompson. *Constant Envelope OFDM Phase Modulation*. PhD thesis, University of California, San Diego, CA, 2005.

[376] R.F. Tigrek, De W.J.A. Heij, and Van P. Genderen. OFDM signals as the radar waveform to solve Doppler ambiguity. *IEEE Transactions on Aerospace and Electronic Systems*, 48(1):130–143, January 2012.

[377] P.G. Tomlinson. Modeling and analysis of monostatic/bistatic space-time adaptive processing for airborne and space-based radar. In *Proceedings of the IEEE Radar Conference*, pages 102–109, Waltham, MA, May 1999.

[378] E.B. Tomme. Balloons in today's military: an introduction to near-space concept. *Air Space Journal*, 19(4):39–50, 2005.

[379] H.L. Trees. Optimum signal design and processing for reverberation-limited environments. *IEEE Transactions on Military Electronics*, 9(3-4):212–229, July-October 1965.

[380] T. Tsao, M. Slamani, P. Varshney, D. Weiner, and H. Schwarzlander. Ambiguity function for a bistatic radar. *IEEE Transactions on Aerospace and Electronic Systems*, 33(3):1041–1051, July 1997.

[381] D. Tse and P. Viswanath. *Fundamentals of Wireless Communication*. Cambridge University Press, Cambridge, UK, 2005.

[382] C.C. Tseng and C. Liu. Complementary sets of sequences. *IEEE Transactions on Information Theory*, 18(6):644–652, September 1972.

[383] R. Turyn and J. Stover. On binary sequences. *Proceeding of American Mathmatics Society*, 12(3):394–399, June 1961.

[384] J. Vankka. *Digital Synthesizers and Transmitters for Software Radio*. Springer, Netherlands, 2005.

[385] D.E. Wahl, P.H. Eichel, D.C. Ghiqlia, and C.V.Jr. Jakowatz. Phase gradient autofocus-a robust tool for high resolution SAR phase correction. *IEEE Transactions on Aerospace and Electronic Systems*, 30(3):827–835, July 1994.

[386] D.W. Wahl, P.H. Eichel, D.C. Ghiglia, and P.A. Thompson. *Spotlight-Mode Synthetic Aperture Radar: A Signal Processing Approach*. Springer, Boston, MA, 1996.

[387] F.L. Walls, C.M. Felton, A.J.D. Clements, and T.D. Martin. Accuracy model for phase noise measurements. In *Proceedings of the 21st Annual Precise Time and Time Interval Planning Meeting*, pages 295–310, Redondo Beach, CA, November 1990.

[388] F.L. Walls, J. Gray, A. O'Gallagher, L. Sweet, and R. Sweet. Time-domain frequency stability calculated from the frequency domain: an update. In *Proceedings of the 4th European Frequency and Time Forum*, pages 197–204, Neuchatel, Switzerland, March 1990.

[389] F.L. Walls and S.R. Stein. Accurate measurements of spectral density of phase noise in devices. In *Proceedings of the 31st Annual Frequency Control Symposium*, pages 335–343, Atlantic City, NJ, June 1977.

[390] F.L. Walls, S.R. Stein, J.E. Gray, and D.J. Glaze. Design considerations in state-of-the-art signal processing and phase noise measurement system. In *Proceedings of the 30st Annual Frequency Control Symposium*, pages 269–274, Atlantic City, NJ, June 1976.

[391] I. Walterscheid, J.H.G. Ender, A.R. Brenner, and O. Loffeld. Bistatic SAR processing and experiments. *IEEE Transactions on Geoscience and Remote Sensing*, 44(10):2710–2717, October 2006.

[392] I. Walterscheid, J. Klare, A.R. Brenner, J.H.G. Ender, and O. Loffeld. Challenges of a bistatic spaceborne/airborne SAR experiment. In *Proceedings of the European Synthetic Aperture Radar Symposium*, pages 1–4, Dresden, Germany, May 2006.

[393] B.C. Wang. *Digital Signal Processing Techniques and Applications in Radar Image Processing*. John Wiley & Sons, Inc., Hoboken, NJ, 2008.

[394] G.Y. Wang and Z. Bao. The minimum entropy crierion of range alignment in ISAR motion compensation. In *Proceedings of the IEEE International Radar Conference*, pages 236–239, Munich, Germany, October 1997.

[395] G.Y. Wang, J.Y. Cai, and Q.C. Peng. Near-space SAR: a revolutionizing remote sensing mission. In *Proceedings of the Asia-Pacific Synthetic Aperture Radar Conference*, pages 127–131, Huangshan, China, November 2007.

[396] R. Wang, O. Loffeld, Y.L. Neo, H. Nies, S. Knedlik, and J. Ender. Chirp-scaling algorithm for bistatic SAR data in the constant-offset configuration. *IEEE Transactions on Geoscience and Remote Sensing*, 47(3):952–964, March 2009.

[397] R. Wang, O. Loffeld, Y.L. Neo, H. Nies, I. Walterscheid, T. Espeter, J. Klare, and J. Ender. Focusing bistatic SAR data in airborne/stationary configuration. *IEEE Transactions on Geoscience and Remote Sensing*, 48(1):452–465, January 2010.

[398] R. Wang, O. Loffeld, H. Nies, and J. Ender. Focusing spaceborne/airborne hybrid bistatic SAR data using wavenumber-domain algorithm. *IEEE Transactions on Geoscience and Remote Sensing*, 47(7):2275–2283, July 2009.

[399] W.Q. Wang. Design one high performance signal generator based on triple tuned algorithm. In *Proceedings of the Asia-Pacific Microwave Conference*, pages 3190–3193, Suzhou, China, December 2005.

[400] W.Q. Wang. An approach for multiple moving targets detection and velocity estimation. In *Proceedings of the IEEE Radar Conference*, pages 749–753, New York, May 2006.

[401] W.Q. Wang. Analysis of waveform errors in millimeter-wave LFMCW synthetic aperture radar. *International Journal of Infrared and Millimeter-Waves*, 27(11):1433–1444, November 2006.

[402] W.Q. Wang. Application of near-space pasive radar for home security. *Sensing and Imaging: An International Journal*, 8(1):39–52, May 2007.

[403] W.Q. Wang. Application of near-space passive radar for homeland security. *Sensing and Imaging: An International Journal*, 8(1):39–52, March 2007.

[404] W.Q. Wang. Applications of MIMO technique for aerospace remote sensing. In *Proceedings of the IEEE Aerospace Conference*, pages 1–10, Big Sky, MT, March 2007.

[405] W.Q. Wang. Approach of adaptive synchronization for bistatic SAR real-time imaging. *IEEE Transactions on Geoscience and Remote Sensing*, 45(9):2695–2700, September 2007.

[406] W.Q. Wang. Approach of multiple moving targets detection for microwave surveillance sensors. *International Journal of Information Acquisition*, 4(1):57–68, March 2007.

[407] W.Q. Wang. GPS-based time & phase synchronization processing for distributed SAR. *IEEE Transactions on Aerospace and Electronic Systems*, 45(3):1040–1051, July 2009.

[408] W.Q. Wang. Near-space wide-swath radar imaging with multiaperturing antenna. *IEEE Antennas and Wireless Propagation Letters*, 8(1):461–464, 2009.

[409] W.Q. Wang. Moving target indication via three-antenna SAR with simplified fractional Fourier transform. *EURASIP Journal on Advances in Signal Processing*, 117:1–10, December 2011.

[410] W.Q. Wang. Near-space vehicles: supply a gap between satellites and airplanes for remote sensing. *IEEE Aerospace and Electronic System Magazine*, 25(4):4–9, April 2011.

[411] W.Q. Wang. Space-time coding MIMO-OFDM SAR for high-resolution imaging. *IEEE Transactions on Geoscience and Remote Sensing*, 49(8):3094–3104, August 2011.

[412] W.Q. Wang. Near-space vehicle-borne SAR with reflector antenna for high-resolution wide-swath remote sensing. *IEEE Transactions on Geoscience and Remote Sensing*, 50(2):338–348, February 2012.

[413] W.Q. Wang. OFDM waveform diversity design for MIMO SAR imaging. In *Proceedings of the IEEE International Geoscience and Remote Sensing Symposium*, Munich, Germany, July 2012.

[414] W.Q. Wang. MIMO SAR: potentials and challenges. *IEEE Aerospace and Electronic Systems Magazine*, 2013. in press.

[415] W.Q. Wang and J.Y. Cai. An approach of developing high performance millimeter-wave frequency synthesizer. *International Journal of Infrared Millimeter Waves*, 27(7):931–940, July 2006.

[416] W.Q. Wang and J.Y. Cai. A technique for jamming bi- and multistatic SAR systems. *IEEE Geoscience and Remote Sensing Letters*, 4(1):80–82, January 2007.

[417] W.Q. Wang and J.Y. Cai. Antenna directing synchronization for bistatic synthetic aperture radar systems. *IEEE Antennas and Wireless Propagation Letters*, 9:307–310, 2010.

[418] W.Q. Wang and J.Y. Cai. MIMO SAR using chirp diverse waveform for wide-swath remote sensing. *IEEE Transactions on Aerospace and Electronic Systems*, 48(4), October 2012.

[419] W.Q. Wang, J.Y. Cai, and Q.C. Peng. Conceptual design of near-space SAR for high-resolution and wide-swath imaging. *Aerospace Science and Technology*, 13(6):340–347, September 2009.

[420] W.Q. Wang, J.Y. Cai, and Q.C. Peng. *Passive Ocean Remote Sensing by Near-Space Vehicle-borne GPS Receiver*, chapter in Remote Sensing of The Changing Oceans, pages 77–99. Springer-Verlag, Berlin Heidelberg, 2011.

[421] W.Q. Wang, J.Y. Cai, and Y.W. Yang. Extracting phase noise of microwave and millimeter-wave signals by deconvolution. *IEE Proceedings-Science and Measurement Technology*, 153(1):7–12, January 2006.

[422] W.Q. Wang, C.B. Ding, and X.D. Liang. Time and phase synchronization via direct-path signal for bistatic synthetic aperture radar systems. *IET Radar Sonar and Navigation*, 2(1):1–11, February 2008.

[423] W.Q. Wang, X.D. Liang, and C.B. Ding. A phase synchronization approach for bistatic SAR systems. In *Proceedings of the European Synthetic Aperture Radar Conference*, pages 1–4, Dresden, Germany, May 2006.

[424] W.Q. Wang, Q.C. Peng, and J.Y. Cai. Conceptual design of near-space SAR for high-resolution and wide-swath imaging. *Aerospace Science and Technology*, 13(6):340–347, September 2009.

[425] W.Q. Wang, Q.C. Peng, and J.Y. Cai. Diversified MIMO SAR waveform analysis and generation. In *Proceedings of the International Synthetic Aperture Radar Conference*, pages 270–273, Xi'an, China, October 2009.

[426] W.Q. Wang, Q.C. Peng, and J.Y. Cai. Waveform diversity-based millimeter-wave UAV SAR remote sensing. *IEEE Transactions on Geoscience and Remote Sensing*, 45(3):691–700, March 2009.

[427] W.Q. Wang and H.Z. Shao. A flexible phased-MIMO array antenna with transmit beamforming. *International Journal of Antennas and Propagation*, 2012:1–10, 2012.

[428] X.Q. Wang, X.Q. Sheng, and M.H. Zhu. Variable pulse repetition frequency in the wide-swath SAR. *System Engineering and Electronics*, 26(11):1631–1634, November 2004.

[429] Y.P. Wang, B. Wang, W. Hong, L. Du, and Y.R. Wu. Imaging geometry analysis of 3D SAR using linear array antennas. In *Proceedings of the IEEE International Geoscience and Remote Sensing Symposium*, volume III, pages 1216–1219, Boston, MA, July 2008.

[430] Z.B. Wang, F. Tigrek, O. Krasnov, F. Van Der Zwan, P. Van Genderen, and A. Yarovoy. Interleaved OFDM radar signals for simultaneous polarimetric measurements. *IEEE Transactions on Aerospace and Electronic Systems*, 48(3):2085–2099, October 2012.

[431] Adaptive LFM waveform diversity. Mitigation of autocorrelation sidelobe peaks of orthogonal discrete frequency-coding waveform for MIMO radar. In *Proceedings of the IEEE Radar Conference*, pages 1–6, Rome, Italy, May 2008.

[432] D.R. Wehner. *High Resolution Radar, 2nd edition*. Artech House, Norwood, MA, 1994.

[433] M. Weiβ. Synchronization of bistatic radar systems. In *Proceedings of the IEEE Geoscience and Remote Senisng Symposium*, pages 1750–1753, Anchorage, July 2004.

[434] M. Weiβ. Time and phase synchronization aspects for bistatic SAR systems. In *Proceedings of the Europe Synthetic Aperture Radar Symposium*, pages 395–398, Ulm, Germany, May 2004.

[435] M. Weiβ. Transponder for calibrating bistatic SAR systems. In *Proceedings of the Europe Synthetic Aperture Radar Symposium*, pages 925–928, Ulm, Germany, May 2004.

[436] M. Weiβ. Determination of baseline and orientation of platforms for airborne bistatic radars. In *Proceedings of the IEEE Geoscience and Remote Senisng Symposium*, pages 1967–1970, Seoul, Korea, July 2005.

[437] M. Weiβ and P. Berens. Motion compensation of wideband synthetic aperture radar with a new transponder technique. In *Proceedings of the IEEE International Geoscience and Remote Sensing Symposium*, pages 3649–3651, Toronto, Canada, June 2002.

[438] M. Weiβ, J. Ender, O. Peters, and T. Espeter. An airborne radar for three dimensional imaging and observation-technical reaslisation and

status of ARTINO. In *Proceedings of the Europe Synthetic Aperture Radar Symposium*, pages 1–4, Dresden, Germany, May 2006.

[439] M. Weiβ, O. Peters, and J. Ender. A three dimensional SAR system on an UAV. In *Proceedings of the IEEE International Geoscience and Remote Sensing Symposium*, pages 5315–5318, Barcelona, Spain, July 2007.

[440] M. Weiβ, O. Peters, and J. Ender. First flight trials with ARTINO. In *Proceedings of the European Conference on Synthetic Aperture Radar*, pages 1–4, Friedrichshafen, Germany, June 2008.

[441] L.G. Weiss. Wavelets and wideband correlation processing. *IEEE Signal Processing Magazine*, 11(1):13–32, January 1994.

[442] M.C. Wicks and P. Antonik. Frequency diverse array with independent modulation of frequency, amplitude, and phase, January 15 2008.

[443] M.C. Wicks and P. Antonik. Method and apparatus for a frequency diverse array, March 31 2009.

[444] M.C. Wicks, E.L. Mokole, S.D. Blunt, R.S. Schneible, and V.J. Amuso. *Pinciples of Waveform Diversity and Design*. Scitech Publishing Inc., Raleigh, NC, 2010.

[445] D. Wiley, S. Parry, C. Alabaster, and R. Hughes. Performance comparison of PRF schedules for medium PRF radar. *IEEE Transactions on Aerospace and Electronic Systems*, 42(2):601–611, April 2009.

[446] N.J. Willis. *Bistatic Radar*. Artech House, Norwood, MA, 1991.

[447] F. Witte. Forward looking radar ((coherent), US patent 1993.

[448] F.H. Wong, I.G. Cumming, and Y.L. Neo. Focusing bistatic SAR data using the nonlinear chirp scaling algorithm. *IEEE Transactions on Geoscience and Remote Sensing*, 46(9):2493–2505, September 2008.

[449] F.H. Wong and T.S. Neo. New application of nonlinear chirp scaling in SAR data processing. *IEEE Transactions on Geoscience and Remote Sensing*, 39(5):946–953, May 2001.

[450] F.H. Wong, T.S. Neo, and I.G. Cumming. Focusing bistatic SAR data using the nonlinear chirp scaling algorithm. *IEEE Transactions on Geoscience and Remote Sensing*, 46(9):2493–2505, September 2008.

[451] S.K. Wong. High range resolution profiles as motion-invariant features for moving ground targets indification in SAR-based automatic target recognition. *IEEE Transactions on Aerospace and Electronic Systems*, 45(3):1017–1039, July 2009.

[452] P.M. Woodward. *Probability and Information Theory, with Applications to Radar.* Pergamon Press. Reprinted by Artech House, 1980.

[453] L. Xu and Y.K. Deng. Multichannel SAR with reflector antenna for high-resolution wide-swath imaging. *IEEE Antennas and Wireless Propgation Letters,* 9(1):1123–1126, 2010.

[454] L. Xu and J. Li. Iterative generalized-likelihood ratio test for MIMO radar. *IEEE Transactions on Signal Processing,* 55(6):2375–2385, June 2007.

[455] L. Xu, J. Li, and P. Stoica. Target detection and parameter estimation for MIMO radar systems. *IEEE Transactions on Aerospace and Electronic Systems,* 44(3):927–939, July 2008.

[456] J. Yang and T.K. Sarkar. A novel Doppler-tolerant polyphase codes for pulse compression based on hyperbolic frequency modulation. *Signal Processing,* 17:1019–1029, 2007.

[457] J.G. Yang, X.T. Huang, J. Tian, J. Thompson, and Z. Zhou. New approach for SAR imaging of ground moving targets based on keystone transform. *IEEE Geoscience and Remote Sensing Letters,* 8(4):829–833, April 2011.

[458] L. Yang, T. Wang, and Z. Bao. Ground moving target indication using an In-SAR system with a hybrid baseline. *IEEE Geoscience and Remote Sensing Letters,* 5(3):373–377, July 2008.

[459] Y. Yang and R.S. Blum. MIMO radar waveform design based on mutual information and minimum mean-square error estimation. *IEEE Transactions on Aerospace and Electronics Systems,* 43(1):330–343, January 2007.

[460] Y. Yang and R.S. Blum. Minimax robust MIMO radar waveform design. *IEEE Journal of Selected Topics in Signal Processing,* 1(1):147–155, June 2007.

[461] Y. Yang and R.S. Blum. Phase synchronization for coherent MIMO radar: algorithms and their analysis. *IEEE Transactions on Signal Processing,* 59(11):5538–5557, November 2011.

[462] Y. Yang, R.S. Blum, Z.S. He, and D.R. Fuhrmann. MIMO radar waveform design via alternating projection. *IEEE Transactions on Signal Processing,* 58(3):1440–1445, March 2010.

[463] H.J. Yardley. Bistatic radar based on DAB illuminators: the evolution of a practical system. *IEEE Aerospace and Electronic Systems Magazine,* 22(11):13–16, November 2007.

[464] C.E. Yarman, B. Yazc, and M. Cheey. Bistatic synthetic aperture radar imaging for arbitrary flight trajectories. *IEEE Transactions on Image Processing*, 17(1):84–93, January 2008.

[465] J.F. Yin, D.J. Li, and Y.R. Wu. Research on the method of moving target detection and location with three-frequency three-aperture along-track spaceborne SAR. *Journal of Electronic and Information Technology*, 32(4):902–907, April 2010. in Chinese.

[466] M. Younis. *Digital Beam-forming for High-resolution Wide Swath Real and Synthetic Aperture Radar*. PhD thesis, University of Karlsruhe, Karlsruhe, Germany, 2004.

[467] M. Younis, C. Fischer, and W. Wiesbeck. Digital beamforming in SAR systems. *IEEE Transactions on Geoscience and Remote Sensing*, 41(7):1735–1739, July 2003.

[468] M. Younis, R. Metzig, and G. Krieger. Performance prediction of a phase synchronization link for bistatic SAR. *IEEE Geoscience and Remote Sensing Letters*, 3(3):429–433, July 2006.

[469] H.A. Zebker, T.G. Farr, and R.P. Salazar. Mapping the world's topgraphy by using radar interferometry: the TOPSAT mission. *Proceedings of the IEEE*, 82(12):1774–1786, December 1994.

[470] H.A. Zebker and J.V. Villasenor. Decorrelation in interferometric radar echoes. *IEEE Transactions on Geoscience and Remote Sensing*, 30(5):950–959, May 1998.

[471] T. Zeng, M. Cherniakov, and T. Long. Generalized approach to resolution analysis in BSAR. *IEEE Transactions on Aerospace and Electronic Systems*, 41(2):461–474, April 2005.

[472] H. Zhang, M.D. Xing, J.S. Ding, and Z. Bao. Focusing parallel bistatic SAR data using the analytic transfer function in the wavenumber domain. *IEEE Transactions on Geoscience and Remote Sensing*, 45(11):3633–3645, November 2009.

[473] X.L. Zhang, H.B. Li, and J.G. Wang. The analysis of time synchronization error in bistatic SAR system. In *Proceedings of the IEEE International Geoscience and Remote Sensing Symposium*, pages 4615–4618, Seoul, Korea, July 2005.

[474] Y.S. Zhang, D.N. Liang, and Z. Dong. Analysis of frequency synchronization error in spaceborne parasitic interferometric SAR system. In *Proceedings of the IEEE International Geoscience and Remote Sensing Symposium*, pages 1–4, Dresden, Germany, May 2006.

[475] F. Zhou, R. Wu, M. Xing, and Z. Bao. Approach for single channel SAR ground moving target imaging and motion parameter estimation. *IET Radar, Sonar and Navigation*, 1(1):59–66, February 2007.

[476] H. Zhou and X.Z. Liu. An extended nonlinear-scaling algorithm for focusing large baseline azimuth-invariant SAR data. *IEEE Geoscience and Remote Sensing Letters*, 6(3):548–52, July 2009.

[477] J.M. Zhou. Spurious reduction techniques for DDS-based synthesizers. *IEICE Transactions on Electronics*, E92-C(2):252–257, February 2009.

[478] D.Y. Zhu, Y. Li, and Z.D. Zhu. A keystone transform without interpolation for SAR ground moving target imaging. *IEEE Geoscience and Remote Sensing Letters*, 4(1):18–22, January 2007.

[479] S.Q. Zhu, G.L. Liao, Y. Qu, Z.G. Zhou, and X.Y. Liu. Ground moving targets imaging algorithm for synthetic aperture radar. *IEEE Transactions on Geoscience and Remote Sensing*, 49(1):462–477, January 2011.

[480] L. Zhuang and X.Z. Liu. Application of frequency diversity to suppress grating lobes in coherent MIMO radar with separated subapertures. *EURASIP Journal on Advances in Signal Processing*, 2009:1–10, 2009.

Index

accumulator, 286
adaptive matched filtering, 231
airborne, 227
algorithm
    chirp scaling, 54
    range-Doppler, 52
aliasing, 195
Allan variance, 259
along-track, 2
along-track interferometry, 95
ambiguities
    azimuth, 44
    range, 42, 46, 180
analog-digital-converter, 282
analog-to-digital (A/D), 40
angle
    incidence, 114, 204
    looking-down, 114
    off-boresight, 193
    off-nadir, 193
angle-Doppler coupling, 229
angular, 190
annealing algorithm, 141
antenna
    length, 4, 246
    separation, 246
array
    elements, 182
    linear, 187
    manifold, 194
    nonuniform, 187
    phased, 184
    quad-element, 179
    virtual, 229
ATI
    three-antenna, 98
    two-antenna, 95

ATI SAR
    three-channel, 234
    two-channel, 228
atmospheric turbulence, 353
autofocus, 280
average, 254
azimuth only, 260
azimuth phase coding, 180
azimuth scanning, 93
azimuth-invariant, 335
azimuth-variant, 335

backscatter coefficient, 204
bandwidth, 12
baseline, 330
beamformer
    weighting vector, 193
beamforming
    analog, 179
    digital, 179
beampattern, 197, 376
beamsteering, 192
    intrapulse, 192
beamwidth
    diffraction-limited, 226
    elevation, 180
bending-beam, 376
blind-range, 66, 121
Boltzmann constant, 42

calibration, 285
carrier frequency, 115, 376
center-to-edge, 355
chirp scaling factor, 124
clutter suppression, 229
clutter tuning, 251
coherent processing interval, 180

compressive sensing, 365
configuration
    azimuth-invariant, 5
    azimuth-variant, 5
constant-envelope, 173
constrained optimization, 388
convolution
    theorem, 20
convolution integral, 19
correlation, 254
correlation function, 21
Costas codes, 145
cross-track, 2
curvature equalization, 55

degrees-of-freedom, 8, 240
digital beamforming, 179
digital-earth model, 335
dihedral, 251
direct digital synthesizer, 169
direct-path, 275
displaced phase center antenna, 63
diversity
    frequency, 183
    spatial, 8
    waveform, 11, 138
diversity gain, 182
    array, 183
    spatial multiplexing, 182
Doppler
    bandwidth, 41, 65
    chirp rate, 101
    frequency center, 68
    frequency shift, 36
    oversampling, 180
    parameters, 243
    resolution, 36
    zero, 35
Doppler tolerance, 135
double-balanced mixer, 311
double-interferometry, 227
downward-looking, 365
DPCA, 3, 234
    in azimuth, 3
duty cycle, 192

early-gate, 297
Earth radius, 193
Envisat, 344
epoch, 287

footprint chasing, 326
forward-looking, 364
fractional Fourier, 102, 243
    peak, 103
    simplified, 102
frequency diverse array, 376
frequency increment, 376
frequency scanning, 378
frequency stability, 252
frequency-coding, 382
frequency-offset, 378
Fresnel approximation, 349
fully active, 5, 336

Gaussian noise, 217
GMTI, 8, 227
GMTI detector
    clutter-limited, 97, 228
    noise-limited, 108, 228
Gold code, 331
grating lobes, 376
guard bands, 66

high-resolution wide-swath, 177
homogenous, 191

imaging
    three-dimensional, 10, 364
    two-dimensional, 10, 365, 382
infrared, 1
initial phase, 276
InSAR, 321
inter-pulse, 180
interferences
    cross, 233
interferometric phase, 102, 228, 242
intermediate frequency, 286
intermediate-frequency, 308
interpolation, 301
interpolation function, 25
Ipatov, 180

ISLR, 87, 270, 361
isolation, 183

jitter, 253, 270
    time-domain, 253

Kronker product, 185

Lagrangian, 350
Laplace transform, 310
least-square, 292
LFM signal, 29
    down-chirp, 29
    up-chirp, 29
line-of-sight, 338
linear, 17
link
    duplex, 304
    one-way, 306
    pulsed alternate, 305
    pulsed duplex, 305
    two-way, 303
link reliability, 183
local oscillator, 12

Mapdrift, 299
matched filtering, 30, 185, 382
    compression gain, 38
    compression ratio, 38
maximum likelihood, 215
MIMO radar, 182
MIMO SAR, 135
    degrees-of-freedom, 135
    multiple phase center, 183
    single phase center, 183
    spatial diversity, 135
    waveform diversity, 135
minimum antenna area, 2
minimum antenna area constrain, 65
minimum detectable speed, 227
minimum entropy alignment, 276
mismatch, 274
modified Allan variance, 259
moving target indication, 66
multi-antenna SAR
    MIMO, 5

SIMO, 5
multi-baseline, 321
multibeam, 68
    multiple-phase center, 76
    single-phase center, 68
multichannel, 3
    in azimuth, 3
    in azimuth and elevation, 4
    in elevation, 2
multidimensional encoding, 177
multidimensional waveform, 188
mutual interferences, 196

nadir
    interferences, 118
    returns, 66
near-space, 344
NESZ, 43
noise
    amplifier, 281
    amplitude, 253
    bandwidth, 42
    figure, 42
    Gaussian, 264
    phase, 253
    temperature, 42
    thermal, 43
nonuniform sampling, 78
    reconstruction, 81
    subsampled, 81
null steering, 112
Nyquist interval, 23
Nyquist rate, 23

OFDM chirp waveform, 156
    chirp diverse, 156
    correlation, 156
    cross-correlation, 159
OFDM waveform, 150
    ambiguity function, 154
    bandwidth, 156
    subcarriers, 151
optical, 1
orbit radius, 193
orientation, 330

oscillator
    crystal, 252
    quartz, 252
    sinusoidal, 252

paired echoes, 321
PAPR, 173
parameter identifiability, 197
passive imaging, 1
peak-to-sidelobe, 141, 233
phase center, 67
phase centers
    effective, 186
    equivalent, 185
    virtual, 186
phase noise, 251
    flicker, 259
    frequency-domain, 253
    white, 259
    Wiener, 262
phase shifter, 191
phased-array, 177
phased-locked loop, 169
pitch, 353
point spread function, 7, 47
polyhedral, 251
polyphase code
    Frank, 139
    P1, 139
    P2, 139
power-law, 256
PSD, 254
PSLR, 361
Pulse chasing, 325
pulse compression, 32
pulse duration, 246
pulse repetition frequency (PRF), 4
pulse repetition interval, 121, 180
pulse-per-second, 286

quadratic phase error, 87
quantization, 23, 282
quasi-monostatic, 349

radar
    bistatic, 4

distributed, 5, 251
monostatic, 2
multistatic, 4
radar ambiguity function, 33
    range, 35
    velocity, 36
radar cross section (RCS), 7
range alignment, 276
    frequency domain, 276
    misalignment, 276
    spatial domain, 276
range migration algorithm, 365
range-azimuth, 336
range-dependent, 376
RASR
    multi-antenna, 127
    single-antenna, 127
ratio
    peak-to-sidelobe, 233
RCMC, 54
receive-only, 251
receiver noise, 281
recursive alignment, 276
register, 286
remote sensing, 1
    high-resolution, 9
resolution, 3
    angular, 197
    azimuth, 3, 8, 65
roll, 353

sample variance, 258
sampling, 23
    aliasing, 26
    spatially, 72
    theorem, 22
SAR
    multi-antenna, 1
    single-antenna, 1
SAR imaging
    wide-swath, 89
scan-on-scan, 323
scanning on receive, 179
Schwartz inequality, 276, 297
scintillations, 182

SCR, 97
second range compression, 59
semi-active, 5, 336
sensing matrix, 117
    condition number, 118
    ill-conditioned, 118
    well-conditioned, 118
shadowing effects, 364
side-looking, 193, 363
sidelobe, 7
signal-to-noise ratio (SNR), 8
slant range, 4
sliding spotlight, 317
small signal theory, 310
space-frequency, 194
space-time coding, 214
space-time encoding, 177
    Alamouti, 177
spaceborne, 227
spatial aliasing, 378
spatial resolution, 1
spotlight, 1
spurious level, 288
standard deviation, 254
STAP, 182
staring spotlight, 327
stationary phase, 27
STBC, 214
    higher-order, 216
    quasi-orthogonal, 216
steering vector, 385
    receiver, 385
    transmitter, 385
    virtual, 386
stop-and-go, 41, 221
stripmap, 1
subaperturing
    AASR, 203
    beampattern, 203
    NTNR MIMO, 200
    NTWR MIMO, 200
    RASR, 202
subband, 191
superheterodyne receiver, 308
swath, 2

wide, 2
width, 2, 63
synchronization
    antenna, 317
    phase, 12, 252
    spatial, 12, 252
    time, 12, 252, 270
synchronization link, 301
synthetic aperture radar (SAR), 1
system
    time-invariant, 17
system sensitivity, 42

Tandem, 335
target
    aspect, 182
    RCS, 251
Taylor series, 28, 77
TerraSAR-X, 344
time multiplexing, 324
time-bandwidth product, 38
time-varying, 309
tomography, 364
translational invariant, 335
transponder, 312
triple tuned algorithm, 171
true variance, 258

variable frequency divider, 288
variance, 254
vector
    beamforming, 199
    steering, 199
velocity
    along-track, 227
    radial, 9, 228
virtual aperture, 6, 227

waveform
    basic RSF, 145
    DFCW, 141
    OFDM, 149
    order, 218
    orthogonality, 203, 246
    phase-modulated RSF, 148
    polyphase-coded, 139

  RSF-LFM, 147
  stepped-frequency, 144
waveform diversity, 6
waveform optimization, 388
waveform repetition interval, 181
wide-sense stationary, 254
Wiener-Khintchine Theorem, 254
Wigner-Ville distribution, 105

yaw, 353